printreading

based on the

1999 NEC®

National Electrical Code®

AMERICAN TECHNICAL PUBLISHERS, INC.
HOMEWOOD, ILLINOIS 60430-4600

R. T. Miller

1 2 3 4 5 6 7 8 9 – 98 – 9 8 7 6 5 4 3 2 1

Printed in the United States of America

ISBN 0-8269-1561-2

PREFACE

This edition of *National Electrical Code® Printreading* has been revised to reflect many of the important changes that appear in the 1999 National Electrical Code®. The changes are presented as they pertain to One-Family Dwellings, Multifamily Dwellings, Commercial Locations, Industrial Locations, and Hazardous Locations. In addition to these appropriate changes to the text and illustrations, the Trade Competency Tests and two Final Tests have been thoroughly updated. A new Chapter 1 has been added to more thoroughly cover printreading. A new Glossary has been added to provide definitions for electrical and printreading terms used throughout the text.

National Electrical Code® Printreading is designed to enable the student to learn electrical printreading and become familiar with applicable sections of the 1999 National Electrical Code®. Complete references to the 1999 NEC® are presented throughout the text. Trade Competency Tests appear at the end of each Chapter to help students check their understanding of the text material and the NEC®. Two Final Tests are included in Chapter 7 of the text. Answers, with NEC® validation for the Trade Competency Tests and Final Tests are included in the *Instructor's Guide for National Electrical Code® Printreading*.

NATIONAL ELECTRICAL CODE®

The National Electrical Code® is sponsored and controlled by the National Fire Protection Association (NFPA). The primary function of the NEC® is to safeguard people and property against electrical hazards. It is mandatory that the *National Electrical Code® Printreading* be used only in conjunction with the 1999 NEC®. Copies of the 1999 NEC® (NFPA No. 70) can be ordered directly from its publisher:

> National Fire Protection Association, Inc.
> 1 Batterymarch Park
> Quincy, MA 02269

WATTS AND VOLT-AMPERES

In general, within the NEC®, the term *watts* (W) has been superseded by the term *volt-amperes* (VA) for the computation of loads. However, references to nameplate ratings still reflect the term watts on certain loads.

CALCULATIONS

When total wattage or VA is to be divided by phase-to-phase (3φ) voltage times 1.732, the following values may be substituted:

> for 208 V × 1.732, use 360
> for 230 V × 1.732, use 398
> for 240 V × 1.732, use 416
> for 440 V × 1.732, use 762
> for 460 V × 1.732, use 797
> for 480 V × 1.732, use 831

MANDATORY USE OF SHALL

Section 90-5(a) states that mandatory rules use the word *shall*. Always refer to the NEC® for mandatory rules.

FRACTIONS OF AN AMP

Section 220-2(b) states that fractions less than .5 may be dropped.

THW Cu

Unless otherwise specified, copper conductors are sized on THW per Table 310-16.

TRADE COMPETENCY TESTS AND FINAL TESTS

National Electrical Code® Printreading contains 23 Trade Competency Tests that should be completed after studying the corresponding chapter. The two Final Tests include questions that cover the content of the entire text. The Final Tests should be taken after completing the text and all of the Trade Competency Tests.

QUESTIONS

Question formats include True-False, Multiple Choice, Completion, and Matching. Always record the answer in the space provided in the Answer column and provide Code validation, as required, in the NEC® column. XXXXX in the NEC® column indicate that no answer is required.

Contents

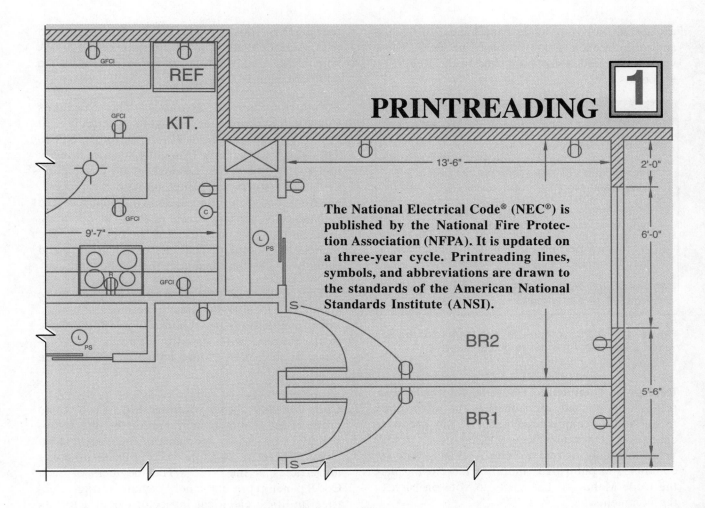

The National Electrical Code® (NEC®) is published by the National Fire Protection Association (NFPA). It is updated on a three-year cycle. Printreading lines, symbols, and abbreviations are drawn to the standards of the American National Standards Institute (ANSI).

THE NATIONAL ELECTRICAL CODE®

The National Electrical Code® (NEC®) is one of the most widely used and recognized consensus standards in the world today. It is a true consensus standard because members from throughout the electrical industry contribute to its development. The NEC® is updated and revised every three years to reflect current trends in the electrical industry. The purpose of the NEC® is to protect people and property from hazards that arise from the use of electricity per 90-1(a).

NEC® Process

The National Fire Protection Association (NFPA) sponsors the development of the NEC®. The NFPA publishes guidelines in the Regulations Governing Committee Projects for the procedures for all of the standards it publishes. For the NEC®, the procedures call for essentially a four-step process:

1. Receipt of Proposals (ROP)
2. Receipt of Comments (ROC)
3. NFPA Annual Meeting
4. Standards Council Issuance

The schedule for the NEC® process is given following the Index. The methods of submitting a proposal and a proposal form also follow the Index.

Receipt of Proposals. The first step in the Code process is the receipt of proposals. Anyone can submit a proposal to change the NEC® provided it contains the required information. See Figure 1-1.

The key to successful Code proposals is proper substantiation for the proposed changes. The time window to submit proposals is very small. For example, the last date to submit changes for the 1999 NEC® was November 8, 1996. After the last date to receive proposals, code panels meet to discuss and vote on each of the proposed changes. These votes are recorded and published in the Report on Proposals.

NEC® PROPOSALS
NEC® proposals shall contain:
1. Identification of the submitter. (name and organization or company affiliation)
2. Identification of the specific Code section and type of revision that is proposed.
3. A statement of the problem and substantiation for why the change is necessary.
4. The actual wording of the revised text or the wording to be deleted.

Figure 1-1. NEC® proposals shall contain the required information in order to be considered.

Receipt of Comments. Once the Report on Proposals is published, everyone has the opportunity to submit a comment on each of the proposed changes whether the Code-Making Panel (CMP) voted to accept or reject them. A closing date for comments is published, and after the closing date the Code-Making Panel meets to vote on each of the comments.

The Report on Comments contains all comments and the votes of the respective code panels. If a proposal or comment is not accepted by the Code-Making Panel, the statement must state the reason the proposal or comment was not accepted. New proposals cannot be submitted during this stage of the Code cycle. The Code-Making Panel can only take action on proposals that have received adequate public review during the proposal stage.

NFPA Annual Meeting. Once the Code-Making Panels have reviewed the proposals and comments, the next step is to present the changes at the NFPA Annual Meeting. Floor action on individual proposals can occur during this meeting. Usually there are very few of these actions on the floor of the NFPA Annual Meeting, because most members choose to support the actions of the Code panels during the initial stages of the Code process.

Voting on floor actions at the NFPA Annual Meeting is limited to NFPA members only. A simple majority vote is required for a floor action to pass. Actions that occur on the floor of the NFPA Annual Meeting are still subject to review by the NFPA Standards Council.

Standards Council Issuance. The Standards Council has the responsibility for overseeing all of the codes and standards developed for the NFPA. The NEC® Correlating Committee works directly under the Standards Council. The Correlating Committee steers the panels through the process, ensuring that each proposal and comment received is handled according to an established operating procedure. Once the process is complete, the Standards Council reviews the entire process and actually issues the document for publication.

The NEC® is a legal document designed to be adopted by local and/or state governmental bodies. Local jurisdictions may choose to adopt the Code in its entirety, with specific additions or exceptions, or they may choose not to adopt the Code at all.

Code-Making Panel Membership. Each Code proposal or comment is reviewed by the representatives of various segments of the electrical industry. There are 20 Code-Making Panels to cover the articles in the NEC®. The members of each Code panel represent labor, manufacturing, electrical utilities, electrical inspectors, contractor associations, and testing laboratories.

NFPA Operating Procedures for the National Electrical Code® Committee state that no interest group shall comprise more than $\frac{1}{3}$ of the total voting panel membership. Membership is designated as either principal or alternate. Alternate members assume the participation and voting rights only when the principal member is not present.

The official vote on panel proposals and comments occurs on a written ballot after the meetings conclude. Although each proposal is voted on during the Code-Making Panel discussions, the official vote on each proposal occurs when the Code-Making Panel member returns a written ballot which was mailed after the conclusion of the Code-Making Panel meetings. All members who vote against a panel action must state the reasons for doing so, and their comments are recirculated to each panel member.

Using the NEC®

The NEC® is available in soft cover, loose-leaf, and PC versions. It may be ordered directly from its publisher:

National Fire Protection Association, Inc.
1 Batterymarch Park
Quincy, MA 02269

While the NEC® is not difficult to use, it does have its own method of organization which must be understood for ease of use.

Revisions, Deletions, and Extracted Text. Any material in the current edition of the text that has changed from the previous edition is identified by a vertical line in the margin. See Figure 1-2. Text that has been deleted from the previous edition is shown as a bullet (•) that also appears in the margin. Text that has been extracted from another NFPA document is designated by the use of the superscript letter (^{X}). These editorial markings help trace the development of Code requirements.

Outline Format. The NEC® is arranged in a simple outline format. See Figure 1-3. Chapters are divided into articles. With the exception of Article 90, these are 3-digit numbers. For example, Article 210 covers Branch Circuits. The Contents in the front matter of the Code contains a list of the chapters with the article numbers covered in each chapter.

Section numbers are designated by a dash following the article number. For example, 210-8 indicates Article 210, Section 8. Section numbers often proceed in a normal numbering sequence, but sometimes sections are not used in numerical sequence. This occurs when sections have been deleted from the NEC® or perhaps if the Code panel wanted to leave room for future Code sections.

Subsections are part of the sections and are designated by lowercase letters set off in parentheses. For example, 210-8(a) is a subsection of 210-8. Subsections are further broken down by numbers to indicate parts of subsections. For example, 210-8(a)(4) is a part of subsection 210-8(a). In a few instances, the NEC® has further subsections of subsections. In most cases, the format remains in this number-letter-number sequence. The only time the NEC® deviates from this fashion is when sections do not contain any subsections, but do contain numbered items. For example, 230-43(4) is a numbered item of Section 230-43.

Exceptions. Exceptions permit alternate methods to the general (main) rule. Exceptions are permitted in lieu of the main rule when all of the provisions of the exception are met. For example, 210-8(a)(5) requires that, in general, all receptacles in unfinished basements of dwelling units shall be GFCI-protected. Exception 2 allows receptacles to be installed without GFCI protection provided that a single receptacle is used. This ensures that the receptacles cannot be easily accessed for other uses. A duplex receptacle is also permitted to be used if two appliances, located in dedicated spaces and not easily moved, are cord- and plug-connected per 400. The purpose of this exception is to allow appliances, such as freezers or refrigerators which may be located in unfinished basements to be installed without GFCI protection provided there is no means for connecting portable cords or other appliances to the same receptacle.

Because the NEC® is organized in an outline format, exceptions are applied only to the section they follow or as noted. For example, the exception for receptacles that are not readily accessible follows 210-8(a)(2) and cannot be applied to 210-8(a)(1).

	NEC® EDITORIAL SYMBOLS	
I	**VERTICAL LINE**	Vertical lines appearing in the margins of the NEC® represent sections where the text has been revised from the previous edition of the NEC®. For example, see 100 where the definition of bathroom has been relocated.
•	**BULLET**	A bullet is used to show where text has been deleted from the previous edition of the NEC®. For example, see 230-65 where the previous requirement has been deleted from the NEC®.
x	**SUPERSCRIPT "X"**	The superscript letter "X" is used whenever text that appears in the NEC® has been extracted from another NFPA document. For example, see 517-25 where text that has been extracted from NFPA 99, *Health Care Facilities*, appears.

Figure 1-2. NEC® editorial symbols are useful tools for studying the latest NEC® changes.

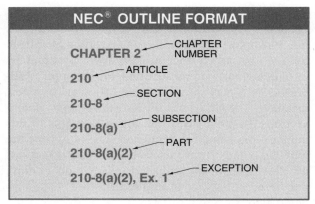

Figure 1-3. The NEC® is arranged in a simple outline format.

Applying the NEC®

The National Electrical Code® is adopted for use by local (village, town, city, etc.), county or parish, and state authorities. These authorities are generally represented by building officials who issue permits for jobs and make periodic inspections of work in progress. Additionally, these authorities certify that the completed building meets all applicable codes before occupancy occurs.

Purpose and Intent – 90-1. The two-fold purpose of the NEC® is to protect people and property from the dangers associated with the use of electricity. This two-fold purpose should be kept in mind whether applying the NEC® or submitting new proposals to modify the NEC®.

The NEC® is not intended to be used as an instruction manual per 90-1(c). The rules are stated without any reasons for their existence. For example, 210-8(a)(1) states that all 15 A and 20 A receptacles installed in dwelling bathrooms shall have GFCI protection. In order to understand the reason for the rule, the Report on Proposals would have to be read to determine the substantiation when the Code panel first accepted the change.

The NEC® is also not a design specification per 90-1(c). It contains the rules and necessary provisions for electrical systems, but leaves the design and layout of the electrical systems to others.

Scope – 90-2. Before applying the NEC®, the scope should be reviewed to determine if the electrical installation falls within the jurisdiction of the NEC®. Installations which are not covered are given in 90-2(b). Often, the distinction between what is covered and what is not covered is very fine and can lead to some difficult interpretations. For example, installations for electrical utilities are not covered if they involve the metering, generation, transmission, or distribution of electric energy. Those installations, however, that are used by an electric utility, but are not an integral part of the electrical energy transmission or distribution are covered. Another example is watercraft. Installations on ships are not covered, but installations on floating buildings are.

Code Arrangement – 90-3. The arrangement of the NEC® is specified in 90-3. The NEC® contains an Introduction and nine chapters. The first four chapters apply generally, unless they are modified by the latter chapters. For example, 300-14 requires that, in general, a minimum of 6″ of free conductor be provided at each outlet for splices or terminations. This applies throughout the NEC® unless modified.

Chapter 8 is independent of the other chapters. None of the NEC® provisions apply unless they are directly referenced. Chapter 8 is independent of the other chapters because it covers communication circuits. These circuits are part of the premises wiring, but are treated separately because they do not, in general, pose a threat of electrical shock to persons. None of the NEC® provisions therefore apply, unless they are directly referenced in Chapter 8.

Chapter 9 contains tables and examples which are referenced throughout the entire NEC®. The Appendix, which appears at the end of the Code, does not contain any mandatory provisions, but it does contain information that may be helpful in designing or installing electrical installations. For example, Appendix C contains many useful tables for determining the maximum number of conductors permitted in a raceway.

A comprehensive Index follows Chapter 9. The Index is copyrighted separately from the NEC®. The alphabetical entries are boldfaced with Code numbers giving the location of the entry as it applies to particular subjects. Page numbers are not referenced in the Index.

Enforcement – 90-4. All applications of the NEC® are ultimately subject to approval by the Authority Having Jurisdiction (AHJ). See Figure 1-4.

The NEC® is designed to be used as a legal document. Any interpretations and approval of equipment rest with the Authority Having Jurisdiction. In many cases, the AHJ is a local or municipal building official.

AUTHORITY HAVING JURISDICTION

Duties and responsibilities:

1. Make interpretations of NEC® rules.

2. Approve equipment and materials.

3. Grant special permission.

4. Waive specific requirements.

5. Permit alternate methods.

Figure 1-4. The AHJ is responsible for enforcement of the NEC® per 90-4.

Some localities permit the use of independent third party electrical inspection agencies. In other cases, the AHJ may be a governmental official, local fire marshal, or an insurance official with jurisdiction over the installation. In any case, whoever the AHJ is, they can permit alternate methods or even waive specific requirements if they believe the installation is equally effective at meeting the intended objective.

Mandatory Rules and Explanatory Material – 90-5. The use of the word "shall" indicates mandatory rules of the Code. For example, per 220-3(a), lighting loads for occupancies shall be figured using the values in Table 220-3(a).

Fine Print Notes (FPNs) are used throughout the NEC® to explain information that is contained in the Code. For the purposes of enforcement, FPNs are not mandatory and do not contain any mandatory provisions. For example, 220-3(a) states that lighting loads shall not be less than the unit load per square foot in Table 220-3(a). The FPN clarifies that the unit values are based on minimum load conditions and 100% power factor, and may not be sufficient for the installation.

Future Use – 90-8. Sections 90-1(b) and 90-8(a), taken together, indicate that installations completed in accordance with the NEC® merely meet a minimum established standard. The installations may not be adequate for future use, and while it is not required, allowance for future expansion is generally in the best interest of all concerned. Essentially, the NEC® is a minimum standard. The designer or installer can choose to provide for future use if desired. Installations in accordance with the NEC® are not necessarily efficient and may not meet the future needs and demands for electricity.

PRINTS

A *print* is a detailed plan of a building drawn orthographically using a conventional representation of lines, and including abbreviations and symbols. See Figure 1-5. Prints were formerly known as blueprints because the method used to produce them yielded white lines on a blue background. Today, most prints are black lines on white paper. Conventional representation of lines refers to prescribed line types as specified in ANSI Y14.2M.

Figure 1-5. A print is a detailed plan of a building drawn orthographically using a conventional representation of lines, and including abbreviations and symbols.

By drawing lines to ANSI standards, an "alphabet of lines" is established and easily read. See Appendix. For example, a line that is always drawn with a series of dashes of equal length is immediately identified as a hidden line. Lines used on drawings include object lines, hidden lines, centerlines, dimension lines, extension lines, leaders, break lines, and cutting plane lines. Other types of lines are also used on drawings. Refer to ANSI Y14.2M.

Prints are first drawn as working drawings and copies are made to produce the prints. Working drawings may be drawn with pencil or ink and drafting instruments, although the common practice today is to use a computer software program such as CAD (Computer Assisted Drafting). Some computer software programs are available which automatically locate electrical receptacles per the NEC®. See Figure 1-6.

Figure 1-6. Some computer software programs automatically locate electrical receptacles per the NEC®.

Shape Description

Each print of a set of working drawings represents a part of a building drawn as an orthographic projection. This method of drawing allows the individual parts of the building to be shown on a flat plane in their true shape. A substantial number of individual prints may be required to clearly show the shape and size of all components. The most essential plans required to show a building include plot plans, floor plans, elevations, section views, and detail views. Of these, the floor plan is the primary plan in that it gives the most information about the building.

Plot Plans. A *plot plan* is an orthographic drawing that shows the location and orientation of the structure on the lot and the size of the lot. See Figure 1-7. A legal survey of the property made by a licensed surveyor establishes the corners of the lot in relation to official points of measurement in the vicinity. The compass direction and the angles at the corners of the lot are included. The survey locates existing buildings, shows the contour of the land, large trees, and other physical features. Other information on the plot plan includes property dimensions, grade elevations, and the location of water, sewer, gas, and electrical utilities. A utility pole indicates that the electrical service is overhead. A lateral service is underground.

The *point of beginning* is a fixed point from which all measurements, both vertical and horizontal, are made to show the building lot. The point of beginning, in turn, is related to a bench mark or city datum point. These datum points may be concrete markers within sighting distance of the lot, a mark on a fire hydrant, the top of a sewer in the street, or other points established by the city as a local point of reference. The datum points are not only points from which the point of beginning is located by horizontal measurement, but are also the basic points for vertical dimensions.

Floor Plans. A *floor plan* is an orthographic drawing of a building as though cutting planes were made through it horizontally. See Figure 1-8. The cutting plane is generally taken 5'-0" above the floor being shown. Objects below the 5'-0" elevation are shown with object lines. Objects above the 5'-0" elevation are shown with hidden lines.

Figure 1-7. A plot plan is an orthographic drawing that shows the location and orientation of the structure on the lot and the size of the lot.

Floor plans are the most frequently used drawings in a set of plans. They are drawn to scale to show the size relationship of various components. A one-story building requires one floor plan. A one-story building with a basement requires two floor plans. A building with more than one story generally requires a floor plan for each level.

The location of all exterior walls and interior partitions are shown and dimensioned on the floor plans. Overall size dimensions and specific location dimensions are given for structural elements. Door and window openings are shown. Reference letters or numbers refer to door and window schedules showing the type and size of door or window required at each opening.

On smaller sets of prints, electrical information is shown directly on the floor plans. On larger sets of residential plans and on commercial plans, electrical information is usually shown on separate electrical plans. The general location of receptacles, lights, and switches is shown. The electrician determines the specific location in conformance to applicable NEC® articles. For example, per 210-52(a), no point of unbroken wall space 2′ or more in length measured horizontally shall be more than 6′ from a receptacle in kitchens, family rooms, dining rooms, living rooms, parlors, libraries, dens, sun rooms, bedrooms, recreation rooms, or similar rooms of dwelling units.

Plumbing fixtures such as water closets, sinks, shower stalls, and bathtubs are shown on residential floor plans. Information pertaining to heating, ventilating, and air conditioning (HVAC) systems may also be shown on floor plans. Registers connecting to the system are located on the floor plan. On larger sets of plans, HVAC information is usually shown on mechanical plans.

Elevations. An *elevation* is an orthographic drawing showing vertical planes of a building. See Figure 1-9. Depending on building size, shape, and complexity, four or more elevations may be required to clearly show all exterior walls. Generally, however, four exterior views are sufficient. These are the front, rear, right side, and left side or North, South, East, and West. The North Elevation of a building is the elevation facing North, not the direction a person faces to see that side of the house. Elevations are identified and the scale is given on elevation drawings. Interior elevations are used to clarify locations of such items as cabinets, mantels, and bookcases.

Sections. A *section* is an orthographic drawing created by passing a cutting plane (either horizontal or vertical) through a portion of a building. See Figure 1-10. Section views are shown wherever an important part of a building is not apparent on the plan views or elevations views. Cutting planes are passed through the building at the most advantageous point to provide clear information.

The most common section view is taken through an outside wall. It shows information such as details on the foundation wall and footing, details of wall and floor framing, height of windows above the floor, eave construction, and roof construction.

Cutting planes for section views are shown on the floor plan. Direction arrows indicate line of sight and a code reference shows the sheet number on which the identified section is found.

Several section views may be required if the construction varies from one part of the building to another part of the building. In some instances, cutting planes are taken lengthwise or crosswise in a building. A *longitudinal section* is a section taken lengthwise. A *transverse section* is a section taken crosswise.

Rodger A. Brooks, Architect

Figure 1-8. A floor plan is an orthographic drawing of a building as though cutting planes were made through it horizontally.

VERTICAL PLANES
OF BUILDING

Rodger A. Brooks, Architect

Figure 1-9. An elevation is an orthographic drawing showing vertical planes of a building.

Details. A *detail* is a part of a plan, elevation, or section view drawn at a larger scale. See Figure 1-11. Often the plan, elevation, and section views do not show all the information the builder must have to construct the building. The basic set of prints may be drawn at a small scale such as $\frac{1}{8}''$ = 1′ or $\frac{1}{4}''$ = 1′ and not show small details clearly. When this occurs, a small part of the building is drawn at a larger scale. Details may be drawn of any part of the building that cannot be shown clearly and conveniently on the working drawings. The scale is always given for the detail.

Written Description

Working drawings show all walls, partitions, and other features of a building by means of lines, symbols, and conventional representations. With few exceptions, each part is located by precise dimensions. However, in order to develop accurate cost estimates and to quickly solve problems that may arise during construction, information is shown on the prints in the form of notations. Also, additional information is included in the specifications, which may consist of a few typed sheets or a book of several hundred pages, depending on the complexity of the project.

Written information is shown on prints in four ways:

1. Some information is found in the title block.
2. Descriptive titles are placed near a specific item and connected with a leader line terminated by an arrowhead or a dot.
3. Specific information that refers to only one situation may be placed near the situation.
4. General information applying to several sheets in the prints may be placed in any convenient space.

Specifications. *Specifications* are typed sheets gathered together into a pamphlet (or into several volumes for a large building) covering a number of subjects in detail. These subjects include information, in paragraph form, that describes the building to be constructed, the materials to be used, and the responsibilities of those involved.

Figure 1-10. A section is an orthographic drawing created by passing a cutting plane (either horizontal or vertical) through a portion of a building.

The simplest type of construction job should have specifications. They are included with prints and a written contract to become the basis for agreement between the owner and the contractor.

Complete specifications are written to give detailed information about the job. This information cannot conveniently be shown on a set of prints because it is too lengthy. The general purpose of specifications includes the following:

- Specifications provide the general conditions giving the broad provisions of the contract and outline the responsibilities of the owner, architect, contractor, and subcontractors. Performance guarantees are included.

- Specifications supplement the working drawings with detailed technical information about the work to be done, specifying the material, equipment, and fixtures to be used.

• Specifications serve, with the contract agreement and the prints, as the legal basis for the transaction of erecting the building from start to finish.

Figure 1-11. A detail is a part of a plan, elevation, or section view drawn at a larger scale.

Specifications are intended to supplement the set of prints with data that will spell out the job in a general sense. They also include the details of the job to be done by each individual trade. The set of prints and the specifications are intended to be in agreement. In the event of any discrepancy or conflict between the two, the specifications take precedence.

Specifications provide the contractor and sub-contractors with detailed information for precise and competitive bidding when estimating the costs of labor and material. Subcontractors bid their work based upon codes referenced in the specifi-

cations. For example, electricians comply with the requirements of the National Electrical Code® when estimating cost for specific materials.

The building department of the town or city in which the building is to be constructed uses the specifications and the prints to determine their compliance with all applicable building codes and zoning ordinances to meet structural, fire, and health standards. Banks and loan agencies use the same information to help appraise the building to determine its value. When governmental agencies such as the Veterans Administration (VA) or the Federal Housing Administration (FHA) provide part of the financing, they require a copy of the specifications for their approval. Certain sections of the specifications must be written on forms supplied by the individual agency when either VA or FHA financing is to be used.

The Construction Specifications Institute (CSI) has developed standard procedures that are helpful in writing specifications, particularly in writing in the CSI format. Specification writing is divided into 16 technical divisions, each with several broad (smaller subheads) section headings. See Figure 1-12. See Appendix.

Figure 1-12. The CSI format contains 16 divisions.

Title Block. The title block is the logical place to begin reading a set of prints. See Figure 1-13. Information in the title block includes the name of the project, project location, architect's name and office location, date on which plans were completed, initials of the drafter, number of the sheet, number of sheets in the set of prints, name of the sheet, and other information as determined by the architect. Title blocks are commonly located in the lower right-hand corner of the sheet.

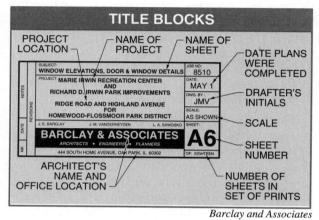

Barclay and Associates

Figure 1-13. The title block contains specific information related to the prints.

Abbreviations

An *abbreviation* is the shortened form of a word or phrase. See Figure 1-14. Abbreviations are used throughout a set of plans to describe materials and processes while conserving space on the drawing sheets. They save drafting time while presenting information in a standard manner.

Uppercase letters are used for most abbreviations. All abbreviations that make an actual word end in a period in order to distinguish the abbreviation from the actual word. For example, IN. is the abbreviation for inch.

While many abbreviations are accepted by virtue of their common usage, only standardized abbreviations should be used on drawings. Standardized abbreviations eliminate confusion and misinterpretation. Although some different words use the same standardized abbreviations, an interpretation based upon the context in which they are used helps clarify the specific use. Examples of the same abbreviations for different words are A and V. A is the

abbreviation for amp and area. V is the abbreviation for volts, valve, and vent. See Appendix.

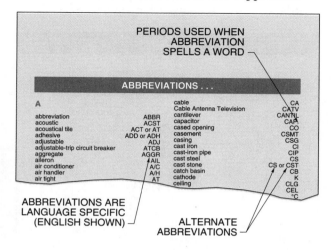

Figure 1-14. An abbreviation is the shortened form of a word or phrase.

Symbols

A *symbol* is a graphic representation of an object. See Figure 1-15. Symbols provide a uniform representation of building materials, fixtures, and structural parts that are easily recognizable. Symbols are universal in that the symbol can be understood no matter what language a person speaks. Symbols are shown on plan views (plot, floor, framing, etc.), elevation views, sections, and details. See Appendix.

Architectural symbols show building materials such as wood, brick, concrete, glass, roofing, and structural steel. See Appendix. Electrical symbols show various types of outlets, receptacles, lights, wiring, signaling systems, etc. See Appendix.

PRINTREADING

Prints for the one-family dwelling include the plot plan, foundation and unfinished basement plan, floor plan, elevations, sections, and details. A door schedule and a window schedule are shown. Specifications are not included. The electrical plans are shown as part of the foundation and basement floor plans and the floor plan.

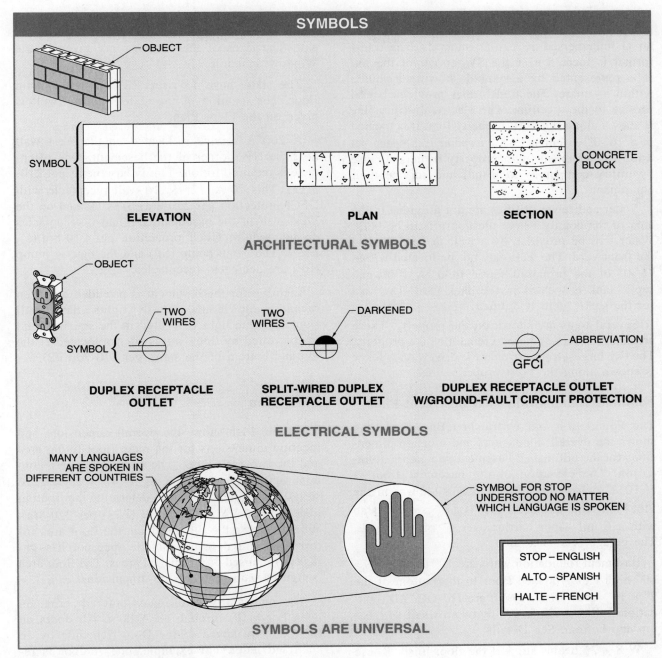

Figure 1-15. A symbol is a graphic representation of an object.

Plot Plan

The Plot Plan shows a one-family dwelling located on S. Juniper Dr. From the point of beginning on the NE corner (which is assigned an elevation of ±0'-0"), the property line parallels S. Juniper Dr. to the SE corner for 192'-6". From that point to the SW corner is 180'-0". The four corners of the property are at 90° angles. The lot contains 34,650 sq ft (192.5' × 180' = 34,650 sq ft). Contour lines are not shown as the lot is relatively flat.

The house is set back 58'-5" from S. Juniper Dr. and 63'-8" from the North property line. This is well beyond the 30'-0" minimum building line. The main floor is 2'-10" above the POB and the garage floor is 2'-0" above the POB to provide good drainage away from the house. A 4'-0" concrete sidewalk is shown from the blacktop drive to the front entrance.

The gas line is on the East side of the street. It is represented by the long-dashed line with the letter G interspersed at regular intervals. The transformer is located near the SW corner of the lot. It is represented by a symbol showing a square within a square. The transformer provides lateral service to the dwelling. The electrical utility line is shown along the West property line. It is located in a 10'-0" easement. An *easement* is a space set aside by agreement with property owners for use by utility companies to install and service their equipment.

Water and sewer utilities are not provided in the area of the county where the house is to be built. Water will be provided by a well to be drilled in the front yard. The well may not be located within 75'-0" of the proposed septic field. A 1000 gal. septic tank is located in the back yard. The area for the septic field is defined.

Several trees are shown on the property. These are existing trees and are to remain on the property. The two largest trees are near the driveway. A fence is shown along the West property line.

Foundation and Unfinished Basement Plan

The Foundation and Unfinished Basement Plan shows the overall dimensions and location dimensions for the unfinished basement and garage foundation. The elevation of the basement floor is 5'-10¾" below the POB, which is shown on the Plot Plan. The basement floor is a 4" concrete slab with a 6 mil vapor barrier over 4" of gravel fill. The garage is unexcavated.

Basement foundation walls are 10" thick and 7'-10" high from finished floor to the bottom of the 2" × 10" floor joists, which are 16" OC. All foundation walls are reinforced by two rows of #4 rebar top and bottom. See Details.

W 8 × 21 beams support the floor joists. These wide flange steel beams are 8" deep and weigh 21 lb per foot. The beams are supported by 8" wide × 8½" high beam pockets 4" deep and 3" diameter lally columns, which are placed in 3' × 3' × 1' deep concrete footings. The lally columns are 17'-11¹⁄₁₆" OC from North to South and 14'-7" OC East to West.

Two windows are set in window wells in the South wall. The larger window well, in the SE corner, is designed for egress in the event of an emergency. The glass can be easily removed and the window well is large enough to allow a person to exit. The sizes of the windows are given in the Window Schedule.

The stairs have 13 risers (R) up to the main floor. The location of the stairs and stairwell is given on the Floor Plan.

A 200 A service is located on the West wall near the NW corner of the basement. One receptacle is required for one-family basements per 210-52(g). This 20 A, 125 V, 1φ wall receptacle with GFCI protection per 210-8(a)(5) is located on the West wall of the basement. Dedicated receptacles do not require GFCI protection per 210-8(a)(5), Ex. 2. The sump pump (SP) and the ejector pump (EP) are dedicated receptacles.

Lighting for the basement is provided by seven incandescent lamps. These lights have pull switches. One light is located in the stairwell and is controlled by three-way switches located at the top and bottom of the stairs per 210-70(a)(2).

Floor Plan

The Floor Plan shows the overall dimensions and location dimensions for the main floor living area and the two-car garage. The main living area contains approximately 2226 sq ft (42 × 53 = 2226 sq ft). This figure is used to determine the lighting load of 6678 VA (2226 × 3 = 6678) per 220-3(a). When figuring the lighting load, the floor area for dwelling units does not include open porches, garages, or unused or unfinished areas. The floor area is figured using the outside dimensions.

The one-family dwelling is entered from an open Porch (P) through an A door. All doors on this plan are keyed to the Door Schedule by an encircled letter. For example, one A door is required. It is a 3'-0" × 6'-8", right-hand, Therma-Tru, Model # CC45 door. The porch has an incandescent lighting fixture on the wall to the left of the front door. The symbol for a wall-mounted incandescent lighting fixture is a circle with one leg. This lighting fixture, and the two garage front wall exterior lighting fixtures, are controlled by three-way switches near the front door and in the Garage (G) at the B door near the Dining Room (DR). S₃ is the symbol for a three-way switch. A three-way switch used with another three-way switch provides control from two locations. Three-way switches have no OFF-ON markings.

The weatherproof (WP) receptacle on the porch has ground-fault circuit interrupter (GFCI) protection. All outdoor 125 V, 1φ, 15 A and 20 A receptacles shall have ground-fault circuit interrupter protection for personnel per 210-8(a)(3).

Two single-pole switches near door A control ceiling-mounted lighting fixtures in the entry area and the closet. S is the symbol for single-pole switches. The symbol for a ceiling-mounted incandescent lighting fixture is a circle with four legs.

The entry area contains a coat closet with a shelf. The shelf is represented by hidden lines because it is over 5'-0" above the floor. The coat closet has a 6'-0" × 6'-8" bypass door. This model door is mirrored. See the M door in the Door Schedule. The entry area is separated from the Living Room (LR) by a kneewall of glass block (GLB). Details are referenced on this floor plan by an encircled number with a darkened arrow indicating the direction of sight. See Detail 3 for an elevation view of the kneewall.

An F door leads to the unfinished basement. This is a left-hand reverse door (LHR). See the Door Schedule. There are 13 risers (R) down to the basement. A ceiling-mounted lighting fixture is controlled by three-way switches located at the top of the stairs and at the foot of the stairs in the unfinished basement. See the Foundation and Unfinished Basement Plan.

The Living Room (LR) has an A window on the front (East) wall and a B window on the North wall. The Window Schedule gives the quantity, sash size, rough openings, and other information for all windows. See the Window Schedule.

Two ceiling-mounted incandescent lighting fixtures are shown in the Living Room. The B in one of the symbols indicates that the fixture is blanked out. This fixture can be used later when a lighting fixture is hung. The active ceiling-mounted incandescent lighting fixture is controlled by three-way switches located near the stairs and near the Pantry (PAN.) in the Kitchen (KIT.).

Receptacle outlets are placed so that no point along the floor line in any unbroken wall space is more than 6'-0" measured horizontally from an outlet per 210-52(a)(1). These duplex receptacle outlets are shown by a circle with two legs. The split-wired duplex receptacle outlet on the East

wall beneath Window A is split-wired as shown by the darkened area in the symbol. One receptacle is controlled by a wall-mounted switch and the other receptacle remains hot. A television outlet is shown on the North Wall. The symbol for a television outlet is a square with the letters TV inscribed.

The Kitchen (KIT.) and Dining Area (DR) are entered from the Living Room (LR), the Garage (G), the laundry room (LAU), or through sliding doors on the West side of the dwelling. See Code E of the Door Schedule. The Dining Area has a ceiling-mounted incandescent lighting fixture controlled by three-way switches located near the B door to the Garage and on the wall near the Pantry (P). Receptacle outlets are provided as required.

The Kitchen (KIT.) is L-shaped with an island counter and a Pantry (PAN.). The Pantry contains a ceiling lampholder with a pull switch (PS). The ceiling-mounted incandescent lighting fixture in the Kitchen is controlled by three-way switches on both sides of the Pantry door. See Code N of the Door Schedule. A receptacle outlet and a telephone jack are located in the Pantry wall near the GLB kneewall. This area is designed for a small desk. The symbol for a telephone jack is a triangle with a base line.

The island counter holds a 5.8 kW, 240 V, counter-mounted electric cooktop. This appliance is on a minimum branch circuit of 30 A per 210-19(a). The 240 V range receptacle outlet is shown by a circle with three legs and an R. A receptacle outlet is shown on each end of the island counter. These receptacle outlets serve the countertop of the island per 210-52(c)(2); 210-52(c)(3); and 210-52(c)(5), Ex. (b). Because these receptacles serve kitchen countertops, they have GFCI protection per 210-8(a)(6).

The South wall of the Kitchen shows base cabinets, wall cabinets, and the refrigerator. See Detail 1, South Wall Elevation. The receptacle outlet for the refrigerator does not have GFCI protection because it is a dedicated receptacle and does not serve the countertop. Two receptacle outlets along the South wall are intended for countertop use and have GFCI protection per 210-8(a)(6). One receptacle outlet does not have GFCI protection. This receptacle outlet is located in a wall cabinet, is dedicated to an under-the-cabinet light, and is over 5½' above the floor. This receptacle outlet does not count as part of the required countertop receptacles. Two soffit-mounted incandescent lighting fixtures are controlled by switches located in the West wall.

The West wall of the Kitchen shows base cabinets, wall cabinets, sink, and dishwasher. See Detail 2, West Kitchen Elevation. The sink is centered on a D window. See the Window Schedule. The dishwasher (DW) is located to the right of the sink. The symbol for a dishwasher is a triangle within a circle with one leg. The letters DW are placed beside the symbol. Two receptacle outlets with GFCI protection serve the countertop. One soffit-mounted incandescent lighting fixture is controlled by a switch located in the West wall. This is the same switch that controls the two soffit-mounted incandescent lighting fixtures on the South wall. Individual switches near the E door control three exterior wall-mounted incandescent lighting fixtures on the West wall of the dwelling. See the Door Schedule.

The Laundry (LAU) is conveniently located as a pass-through from Bedroom 1 (BR1) and the Kitchen. Plumbing economy is realized by the proximity of the Laundry, Baths 1 and 2, and the Kitchen. The K doors of the Laundry are pocket doors. See the Door Schedule. The ceiling-mounted incandescent lighting fixture is controlled by three-way switches near each door. A wall receptacle outlet is provided near the shelving on the West wall per 210-52(f). The washing machine (WM) and gas dryer (D) are located on the East wall next to the tubs. The appliance outlet is within 6'-0" of the laundry equipment per 210-50(c).

Bathroom 1 (B1) has an I door and B2 has an H door. See the Door Schedule. B1 has a small bumpout to the rear of the house. GLB provides natural daylight in B1. B1 has a basin, toilet, and shower. B2 has a basin, toilet, and tub. Each of these meets the definition of a bathroom. See 100.

Bathrooms 1 and 2 (B1 and B2) have a ceiling-mounted incandescent lighting fixture, a fan, and a heater in their ceilings. The symbol for the fan is the letter F inside a square. The symbol for the heater is the letter H inside a square. All ceiling fixtures are controlled by switches near the bathroom doors. See the Door Schedule for I and H doors. All receptacle outlets in the bathrooms are GFCI-protected per 210-8(a)(1). At least one 20 A branch circuit shall be provided to supply the bathroom receptacle outlets. Incandescent track lighting is shown on the East wall of B1. The symbol

for incandescent track lighting is a rectangle with inscribed circles and one leg. The bathtub (BT) in B2 has a dedicated receptacle for a whirlpool. This is on a separate 15 A circuit. See the Panel Schedule.

BR1, BR2, and BR3 have G doors and C windows. See the Door Schedule and the Window Schedule. A large walk-in closet (WIC) is entered through an I door from BR1. Dashed lines indicate shelving on three walls. A Scuttle (S) is shown in the ceiling with a Lampholder (L) and a Pull Switch (PS) to light the attic area per 210-70(a). A ceiling-mounted incandescent lighting fixture for the walk-in closet is controlled by a wall switch in BR1.

The three bedrooms have ceiling-mounted incandescent lighting fixtures controlled by wall switches. All receptacle outlets in the bedrooms are installed per 210-52(a)(1). BR1 and BR3 have telephone jacks. BR1 also has a TV jack on the East wall.

A cutting plane line is shown in the East wall of BR3. This refers to Section A-A which shows details of the foundation footing, foundation wall, exterior wall, ceiling, and roof. Specific materials required are noted. The 4" perforated drain tile ties into the Sump Pump (SP) shown on the Foundation and Unfinished Basement Plan.

Elevation Views

The Elevation Views show East, West, North, and South elevations. The slope symbol on the East Elevation shows that the roof over the front porch has a rise of 8" for every 12" of run. The main roof has a rise of 6" for every 12" of run. The slope symbol is a right triangle with the run given on the horizontal leg and the rise given on the vertical leg.

Sections and Details

Section A-A is taken in the East wall of the Floor Plan. This section shows the foundation footing through the roof shingles. A trussed roof is specified.

Details of three interior elevations are given. Detail 1 and Detail 2 show details in the kitchen and Detail 3 shows details of the South LR wall. See the Floor Plan for the reference marks for these details.

PLOT PLAN

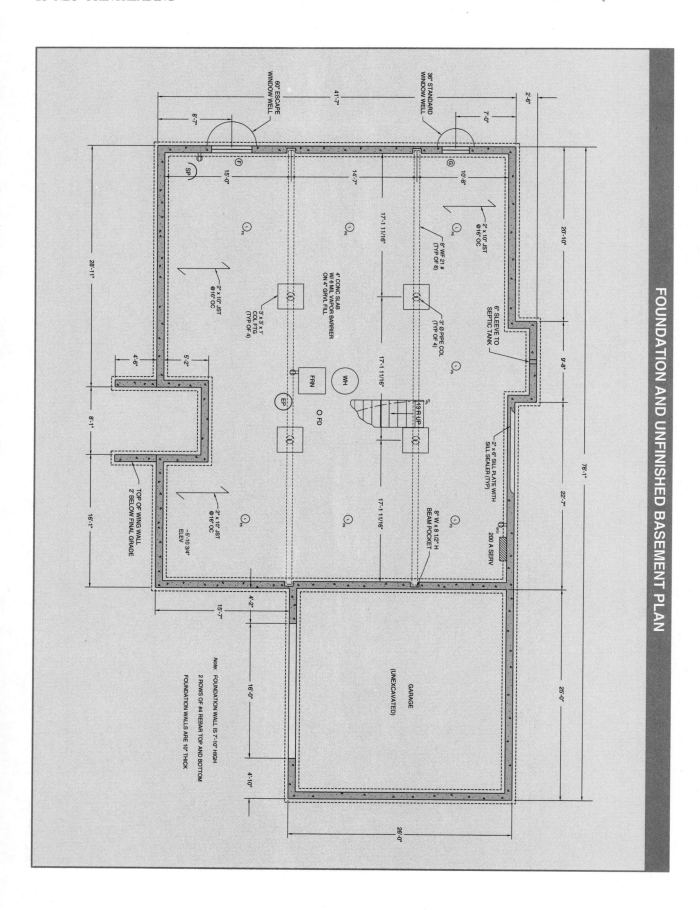

FOUNDATION AND UNFINISHED BASEMENT PLAN

FLOOR PLAN

ELEVATIONS

SECTION AND DETAILS

CELOTEX SHADOWBAND
DIMENSIONAL/25 SHAKE SHINGLES

1/2" PLYWOOD

TRUSSES DESIGNED
BY MANUFACTURER

5/8" DRYWALL

1" x 8" FASCIA
BOARD

ALUMINUM FASCIA

2" x 4"
CHORD

2" x 6" DOUBLE
TOP PLATE

ALUMINUM
SOFFIT PANEL

2" x 2" LEDGER

R19 INSULATION

GEORGIA-PACIFIC
D/4" VINYL SIDING

1/2" DRYWALL

TYVEK HOUSE
WRAP

2" x 6" PRE-CUT
STUD

1" CELOTEX

3/4" T & G PLYWOOD

2" x 6" BOTTOM
PLATE

2" x 10" BOX
JOIST

2" x 6" SILL
PLATE

10" x 7'-10"
FOUNDATION
WALL

GRADE

2" x 10"
FLOOR JOIST

ANCHOR
BOLT

NO. 4 REBAR

10" x 24"
FOOTING

4" PERFORATED
DRAIN TILE

SECTION A-A

DETAIL 1
S KIT. WALL

VAULTED
CEILING

Ⓐ

OAK
TRIM

GLASS BLOCK PATTERN
FOR KNEE WALL

DETAIL 2
W KIT. WALL

DETAIL 3
S LR WALL

SCHEDULE

DOOR SCHEDULE

CODE	QUANTITY	SIZE	SWING	REMARKS
Ⓐ	1	3^0 x 6^8	RH	THERMA-TRU MODEL # CC45
Ⓑ	1	3^0 x 6^8	LH	STEEL
Ⓒ	1	3^0 x 6^8	RH	STEEL
Ⓓ	1	16 x 8	——	WAYNE DALTON MODEL #
Ⓔ	1	6^0 x 6^8	SLIDING	HURD MILLWORK MODEL # CHES-6
Ⓕ	1	2^6 x 6^8	LHR	HOLLOW CORE OAK LAMINATE
Ⓖ	3	2^6 x 6^8	RH	HOLLOW CORE OAK LAMINATE
Ⓗ	1	2^4 x 6^8	RH	HOLLOW CORE OAK LAMINATE
Ⓘ	2	2^4 x 6^8	LH	HOLLOW CORE OAK LAMINATE
Ⓙ	1	2^0 x 6^8	RHR	HOLLOW CORE OAK LAMINATE
Ⓚ	2	2^6 x 6^8	POCKET	HOLLOW CORE OAK LAMINATE
Ⓛ	2	5^0 x 6^8	BYPASS	STANLEY MONARCH MODEL # 1205980AGR
Ⓜ	1	6^0 x 6^8	BYPASS	STANLEY MONARCH MODEL # 50219961
Ⓝ	1	5^0 x 6^8	BI-FOLD	ARROW DOOR OAK LAMINATE MODEL # 413-1166

PANEL SCHEDULE

LEFT	RIGHT
20 A KIT. APPLIANCES	20 A DR RCPT & GAR GFCI
20 A KIT. APPLIANCES	15 A GAR LTS
15 A REF	15 A LR RCPT
15 A DW	15 A KIT. & LR LTS
15 A BR1 & HALLWAY & SD	SPARE
15 A BR1 & BR2	20 A LAU
15 A YARD SPOTS	20 A B1 & B2 GFCI
15 A BSMT S	15 A WHIRLPOOL TUB
20 A B1 HEATER	15 A FUR.
20 A B2 HEATER	15 A EP
30 A OVEN	15 A BSMT N
	SPARE
20 A WELL PUMP	30 A A/C
20 A BSMT GFCI	100 A GAR
SPARE	

WINDOW SCHEDULE

CODE	QUANTITY	SASH SIZE	ROUGH OPENING	REMARKS
◇A	1	2'-4"/5'-8"/2'-4" x 5'-2"	10'-9" x 5'-5"	HURD MILLWORK MODEL # 24-64-2458 24" FLANKERS/CLAD WOOD DOUBLE HUNG PICTURE COMBINATION
◇B	1	3'-8" x 5'-2"	3'-10" x 5'-5"	HURD MILLWORK MODEL # 40x28 CLAD WOOD DOUBLE HUNG
◇C	5	3'-8" x 3'-10"	3'-10" x 4'-1"	HURD MILLWORK MODEL # 40x20 CLAD WOOD DOUBLE HUNG
◇D	1	2'-8" x 3'-2"	2'-10" x 3'-5"	HURD MILLWORK MODEL # 28x16 CLAD WOOD DOUBLE HUNG
◇E	2	2'-0" x 3'-2"	2'-2" x 3'-5"	HURD MILLWORK MODEL # 20x16 CLAD WOOD DOUBLE HUNG
◇F	1	31" x 37"—10" GALV ESCAPE BUCK FRAME	——	KEWANEE WINDOW MODEL # ASV36G STANDARD ESCAPE INSERT
◇G	1	31" x 20"—10" GALV STD BUCK FRAME	——	KEWANEE WINDOW MODEL # AOV-20G STANDARD WINDOW INSERT

Chapter 1

Printreading

Trade Competency Test **1**

NEC®	Answer		

_____   T   F   **1.** The purpose of the NEC® is to protect people and property from hazards that arise from the use of electricity.

xxxxx   T   F   **2.** The key to successful NEC® proposals is proper substantiation for the proposed changes.

xxxxx   _____   **3.** The NEC® is available in _____ versions.
  A. soft cover   C. PC
  B. loose-leaf   D. A, B, and C

xxxxx   _____   **4.** Any material in the current edition of the NEC® that has changed from the previous edition is identified by a(n) _____ line in the margin.

_____   _____   **5.** The NEC® contains an Introduction and _____ chapters.

xxxxx   _____   **6.** A(n) _____ is a detailed plan of a building which is drawn orthographically using a conventional representation of lines, and including abbreviations and symbols.

xxxxx   _____   **7.** A(n) _____ line is always drawn with a series of dashes of equal length.

xxxxx   _____   **8.** The point of _____ is a fixed point from which all measurements are made to show the building lot.

xxxxx   _____   **9.** An elevation is an orthographic drawing showing _____ planes of a building.

_____   T   F   **10.** The NEC® is not intended to be used as an instruction manual.

_____   _____   **11.** All applications of the NEC® are ultimately subject to approval by the _____.

xxxxx   _____   **12.** The most common section view of a building is taken through a(n) _____.
  A. inside partition   C. exterior door
  B. outside wall   D. window

xxxxx   _____   **13.** Cutting planes for section views are shown on the _____.
  A. plot plan   C. both A and B
  B. floor plan   D. neither A nor B

xxxxx   _____   **14.** The Construction Specifications Institute has developed standard procedures that are helpful in writing _____.

xxxxx   _____   **15.** A(n) _____ is the shortened form of a word or phrase.

_____   T   F   **16.** The members of each CMP represent labor, manufacturing, electrical utilities, electrical inspectors, contractor associations, and testing laboratories.

Electrical Symbols

_____ **1.** Ceiling outlet box and incandescent lighting fixture

_____ **2.** Continuous fluorescent fixture

_____ **3.** Duplex receptacle outlet

_____ **4.** Three-way switch

_____ **5.** Home run

_____ **6.** Thermostat

_____ **7.** Motor

_____ **8.** Underground direct burial cable

_____ **9.** Power panel

_____ **10.** Range outlet

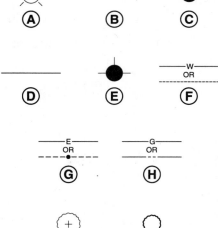

Plot Plan Symbols

_____ **1.** Fire hydrant

_____ **2.** Light standard

_____ **3.** Point of beginning

_____ **4.** Tree

_____ **5.** Bush

_____ **6.** Water line

_____ **7.** Natural grade

_____ **8.** Finish grade

_____ **9.** Electrical service

_____ **10.** Natural gas line

Name _____

Date _____

Trade Competency Test 2

NEC®	Answer	

_____ XXXXX _____   _____   **1.** The first step in the Code process is the receipt of _____.

_____ XXXXX _____   _____   **2.** There are a total of _____ Code-Making Panels.

_____   T   F   **3.** The NEC® is a design specification.

_____ XXXXX _____   _____   **4.** Each print of a set of working drawings represents a part of a building drawn as a(n) _____ projection.

_____ XXXXX _____   _____   **5.** The cutting plane for a floor plan is generally taken _____ above the floor being shown.

    A. 3'-0"   C. 5'-0"
    B. 4'-0"   D. 8'-0"

_____   _____   **6.** No point of unbroken wall space _____ or more in length measured horizontally shall be more than _____' from a receptacle in a living room of a dwelling unit.

    A. 12"; 6   C. 2'; 6
    B. 18"; 12   D. 2'; 12

_____ XXXXX _____   _____   **7.** Division _____ of the CSI format pertains to electrical work.

_____ XXXXX _____   _____   **8.** A(n) _____ is a graphic representation of an object.

_____ XXXXX _____   _____   **9.** A detail is part of a(n) _____ view drawn at a larger scale.

    A. plan   C. section
    B. elevation   D. either A, B, or C

_____ XXXXX _____   _____   **10.** The location of all exterior walls and interior partitions is shown and dimensioned on the _____ plans.

_____   _____   **11.** The first _____ chapters of the NEC® apply generally.

_____ XXXXX _____   T   F   **12.** The Index is copyrighted separately from the NEC®.

_____ XXXXX _____   T   F   **13.** The location of water, sewer, gas, and electrical utilities is given on the plot plan.

_____ XXXXX _____   _____   **14.** A(n) _____ section is taken lengthwise through a building.

_____ XXXXX _____   T   F   **15.** All prints are drawn to ¼" = 1'-0" scale.

_____ XXXXX _____   _____   **16.** The NEC® is _____.

    A. a legal document   C. both A and B
    B. designed to be adopted by   D. neither A nor B
       local and/or state governmental
       bodies

Architectural Symbols

_____ 1. The section symbol for cast iron is shown at _____.

_____ 2. The section symbol for aluminum is shown at _____.

_____ 3. The elevation symbol for ceramic tile is shown at _____.

_____ 4. The section symbol for steel is shown at _____.

_____ 5. The section symbol for bronze or brass is shown at _____.

_____ 6. The section symbol for earth is shown at _____.

_____ 7. The elevation symbol for glass block is shown at _____.

_____ 8. The elevation symbol for concrete is shown at _____.

_____ 9. The plan symbol for loose fill insulation is shown at _____.

_____ 10. The elevation symbol for brick is shown at _____.

Ⓐ Ⓑ Ⓒ Ⓓ Ⓔ

Ⓕ Ⓖ Ⓗ Ⓘ Ⓙ

Abbreviations

_____ 1. The abbreviation for volt is _____.

_____ 2. The abbreviation for specification is _____.

_____ 3. The abbreviation for circuit breaker is _____.

_____ 4. The abbreviation for current is _____.

_____ 5. The abbreviation for dryer is _____.

Name _____

Date _____

NEC®	Answer	

XXXXX   T   F   **1.** Anyone can submit a proposal to change the NEC® provided it contains the required information.

_____   _____   **2.** All 15 A and 20 A receptacles installed in dwelling bathrooms shall have _____ protection.

_____   _____   **3.** Chapter _____ contains tables and examples which are referenced throughout the entire NEC®.

_____   T   F   **4.** The AHJ must be a government employee.

_____   _____   **5.** The use of the word "_____" indicates mandatory rules of the NEC®.

XXXXX   T   F   **6.** Electrical information is shown directly on the floor plans for small sets of prints.

XXXXX   _____   **7.** In the event of a discrepancy between specifications and prints, the _____ take precedence.

XXXXX   _____   **8.** Symbols are _____ in that the symbol can be understood no matter what language a person speaks.

XXXXX   _____   **9.** The abbreviation for overload is _____.

XXXXX   _____   **10.** NEC® _____ numbers are designated by a dash following the article number.

XXXXX   _____   **11.** _____ permit alternate methods to the general rule.
- A. Subsections
- B. FPNs
- C. Exceptions
- D. neither A, B, nor C

XXXXX   T   F   **12.** Page numbers are referenced in the Index to the NEC®.

XXXXX   T   F   **13.** The common practice today is to use a computer software program such as CAD to produce drawings.

XXXXX   _____   **14.** A plot plan _____.
- A. is an orthographic drawing
- B. shows the location of the structure
- C. shows the contour of the land
- D. A, B, and C

XXXXX   _____   **15.** A(n) _____ section is a section taken crosswise.

XXXXX   _____   **16.** On a floor plan, objects _____ the 5'-0" elevation are shown with _____ lines.
- A. above; hidden
- B. below; object
- C. both A and B
- D. neither A nor B

Abbreviations

_____   **1.** The abbreviation for benchmark is _____.

_____   **2.** The abbreviation for basement is _____.

T   F   **3.** A is the abbreviation for amps, anode, and area.

_____   **4.** The abbreviation for dining room and door is _____.

_____   **5.** The abbreviation for ground-fault circuit interrupter is _____.

_____   **6.** _____ is the abbreviation for horsepower.

_____   **7.** The abbreviation for single-pole circuit breaker is _____.

_____   **8.** _____ is the abbreviation for outside diameter.

T   F   **9.** The abbreviation for surface one side is S1S.

T   F   **10.** The abbreviation for transformer is T, TRANS, or XFMR.

_____   **11.** _____ is the abbreviation for watts or west.

_____   **12.** The abbreviation for three-wire is _____.

_____   **13.** The abbreviation for number is _____.

_____   **14.** The abbreviation for overcurrent protection device is _____.

_____   **15.** _____ is the abbreviation for point of beginning.

Alphabet of Lines

_____   **1.** Object line

_____   **2.** Hidden line

_____   **3.** Centerline

_____   **4.** Dimension line

_____   **5.** Extension line

_____   **6.** Leader

_____   **7.** Cutting plane

_____   **8.** Break line

Name _____

Date _____

Trade Competency Test 4

Refer to Floor Plan on page 19.

_____ 1. The walk-in closet entered from BR1 measures 5'-8" × _____.

_____ 2. B1 has a total of _____ duplex receptacle outlets with GFCI protection.

_____ 3. The duplex receptacle outlet on the front porch is _____.

      A. weatherproof   C. both A and B

      B. GFCI-protected   D. neither A nor B

_____ 4. The stairs to the basement are _____ wide.

T   F   5. The ceiling-mounted incandescent lighting fixture in the LR is controlled by a switch to the left of the front door.

T   F   6. Hosebibbs are located on N, E, and W exterior walls.

_____ 7. Cutting plane A-A indicates a _____ section.

      A. transverse   C. wall

      B. longitudinal   D. neither A, B, nor C

_____ 8. C windows on the West wall are spaced _____ OC.

_____ 9. The front of the one-family dwelling faces _____.

_____ 10. The duplex receptacle outlet near the telephone in the KIT. does not require GFCI protection because it does not serve a(n) _____.

_____ 11. The duplex receptacle outlet beneath the A window in the LR is _____.

      A. GFCI-protected   C. 220 V

      B. split-wired   D. neither A, B, nor C

_____ 12. A total of _____ fluorescent fixtures are ceiling-mounted in the GAR.

T   F   13. All rooms in the one-family dwelling contain ceiling-mounted incandescent lighting fixtures.

_____ 14. TV jacks are located in the _____.

      A. LR and KIT.   C. LR and BR1

      B. KIT. and BR2   D. neither A, B, nor C

_____ 15. The ceiling-mounted incandescent lighting fixture in the stairway to the basement is controlled by a _____.

      A. pull switch   C. wall switch at foot of stairs

      B. wall switch at head of stairs   D. both B and C

_____ **16.** The range in the island is _____.

    A. gas-operated   C. hard-wired

    B. connected to a 230 V receptacle   D. connected to a 115 V receptacle

_____ **17.** Hallways in the one-family dwelling are _____ wide.

_____ **18.** The _____ is located in the hallway.

    A. smoke detector   C. doorbell

    B. thermostat   D. A, B, and C

_____ **19.** The garage has an area of _____ sq ft.

    A. 400   C. 600

    B. 450   D. 650

_____ **20.** A total of _____ duplex receptacle outlets are located in BR1.

T   F   **21.** All bedrooms have telephone jacks.

T   F   **22.** The PAN. doors are bi-fold doors.

T   F   **23.** The ceiling fan in BR1 is switch-controlled.

T   F   **24.** All bedroom entrance doors are the same size.

T   F   **25.** The duplex receptacle outlets in the DR are GFCI-protected.

Section A-A

Refer to Section A-A on page 21.

T   F   **1.** Double 4″ vinyl siding is used.

T   F   **2.** The foundation wall is 10″ thick.

T   F   **3.** The drain tile is perforated.

_____ **4.** The R value of insulation in the exterior walls is R _____.

_____ **5.** The 1″ x _____″ fascia board is covered with aluminum.

_____ **6.** The nominal size of the rafters is _____.

_____ **7.** Exterior wall studs are 2″ × _____″.

_____ **8.** The sill plate is attached to the top of the foundation walls with _____.

T   F   **9.** The ceiling and walls are covered with ½″ drywall.

T   F **10.** The soffit panel is aluminum.

ONE-FAMILY DWELLINGS 2

A *one-family dwelling* is a building containing one dwelling unit. The service for a one-family dwelling is commonly 120/240 V, 1ϕ.

ONE-FAMILY DWELLINGS

A *one-family dwelling* is a building consisting solely of one dwelling unit. A *dwelling unit* is one or more rooms used by one or more persons for housekeeping with space for eating, living, and sleeping and having permanent provisions for cooking and sanitation. To meet this definition, kitchen equipment and a bathroom are essential. Kitchen equipment includes appliances for storing food (refrigerators and freezers) and cooking food (stoves, ovens, etc.). A *bathroom* is an area with a basin and a toilet, tub, or shower.

The service for a one-family dwelling is commonly 120/240 V, 1ϕ. It may be either overhead or underground. The utility company installs the service to the dwelling. The *service point* is the connection point between the utility and the premises wiring. *Service equipment* is the necessary equipment to control and cut off the electrical supply. The service equipment consists of CBs or a

switch and fuses normally located near the service point. Electricians install the service equipment when they wire the dwelling.

CALCULATIONS

"Calculation" is a widely used but undefined term in the NEC®. Typically, the term is used to refer to a set of procedures which are followed to arrive at a given value.

The Standard Calculation and the Optional Calculation are the two methods to calculate the load for one-family and multifamily dwellings per 220. Forms designed specifically for determining the Standard Calculation and the Optional Calculation standardize and simplify the operation. See Figure 2-1. In general, the Optional Calculation is less complex and easier to use than the Standard Calculation, and is used with larger loads. In either case, the calculation can be used to determine feeder or service loads.

STANDARD CALCULATION: ONE-FAMILY DWELLING

1. **GENERAL LIGHTING:** *Table 220-3(a)*
   _____ sq ft × 3 VA = _____ VA
   Small appliances: 220-16(a)
   _____ VA × _____ circuits = _____ VA
   Laundry: 220-16(b)
   _____ VA × 1 = _____ VA
   _____ VA

   © 1988 by American Technical Publishers, Inc. This form may be reproduced for instructional use only. It shall not be reproduced and sold.

   Applying Demand Factors: *Table 220-11*
   First 3000 VA × 100% = 3000 VA
   Next _____ VA × 35% = _____ VA   **PHASES**   **NEUTRAL**
   Remaining _____ VA × 25% = _____ VA
   Total _____ VA   _____ VA   _____ VA

2. **FIXED APPLIANCES:** *220-17*
   Dishwasher = _____ VA
   Disposer = _____ VA
   Compactor = _____ VA
   Water heater = _____ VA
   _____ = _____ VA
   _____ = _____ VA
   _____ = _____ VA   (120 V Loads × 75%)
   Total _____ VA × 75% = _____ VA   _____ VA

3. **DRYER:** *220-18; Table 220-18*
   _____ VA × _____% = _____ VA   _____ VA × 70% = _____ VA

4. **COOKING EQUIPMENT:** *Table 220-19; Notes*
   Col A _____ VA × _____% = _____ VA
   Col B _____ VA × _____% = _____ VA
   Col C _____ VA × _____% = _____ VA
   Total _____ VA   _____ VA × 70% = _____ VA

5. **HEATING or A/C:** *220-21*
   Heating unit = _____ VA × 100% = _____ VA
   A/C unit = _____ VA × 100% = _____ VA
   Heat pump = _____ VA × 100% = _____ VA
   Largest Load _____ VA   _____ VA   _____ VA

6. **LARGEST MOTOR:** *220-14*
   φ _____ VA × 25% = _____ VA   _____ VA
   N _____ VA × 25% = _____ VA   _____ VA

   1φ service: PHASES   I = VA/V = _____ A
   NEUTRAL   I = VA/V = _____ A   [_____ VA]   [_____ VA]

   220-22: First 200 A × 100% = 200 A
   Remaining _____ A × 70% = _____ A
   Total _____ A

OPTIONAL CALCULATION: ONE-FAMILY DWELLING

1. **HEATING and A/C:** *220-30(c)(1 – 6)*
   Heating units (3 or less) = _____ VA × 65% = _____ VA
   Heating units (4 or more) = _____ VA × 40% = _____ VA

   © 1988 by American Technical Publishers, Inc. This form may be reproduced for instructional use only. It shall not be reproduced and sold.

   A/C unit = _____ VA × 100% = _____ VA
   Heat pump = _____ VA × 100% = _____ VA   **PHASES**
   Largest Load _____ VA
   Total _____ VA   _____ VA

2. **GENERAL LOADS:** *220-30(b)(1 – 4)*
   General lighting: *220-30(b)(2)*
   _____ sq ft × 3 VA = _____ VA
   Small appliance and laundry loads: *220-30(b)(1)*
   _____ VA × _____ circuits = _____ VA
   Special loads: *220-30(b)(3)(4)*
   Dishwasher = _____ VA
   Disposer = _____ VA
   Compactor = _____ VA
   Water heater = _____ VA
   _____ = _____ VA
   _____ = _____ VA
   _____ = _____ VA
   _____ = _____ VA
   _____ = _____ VA   Total _____ VA

   Applying Demand Factors: *220-30(b)*
   First 10,000 VA × 100% = 10,000 VA
   Remaining _____ VA × 40% = _____ VA
   Total _____ VA   _____ VA

   NEUTRAL (Loads from Standard Calculation)
   1. General lighting = _____ VA
   2. Fixed appliances = _____ VA
   3. Dryer = _____ VA
   4. Cooking equipment = _____ VA
   5. Heating or A/C = _____ VA
   6. Largest motor = _____ VA
   Total [_____ VA]

   1φ service: PHASES   I = VA/V = _____ A
   NEUTRAL   I = VA/V = _____ A   [_____ VA]

Figure 2-1. The Standard Calculation contains six loads. The Optional Calculation is an easier method for calculating larger loads.

Demand is the amount of electricity required at a given time. The concept behind the use of the term is that the total calculated load is rarely placed on the electrical system. This is because all of the electrical loads are not on at the same time. For example, a household electric range may have a nameplate rating of 14 kW. The electric range has four burners, one broiler, one heating element, and accessories such as fans, lights, and timers. Rarely, if ever, would all of these be used at the same time. Therefore, the concept of demand permits the 14 kW range to be considered as a smaller load for the purpose of determining how much load to allow for the range when calculating the total feeder or service load for the building.

Noncoincident loads are loads that are unlikely to be in use at the same time. For example, heating and air conditioning are noncoincident loads. It is permissible to drop the smaller of the two loads when completing the total load of a feeder. Heating or A/C is Load 5 on the Standard Calculation: One-Family Dwelling form and Load 1 on the Optional Calculation: One-Family Dwelling form.

One-Family Dwellings – Standard Calculation

The Standard Calculation for One-Family Dwellings contains six individual calculations that are performed before the minimum size service or feeder conductors required for the calculated load can be determined. Demand factors may be applied to four of these loads. These six loads, their NEC® references, and those to which demand factors may be applied are:

1. General Lighting – *Table 220-3(a)* (Demand Factors)
2. Fixed Appliances – *220-17* (Demand Factors)
3. Dryer – *220-18; Table 220-18* (Demand Factors)
4. Cooking Equipment – *Table 220-19* (Demand Factors)
5. Heating or A/C – *220-21*
6. Largest Motor – *220-14*

One-Family Dwelling – Optional Calculation

The Optional Calculation is designed for dwellings with large electric loads. It is an easier method than the Standard Calculation for calculating feeder and service loads for dwellings. The Optional Calculation is permitted for the ungrounded conductors of 120/240 V, 3-wire or 120/208 V service-entrance or feeder conductors with an ampacity of at least 100 A.

The grounded conductor calculation for the Optional Calculation is determined per 220-22. The Optional Calculation for dwelling units consists of two groups of loads. These two groups of loads and their NEC® references are:

1. Heating and A/C – *220-30(1 – 6)*
2. General Loads – *220-30(b)(1 – 4)*

The Optional Calculation can be used for large-capacity dwelling units. It permits designers of electrical systems to take advantage of the increased diversity and demand factors with these types of dwellings and utilize smaller service-entrance or feeder conductors.

Dwelling A

Dwelling A is a one-story, one-family dwelling. It has a living room (LR), kitchen (KIT.), laundry (LAU), one bathroom (B), three bedrooms (BR), and an entrance foyer leading to a hallway. These abbreviations, and others used on the plan, are standard abbreviations. See Appendix. Note that some architects may use abbreviations which are slightly different. Any abbreviation that forms a word is followed by a period. For example, KIT. is followed by a period.

Standard symbols are used to show building materials, fixtures, etc. For example, the symbol for a duplex receptacle is a circle with parallel lines passing through the circle. See Appendix.

The outside dimensions of the dwelling are 55′ x 28′. These dimensions are multiplied to find the area, which is 1540 sq ft (55′ x 28′ = 1540 sq ft). *Area* is the number of units of length multiplied by the number of units of width. Area is expressed in square units such as square inches, square feet, square yards, etc. Square feet are commonly used to express the area for dwellings.

Table 220-3(a) gives the General Lighting loads by occupancies. The unit load per square foot for dwellings is 3 VA. The General Lighting load is based on the area of the dwelling times the VA (1540 sq ft x 3 VA = 4620 VA), plus the loads for Small Appliances and Laundry loads per 220-3(b)(10). Demand factors are applied to the total per Table 220-11 to determine the total General Lighting load.

By following the Standard Calculation: One-Family Dwelling form, the other five groups of loads can be calculated. The VA for these loads is taken from either the prints or specifications. For example, the loads for Fixed Appliances (220-17), Dryer (220-18; Table 220-18); Cooking Equipment (Table 220-19; Notes); Heating or AC (220-21); and the Largest Motor (220-14) are not given on the print for Dwelling A. For Dwelling A, the loads used to calculate the service are typical for this size dwelling and were taken from specifications which are not shown in this book.

The total VA is divided by the voltage to find the size of the service in amperes (A). The formula used is $I = \dfrac{VA}{V}$, where I = amps, VA = volt-amps, and V = volts. Notice that the Optional Calculation: One-Family Dwelling form can only be completed after the Standard Calculation: One-Family Dwelling form has been completed.

Dwelling B

Dwelling B is also a one-family dwelling. It has a basement in which the den (DEN), laundry room (LAU), and bathroom (B) are finished. The unfinished basement (UNFIN BSMT) area contains the water heater (WH), heating and air conditioning, and the main panelboard. The area of the basement is added to the area of the main floor to determine the area of the dwelling.

Seven window wells are shown for windows in the basement. The windows allow natural light and air to enter the basement and the window wells provide a means of egress in the event of an emergency. The window wells are shown as semicircles on the plan. Again, the loads used to calculate the service are typical for this size dwelling and may be taken from the prints or specifications, depending on where they are given. The specifications for this particular floor plan are not shown in this book.

BRANCH CIRCUIT, FEEDER, AND SERVICE CALCULATIONS

Note: TO AVOID OVERCROWDING OF THE FLOOR PLAN, ALL A-J ARE NOT NOTED.

DWELLING A

ABBREVIATIONS			
B	BATH	**KIT.**	KITCHEN
BR	BEDROOM	**LAU**	LAUNDRY
CLO	CLOSET	**LR**	LIVING ROOM
DR	DINING ROOM	**PNL**	PANEL
FRN	FURNACE	**WH**	WATER HEATER

A. 210-7(a) All receptacles installed on 15 A and 20 A circuits shall be of the grounding type. See Exception.

B. 210-8(a)(3) All 15 A and 20 A, 125 V receptacles installed outdoors of dwelling units shall have GFCI protection.

C. 210-8(a)(1 – 7) Ground-fault protection for personnel in dwellings shall be provided for all 15 A and 20 A, 125 V receptacles in all (1) bathrooms, (2) garages and other buildings used for storage or as workshops, (3) outdoor locations, (4) crawl spaces at or below grade, (5) unfinished basements, (6) kitchen countertops, (7) within 6′ of a wet bar sink intended to serve countertops. See Exceptions.

D. 210-23 Branch circuits may be installed for specific loads (exhaust fan in bath).

E. 210-52(a)(1) All general occupancy rooms in dwellings shall have receptacle outlets installed so that no point along the floor wall line is more than 6′, measured horizontally, from an outlet.

F. 210-52(b)(1) For small appliance load, including all receptacles in kitchen, pantry, breakfast room, dining room, and similar areas, two or more 20 A appliance circuits shall be provided. Such circuits shall have no other outlets. Countertop outlets to be at least two 20 A circuits. Clock outlets and receptacles for supplemental equipment on ranges, ovens, and counter-mounted cooking units are permitted on these circuits. See 210-52(b)(2), Ex. 1, 2. Refrigerator receptacle is permitted on an individual 15 A circuit. See 210-52(b)(1), Ex. 2.

G. 210-52(h) Dwelling hallways more than 10′ in length, measured from the centerline of the hall, without passing through a doorway, shall be provided with at least one receptacle outlet.

H. 210-70(a)(1) Dwelling units shall have at least one wall switch-controlled lighting outlet in every habitable room including bathrooms, hallways, stairways, and at outdoor entrances or exits. In rooms other than kitchen and bathroom, switch-controlled receptacles are permitted in lieu of lighting outlet. See Ex. 1.

I. 220-3(a) Unit lighting load for dwelling occupancies shall be 3 VA per sq ft. See Table 220-3(a). Use outside dimensions of dwelling.

Table 220-3(a) Receptacles other than on the two small appliance circuits are considered part of general illumination and require no allowance for additional load.

J. 210-11(c)(2) At least one 20 A circuit shall be provided for laundry receptacle outlet per 210-52(f).

210-11(b) "VA per sq ft" load shall be apportioned evenly in branch circuits within panelboard.

220-3(a) For general illumination, one 15 A, 120 V branch circuit is required for every 600 sq ft of area. 3 VA per sq ft × 600 = 1800 VA ÷ 120 V = 15 A.

220-16(a) Feeder load for the two small appliance circuits of 210-11(c)(1) is to be taken at 3000 VA (1500 VA for each two-wire circuit). This load may be included with general lighting load in feeder calculations per 220-11.

220-16(b) Feeder load is 1500 VA for each laundry circuit as required by 210-11(c)(2). This load may be included with general lighting load in feeder calculations per 220-11.

230-42(b) In general, service-entrance conductors shall have an ampacity of at least 100 A. See 230-79(c).

422-11(e) Branch circuit to single nonmotor appliance not to exceed OC rating on device. If no appliance rating, 13.3 A permitted on 20 A circuit. If over 13.3 A, the next size OCPD above 150% can be used where it does not correspond to standard OCPD.

DWELLING A – 220, PARTS A AND B

STANDARD CALCULATION: ONE-FAMILY DWELLING

1. GENERAL LIGHTING: *Table 220-3(a)*

<u>1540</u> sq ft × 3 VA = <u>4620</u> VA

Small appliances: *220-16(a)*

<u>1500</u> VA × <u>2</u> circuits = <u>3000</u> VA

Laundry: *220-16(b)*

<u>1500</u> VA × 1 = <u>1500</u> VA

<u>9120</u> VA

Applying Demand Factors: *Table 220-11*

First 3000 VA × 100% = 3000 VA

Next <u>6120</u> VA × 35% = <u>2142</u> VA   **PHASES**   **NEUTRAL**

Remaining <u>—</u> VA × 25% = <u>—</u> VA

Total <u>5142</u> VA   <u>5142</u> VA   <u>5142</u> VA

---

2. FIXED APPLIANCES: *220-17*

Dishwasher = <u>1600</u> VA

Disposer = <u>1200</u> VA

Compactor = <u>900</u> VA

Water heater = <u>1800</u> VA

Exh Fan = <u>350</u> VA

Attic Fan = <u>1200</u> VA

<u>—</u> = <u>—</u> VA

(120 V Loads × 75%)

Total <u>7050</u> VA × 75% = <u>5288</u> VA   <u>5288</u> VA   <u>5288</u> VA

---

3. DRYER: *220-18; Table 220-18*

<u>5800</u> VA × <u>100</u> % = <u>5800</u> VA   <u>5800</u> VA × 70% = <u>4060</u> VA

---

4. COOKING EQUIPMENT: *Table 220-19; Notes*

Col A <u>8000</u> VA × <u>100</u> % = <u>8000</u> VA

Col B <u>—</u> VA × <u>—</u> % = <u>—</u> VA

Col C <u>—</u> VA × <u>—</u> % = <u>—</u> VA

Total <u>8000</u> VA   <u>8000</u> VA × 70% = <u>5600</u> VA

---

5. HEATING or A/C: *220-21*

Heating unit = <u>—</u> VA × 100% = <u>—</u> VA

A/C unit = <u>—</u> VA × 100% = <u>—</u> VA

Heat pump = <u>16,000</u> VA × 100% = <u>16,000</u> VA

Largest Load <u>16,000</u> VA   <u>16,000</u> VA   <u>—</u> VA

---

6. LARGEST MOTOR: *220-14*

φ <u>1200</u> VA × 25% = <u>300</u> VA   <u>300</u> VA

N <u>—</u> VA × 25% = <u>—</u> VA   <u>300</u> VA

1φ service: PHASES $I = \dfrac{40,530 \text{ VA}}{240 \text{ V}} = 168.88$ A

NEUTRAL $I = \dfrac{20,390 \text{ VA}}{240 \text{ V}} = 84.96$ A | 40,530 | VA | 20,390 | VA

220-22; First 200 A × 100% = 200 A

Remaining <u>—</u> A × 70% = <u>—</u> A

Total <u>—</u> A

BRANCH CIRCUIT, FEEDER, AND SERVICE CALCULATIONS

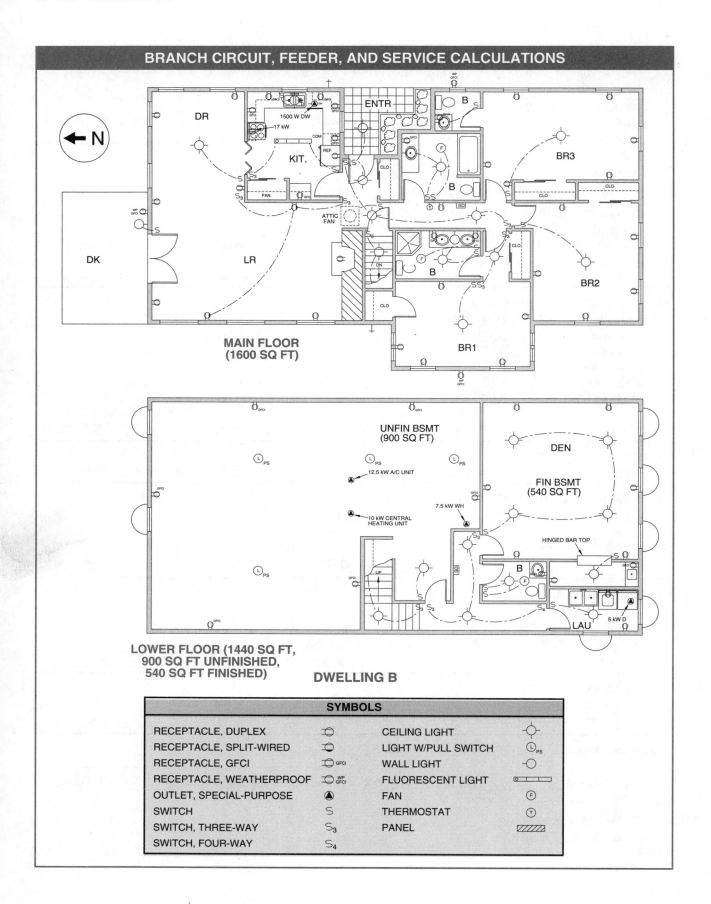

MAIN FLOOR (1600 SQ FT)

LOWER FLOOR (1440 SQ FT, 900 SQ FT UNFINISHED, 540 SQ FT FINISHED)

DWELLING B

SYMBOLS			
RECEPTACLE, DUPLEX		CEILING LIGHT	
RECEPTACLE, SPLIT-WIRED		LIGHT W/PULL SWITCH	
RECEPTACLE, GFCI		WALL LIGHT	
RECEPTACLE, WEATHERPROOF		FLUORESCENT LIGHT	
OUTLET, SPECIAL-PURPOSE		FAN	
SWITCH		THERMOSTAT	
SWITCH, THREE-WAY		PANEL	
SWITCH, FOUR-WAY			

DWELLING B – 220, PARTS A AND B

STANDARD CALCULATION: ONE-FAMILY DWELLING

1. GENERAL LIGHTING: *Table 220-3(a)*

3040 sq ft × 3 VA =   **9120** VA

Small appliances: *220-16(a)*

1500 VA × **2** circuits =   **3000** VA

Laundry: *220-16(b)*

1500 VA × 1 =   **1500** VA

    13,620 VA

Applying Demand Factors: *Table 220-11*

First 3000 VA × 100% =   3000 VA

Next **10,620** VA × 35% =   **3717** VA   **PHASES**   **NEUTRAL**

Remaining **—** VA × 25% =   **—** VA

Total   **6717** VA   **6717** VA   **6717** VA

2. FIXED APPLIANCES: *220-17*

Dishwasher =   **1500** VA

Disposer =   **900** VA

Compactor =   **950** VA

Water heater =   **7500** VA

2 Exh Fans =   **700** VA

Attic Fan =   **1000** VA

    = **—** VA

Total   **12,550** VA × 75% = **9413** VA   **9413** VA   (120 V Loads × 75%) **9413** VA

3. DRYER: *220-18; Table 220-18*

5000 VA × **100** % =   **5000** VA   **5000** VA × 70% = **3500** VA

4. COOKING EQUIPMENT: *Table 220-19; Notes*

Col A **8000** VA × **125** % =   **10,000** VA

Col B **—** VA × **—** % =   **—** VA

Col C **—** VA × **—** % =   **—** VA

Total   **10,000** VA   **10,000** VA × 70% = **7000** VA

5. HEATING or A/C: *220-21*

Heating unit = **10,000** VA × 100% =   **10,000** VA

A/C unit = **12,500** VA × 100% =   **12,500** VA

Heat pump = **—** VA × 100% =   **—** VA

Largest Load   **12,500** VA   **12,500** VA   **—** VA

6. LARGEST MOTOR: *220-14*

φ **1000** VA × 25% =   **250** VA   **250** VA

N **—** VA × 25% =   **—** VA   **250** VA

1φ service: PHASES $I = \dfrac{\textbf{43,880} \text{ VA}}{\textbf{240} \text{ V}} = \textbf{182.83}$ A

NEUTRAL $I = \dfrac{\textbf{26,880} \text{ VA}}{\textbf{240} \text{ V}} = \textbf{112}$ A   **43,880** VA   **26,880** VA

220-22; First 200 A × 100% =   200 A

Remaining **—** A × 70% =   **—** A

Total   **—** A

DWELLING A – 220, PART C

OPTIONAL CALCULATION: ONE-FAMILY DWELLING

1. HEATING and A/C: *220-30(c)(1 – 6)*

Heating units (3 or less) = _____–_____ VA × 65% = _____–_____ VA

Heating units (4 or more) = _____–_____ VA × 40% = _____–_____ VA

A/C unit = _____–_____ VA × 100% = _____–_____ VA

Heat pump = __16,000__ VA × 100% = __16,000__ VA   **PHASES**

Largest Load   __16,000__ VA

Total   __16,000__ VA   __16,000__ VA

2. GENERAL LOADS: *220-30(b)(1 – 4)*

General lighting: *220-30(b)(2)*

__1540__ sq ft × 3 VA   __4620__ VA

Small appliance and laundry loads: *220-30(b)(1)*

__1500__ VA × __3__ circuits =   __4500__ VA

Special loads: *220-30(b)(3)(4)*

Dishwasher = __1600__ VA

Disposer = __1200__ VA

Compactor = __900__ VA

Water heater = __1800__ VA

__Exh Fan__ = __350__ VA

__Attic Fan__ = __1200__ VA

_____–_____ = __–__ VA

_____–_____ = __–__ VA

_____–_____ = __–__ VA

__7050__ VA   __7050__ VA

Total __16,170__ VA

Applying Demand Factors: *220-30(b)*

First 10,000 VA × 100% =   10,000 VA

Remaining __6170__ VA × 40% =   __2468__ VA

Total   __12,468__ VA   __12,468__ VA

NEUTRAL (Loads from Standard Calculation)

1. General lighting =   __5142__ VA

2. Fixed appliances =   __5288__ VA

3. Dryer =   __4060__ VA

4. Cooking equipment =   __5600__ VA

5. Heating or A/C =   __–__ VA

6. Largest motor =   __300__ VA

Total   __20,390__ VA

1φ service: PHASES $I = \dfrac{28,468 \text{ VA}}{240 \text{ V}} = $ __118.62__ A

        NEUTRAL $I = \dfrac{20,390 \text{ VA}}{240 \text{ V}} = $ __84.96__ A   __28,468__ VA

DWELLING B – 220, PART C

OPTIONAL CALCULATION: ONE-FAMILY DWELLING

1. HEATING and A/C: *220-30(c)(1 – 6)*

Heating units (3 or less) = __10,000__ VA × 65% =   __6500__ VA

Heating units (4 or more) = __—__ VA × 40% =   __—__ VA

A/C unit = __12,500__ VA × 100% =   __12,500__ VA

Heat pump = __—__ VA × 100% =   __—__ VA   **PHASES**

Largest Load   __12,500__ VA

Total   __12,500__ VA   __12,500__ VA

2. GENERAL LOADS: *220-30(b)(1 – 4)*

General lighting: *220-30(b)(2)*

__3040__ sq ft × 3 VA   __9120__ VA

Small appliance and laundry loads: *220-30(b)(1)*

__1500__ VA × __3__ circuits =   __4500__ VA

Special loads: *220-30(b)(3)(4)*

Dishwasher = __1500__ VA

Disposer = __900__ VA

Compactor = __950__ VA

Water heater = __7500__ VA

__2 Exh Fans__ = __700__ VA

__Attic Fan__ = __1000__ VA

__—__ = __—__ VA

__—__ = __—__ VA

__—__ = __—__ VA

__12,550__ VA   __12,550__ VA

        Total __26,170__ VA

Applying Demand Factors: *220-30(b)*

First 10,000 VA × 100% =   10,000 VA

Remaining __16,170__ VA × 40% =   __6468__ VA

Total   __16,468__ VA   __16,468__ VA

NEUTRAL (Loads from Standard Calculation)

1. General lighting =   __6717__ VA
2. Fixed appliances =   __9413__ VA
3. Dryer =   __3500__ VA
4. Cooking equipment =   __7000__ VA
5. Heating or A/C =   __—__ VA
6. Largest motor =   __250__ VA

Total   __26,880__ VA

1φ service: PHASES $I = \dfrac{28,968 \text{ VA}}{240 \text{ V}} =$ __120.7__ A

NEUTRAL $I = \dfrac{26,880 \text{ VA}}{240 \text{ V}} =$ __112__ A   __28,968__ VA

CLEARANCES FOR SERVICE DROPS AND SERVICE LATERALS

ROOF SLOPE

230-2 In general, a building or structure shall be supplied by one service. See Exceptions. Several underground sets of No. 1/0 or larger multiple conductors, connected at the source but not at the load, are considered to be only one service. This is allowed to reduce available fault currents. See 230-40, Ex. 2.

230-22 Overhead individual service conductors shall be insulated or covered with extruded thermoplastic or thermosetting insulating material.

230-22, Ex. The grounded conductor (neutral) of multiconductor cables may be bare.

230-24(a) Conductors shall have 8′ minimum clearance above flat roofs and maintain 3′ minimum clearance from roof edge.

230-24(a), Ex. 2 For voltage between service drop conductors of 300 V or less and the roof slope is at least 4″ in 12″, clearances may be reduced to 3′.

230-24(a), Ex. 3 Service-drop conductors of 300 V or less may pass over a roof overhang with a clearance of 18″ or more. Up to 6′ of conductor can

be run in this manner. It shall pass over no more than 4′ of overhang measured horizontally.

230-24(a), Ex. 4 Vertical clearance of 3′ is not required for the conductor span if the service drop is connected to the side of the building.

230-24(b) For service-drops less than 600 V, service entrance or drip loop shall be a minimum of 10′ above finished grade of pedestrian areas or sidewalks. Voltage is limited to 150 V to ground with insulated conductor cabled together and supported with a grounded bare messenger.

For service drops up to 300 V to ground, the minimum clearance over residential property and driveways, and commercial areas not having truck traffic is 12′. For service drops over 300 V to ground, the minimum clearance over residential property and driveways, and commercial areas not having truck traffic, is 15′.

For service drops over public roads, streets, alleys, parking areas with truck traffic, and commercial driveways, the minimum clearance is 18′

230-28 Service mast shall have adequate strength or be supported. All raceway fittings shall be identified for use with service masts. See 230-51.

230-30 Underground service-lateral conductors shall be insulated for the applied voltage.

230-30, Ex. (a) An uninsulated bare copper grounded conductor is permitted in a raceway.

230-30, Ex. (b) An uninsulated bare copper grounded conductor is permitted for direct burial where suitable for soil conditions.

230-30, Ex. (c) An uninsulated bare copper grounded conductor is permitted in any soil if the cable assembly is identified for underground use.

230-30, Ex. (d) An uninsulated bare Al or Cu-clad Al conductor is permitted in a raceway or for direct burial if the conductor is part of a cable assembly identified for underground use.

SERVICE HEADS AND SERVICE-ENTRANCE CONDUCTORS

230-23(b) Service drops shall not be smaller than No. 8 Cu, No. 6 Al, or No. 6 Cu-clad Al.

230-23(b), Ex. For limited loads of a single branch circuit, service drops shall not be smaller than No. 12 hard-drawn Cu or equivalent.

230-41 Service-entrance conductors entering or on a building exterior shall be insulated.

230-41, Ex. Bare Cu conductors is permitted when used:

230-41, Ex. (a) in raceway or as part of service cable assembly.

230-41, Ex. (b) for direct burial if judged suitable for soil conditions.

230-41, Ex. (c) for direct burial in any soil if part of a cable assembly.

230-41, Ex. (d) Bare Al or Cu-clad Al is permitted when used as part of a cable assembly, identified for use in an underground raceway, or for direct burial.

230-42(a) Service-entrance conductors shall have ampacity to carry loads per 220.

230-42(b)(1) Ungrounded conductors shall have an ampacity not less than the sum of the noncontinuous loads plus 125% of the continuous load.

230-42(b)(2) Ungrounded conductors shall have an ampacity of not less than the sum of noncontinuous loads plus the continuous load if the conductors terminate in an OCPD listed for the operation at 100% of the rating.

230-42(c) The grounded conductor shall not be smaller than minimum size per 250-23(b).

230-51(a) Service-entrance cable shall be supported within 12" of service head, gooseneck, or connection to raceway or enclosures and at intervals not exceeding 30".

230-54(a) Service raceways shall have a raintight service head at connection to service-drop conductors.

230-54(b) Service cable shall have a raintight service head. Type SE may be formed into a gooseneck which shall be taped and painted or taped with self-sealing, weather-resistant thermoplastic.

230-54(c) Service heads and goosenecks shall be above point of attachment of service-drop conductors to building.

230-54(c), Ex. Service head is permitted to be located no more than 24" from point of attachment.

230-54(d) Service-drop cables shall be securely fastened in place.

230-54(e) Service conductors of different potential shall be brought out of service head through separately bushed openings.

230-54(e), Ex. Separated bushed openings are not required for jacketed multiconductor service cable without a splice.

230-54(f) Drip loops shall be formed on individual conductors.

230-54(g) Service-drop and service-entrance conductors shall be arranged to prevent water from entering service raceway or equipment.

338-1 Service-entrance cable is a single or multiconductor assembly with or without an overall covering. Type SE has a flame-retardant and moisture-resistant covering. Type USE is identified for underground use. It has a moisture-resistant covering.

Service-drop conductors are required to be insulated or covered per 230-41.

CLEARANCES FOR SERVICE DROPS FROM BUILDING OPENINGS

230-24(b) Service drops shall have a minimum vertical clearance of not less than 10' from a platform provided the conductors are insulated and supported on and cabled with a grounded bare messenger and voltage limited to 150 V to ground.

230-9 Conductors shall have a minimum clearance of not less than 3' from openable windows, doors, porches, balconies, ladders, fire escapes, etc.

230-9, Ex. conductors run above the window are considered as out of reach from the window.

SERVICES — OVERHEAD AND UNDERGROUND

NUMBER OF SERVICES

UTILITY SUPPLY SYSTEM
(OVERHEAD OR UNDERGROUND)

ONE SERVICE
• 230-2

SERVICE FOR
FIRE PUMP
• 230-2(a)(1)

SERVICE-DROP
CONDUCTORS TO
EACH BUILDING

SERVICE FOR
EMERGENCY SYSTEMS
• 230-2(a)(2)

SERVICE FOR
MULTIPLE-OCCUPANCY
• 230-2(b)(1)

SERVICE FOR
LARGE SINGLE
BUILDING
• 230-2(b)(2)

SERVICE FOR
LARGE CAPACITY
• 230-2(c)(1)

ADDITIONAL SERVICE
FOR LARGE BUILDING
• 230-2(b)(2)

ADDITIONAL SERVICE FOR
DIFFERENT CHARACTERISTICS
• 230-2(d)

The service-entrance conductors shall be firmly connected to the mast.

230-2 A building shall be supplied by one service. Service lateral may have several sets of multiple conductors, No. 1/0 and larger, tied together at source but not at load.

230-2(a)(1) For fire pumps.

230-2(a)(2) Additional service for emergency systems as permitted.

230-2(b)(1) More than one service permitted in multiple-occupancy building where no space available to locate service equipment that is accessible to all tenants and then permitted only by special permission.

230-2(b)(2) Additional service shall be permitted for a single building large enough for two services.

230-2(c)(1) Additional services permitted in buildings with large capacity requirements: (a) in excess of 2000 A at 600 V or less; or (b) where 1φ load is in excess of that supplied by serving agency; or (c) by special permission.

230-2(d) Additional service permitted for loads having different voltages, frequencies, phases, or different rate schedules.

230-2(e) Where more than one service, or any combination of branch circuits, feeders, and services supplies a building or structure, a directory or plaque shall be installed denoting all other services, branch circuits, and feeders.

This overhead service is fed through the roof.

230-40, Ex. 2 One set of service-entrance conductors permitted to feed up to 6 disconnects in separate enclosures. Disconnects not required to be all of the same size.

230-71(a) Not more than 6 disconnects in a single enclosure, a group of separate enclosures, or on a switchboard.

230-71(b) Two or three single-pole switches or CBs with "handle ties" or a "master handle" are counted as one disconnect. Not more than six operations of the hand permitted.

UTILITY SUPPLY SYSTEM
(OVERHEAD OR UNDERGROUND)

UP TO 6 DISCONNECTS
IN SEPARATE ENCLOSURES
• 230-40, Ex. 2

SERVICE-
ENTRANCE
CONDUCTORS

NOT OVER 6 DISCONNECTS
IN SINGLE ENCLOSURE
• 230-71(a)

SERVICE-
ENTRANCE
CONDUCTORS

INSTALLATION OF SERVICE-ENTRANCE CONDUCTORS

WIRING
METHODS
• 230-43

METER
SOCKET

230-43 Service-entrance conductors are permitted to be installed as (1) open wiring on insulators, (2) IGS cable, (3) RMC, (4) IMC, (5) EMT, (6) ENT, (7) SE cables, (8) wireways, (9) busways, (10) auxiliary gutters, (11) RNMC, (12) cablebus, (13) MC cable, (14) MI cable, (15) FMC or LTFMC with EBJ per 250-28(a – d) and not over 6′ in length, and (16) LTFNMC.

DISCONNECTING MEANS

POSITION
• 230-77

DISCONNECTING MEANS
• 230-70(a)(b)(c)

READILY ACCESSIBLE
• 230-70(a)

MORE THAN ONE DISCONNECT
• 230-80

230-70(a) Service disconnecting means shall be installed at a readily accessible place nearest entrance of conductors, either inside or outside of building.

230-70(a)(b)(c) Disconnecting means shall be provided for all service-entrance conductors. Each shall be suitable for use as service equipment and shall be permanently marked to identify it as a service disconnecting means. Shall be suitable for prevailing conditions.

230-74 Disconnecting means shall simultaneously disconnect all ungrounded service conductors from premises wiring system.

230-75 Means shall be provided to disconnect the grounded conductor from the premises wiring. A terminal or bus used in conjunction with pressure connectors is permissible.

230-76(1)(2) Disconnecting means shall be a switch or CB and be manual or power operated. If power operated, it must be able to be opened by hand.

230-77 Service disconnecting means shall plainly indicate if it is in open or closed position.

230-79 The service disconnecting means shall have a rating not less than the load as determined by 220 and not smaller than as given in 230-79 (a – d).

230-79(a) For limited loads of one branch circuit it may have a rating of not less than 15 A.

230-79(b) Where not more than two 2-wire branch circuits, it shall have rating of not less than 30 A.

230-79(c) For one-family dwelling where initial computed load is 10 kW or more, or where initial installation has 6 or more 2-wire branch circuits, the service disconnecting means shall have a rating of 100 A, 3-wire.

230-79(d) All other installations require rating of not less than 60 A.

230-80 If more than one disconnect is used per 230-71, the total rating of switches or CBs shall not be less than required by 230-79.

384-16(a) Lighting and appliance panelboard is required to be protected by not more than two main CBs or two sets of fuses.

384-16(a), Ex. 1 Permits feeder overcurrent to also protect panelboard if not larger than rating of panelboard.

384-16(a), Ex. 2 Individual protection for lighting and appliance panelboards not required for existing installation where used as service equipment for a one-family dwelling.

SERVICE DETAILS

CONNECTED AHEAD OF SERVICE DISCONNECTING MEANS
• 230-82(1-7)

SERVICE CONDUCTOR CONNECTIONS
• 230-81

BLADES NOT ENERGIZED IN OPEN POSITION
• 380-6(c)

KNIFE SWITCHES
• 380-6(a)

SUPPLY SIDE
• 230-82
• 230-94, Ex. 1

AMPACITY
• 215-2(c)

215-2(c) Where a feeder carries the total service load, the ampacity of the feeder conductors shall not be less than the ampacity of the service conductors for loads of 55 A or less.

230-75 Means shall be provided for disconnecting the grounded service conductor.

230-81 Service conductors shall be attached to the disconnect by clamps, pressure connectors, or other approved means. Soldered connections are not permitted.

230-82 In general, equipment shall not be connected to supply side of service disconnecting means. See Applications 1 – 7.

230-82(1) Cable limiters or other current-limiting devices are permitted to be connected ahead of service disconnecting means.

230-90(a) Each ungrounded service conductor shall have an overcurrent device in series with and not exceeding ampacity of conductor. See Exceptions.

230-91 Service overcurrent device shall be an integral part of disconnect or located immediately adjacent thereto.

230-94, Ex. 1 Service switch is permitted on supply side of service overcurrent devices. *Note:* This is the usual method.

240-61 Cartridge fuses and fuseholders shall be classified according to voltage and amperage ranges.

380-6(a) Single-throw knife switches shall be installed so they will not tend to close by gravity.

380-6(c) Motor-circuit switches shall be installed so blades are not energized when switch is in open position.

SERVICE GROUNDING

A. **250-50(a)** Water piping grounding electrode.

B. **250-70(1)** Ground clamp listed for the purpose shall be used to connect GEC to grounding electrode.

C. **250-68(b)** Grounding electrode connection shall be made in a permanent manner and shall be accessible. Bonding around water meters, valves, unions, etc. is required.

D. **250-64(d)** If metal enclosure is not electrically continuous, it shall be bonded at each end to grounding conductor.

E. **250-64(d)** GEC bonded to metal enclosure so as to be electrically continuous.

F. **250-64(d)** Metal GEC enclosures shall be electrically continuous from cabinet to grounding electrode.

G. Grounding electrode conductor (GEC)

H. Service-entrance equipment

I. Neutral connection bar

J. **250-24(a); 250-28(a); 250-80** GEC to also ground all metal service-entrance equipment, service raceways, and system grounded conductor (neutral).

EQUIPMENT BONDING JUMPER AND MAIN BONDING JUMPER . . .

See SERVICE GROUNDING AND BONDING.

A. NEUTRAL SERVICE ENTRANCE CONDUCTOR In this drawing, only the identified neutral conductor is shown. The two "ungrounded" or "hot" conductors are not shown.

B. AC CIRCUITS AND SYSTEMS TO BE GROUNDED – 250-20 Service equipment, raceways, cable armor, cable sheaths, etc. and any service conductor that is required to be grounded by 250-20 shall be grounded per 250.

C. METHOD OF BONDING SERVICE EQUIPMENT – 250-94 Electrical continuity shall be assured by the method of bonding used.

D. SERVICE EQUIPMENT ENCLOSURE This contains the service disconnecting means.

E. MAIN and EQUIPMENT BONDING JUMPERS – 250-28; 250-102 The MBJ and EBJ shall be copper or other corrosion-resistant material. They shall be attached per 250-8. Per 250-102(c) and 250-28(d), the EBJ on the supply side of the service and the MBJ are sized per Table 250-66, but not less than 12½% of the largest phase conductor. Where service entrance conductors are parallel in separate raceways as permitted in 310-4, the size of the bonding jumper for each raceway shall be based on the size of the conductors in each raceway or cable.

F. MAIN BONDING JUMPER – 250-28 The MBJ is the connection between the grounded circuit conductor and the EGC at the service. See 100. It shall be used to connect the EGC and service equipment enclosure to the neutral within the service enclosure. The MBJ shall be a wire, bus, screw, or similar suitable conductor without a splice.

G. EQUIPMENT GROUNDING CONDUCTOR CONNECTIONS – 250-130 EGC connections at the service equipment shall be made for grounded and ungrounded systems. For grounded systems, the connection is made by bonding the EGC, grounded service conductor, and the GEC. For ungrounded systems, the connection is made by bonding the EGC to the GEC. See Exception for (a) and (b) and 100.

H. METAL UNDERGROUND WATER PIPE – 250-50(a) When a metal underground water pipe is used as the grounding electrode, a supplemental electrode is required.

I. SYSTEM GROUNDING CONNECTIONS – 250-24(a) Secondary AC systems to be grounded shall have the GEC connected at each service. The GEC is connected to the AC neutral on the supply side of the disconnecting means on the premises, preferably within the service enclosure. This connection is not to be made on the load side of the disconnecting means. See Exceptions.

J. EFFECTIVE GROUNDING PATH – 250-2(d) The path to ground from circuits, equipment, and enclosures shall be permanent and electrically continuous. It shall have the capacity to safely conduct available fault currents and have low impedance so as to limit voltage to ground and to facilitate the circuit overcurrent devices.

K. INSTALLING GROUNDING ELECTRODE CONDUCTOR – 250-64(a) A No. 4 or larger GEC needs no protection if it is not exposed to severe physical damage. A No. 6 needs no protection if it follows the surface of a building and is not exposed to physical damage. A No. 8 shall be in RMC, IMC, RNMC, EMT, or cable armor.

L. GROUNDING ELECTRODE CONDUCTOR MATERIAL – 250-62 The GEC shall be Cu, Al, Cu-clad Al, or other corrosion-resistant material. It shall be solid or stranded, insulated, covered or bare, and shall be continuous without splices or joints. Splices in busbars are permitted per 250-64(b), Ex. Taps are permitted to the GEC where more than one service entrance enclosure is installed if such tap extends inside of such enclosure and is sized per 250-66 for largest conductors entering the enclosure. See 250-64(c).

M. SIZE OF GROUNDING ELECTRODE CONDUCTOR – Table 250-66 The size of the GEC is determined by the size of the largest service entrance conductor or equivalent for parallel conductors. It need not be larger than No. 6 Cu or its equivalent in ampacity when it is the sole connection to made electrodes as permitted in 250-52(c)(d).

N. GROUNDING ELECTRODE CONDUCTOR – 250-24(a); 100 The GEC is used to connect EGCs, service equipment enclosures, and the grounded service conductor (when present) to the grounding electrode.

O. Al or Cu-CLAD Al CONDUCTORS – 250-64 These conductors are not permitted where subject to corrosive conditions or in contact with masonry or earth, and when outside shall not be within 18″ or earth.

P. EGC IN CONDUIT – Ch 9, Note 3 to Tables When in conduit, the EGC shall be included the same as any other conductor when determining conduit size.

Q. CONNECTION TO ELECTRODES – 250-70 The grounding conductor is connected to the grounding fitting by exothermic welding, listed lugs, pressure connectors, clamps, or other listed means. Solder connections are not permitted.

R. TO GROUNDING ELECTRODE – 250-68(a)(b) The connection of the GEC to the grounding electrode shall be accessible and shall ensure a permanent and effective ground.

SIZING LOAD SIDE EBJ

What size Cu EBJ is required for a 4′ FMC with branch-circuit conductors protected by an 80 A CB?

250-102(d); Table 250-122: 80 A = No. 8 Cu

EBJ = **No. 8 Cu**

SIZING MAIN AND SUPPLY SIDE MBJs

What size Cu MBJ is required for a 120/208 V, 3φ, 4-wire service with three paralleled No. 500 kcmil Cu conductors?

Table 250-66: No. 500 kcmil x 3 = No. 1500 kcmil

250-28(d): No. 1500 kcmil x 12½ % = 187,500 cm

Ch 9, Table 8: 187,500 CM = No. 4/0 Cu

MBJ = **No. 4/0 Cu**

. . . EQUIPMENT BONDING JUMPER AND MAIN BONDING JUMPER

S. **NEUTRAL CONDUCTOR** – 250-24(a) The neutral shall be connected to the GEC. Per 230-75, means in the service equipment shall disconnect the neutral from premises wiring. In this case, the terminal or bus to which all grounded conductors are attached by means of pressure connectors is permitted.

T. **NEUTRAL CONDUCTOR** – 250-26 2. The neutral conductor shall be grounded for a 1φ, 3-wire system.

U. **BONDING OF SERVICE EQUIPMENT** – 250-92(a) Metal service equipment enclosures containing service entrance conductors, meter fittings, boxes, etc. interposed in the service raceway of armor and any conduit or armor that forms part of the grounding conductor to the service raceway shall be effectively bonded together, except as permitted by 250-84(a) for underground service cables.

Per 250-94, the grounding continuity of service equipment shall be assured by (1) grounded service conductor, (2) threaded connections, (3) threadless couplings and connections, or (4) other approved devices or bonding jumpers.

V. **SERVICE RACEWAYS AND ENCLOSURES** – 250-80 Metal service equipment and enclosures shall be grounded.

W. **GROUNDED CIRCUIT CONDUCTOR** – 250-142(a) The grounded circuit conductor can be used to ground meter enclosures, service raceways, etc. on the supply side of service disconnecting means.

X. **METER SOCKET** The utility company installs the meter into the meter socket.

Y. **THREADED CONNECTIONS** – 250-94-2 Threaded bosses or hubs are acceptable means of bonding the service conduit to the meter socket.

SERVICE GROUNDING AND BONDING

DEFINITIONS

Service Equipment – All of the necessary equipment to control the supply of electrical power to a building or a structure.

Raceway – A metal or nonmetallic enclosed channel for conductors.

Grounding Conductor – The conductor that connects electrical equipment or the grounded conductor to the grounding electrode.

Grounding Electrode Conductor (GEC) – The conductor that connects the grounding electrode(s) to the grounded conductor and/or the EGC.

Equipment Grounding Conductor (EGC) – An electrical conductor that provides a low-impedance path between electrical equipment and enclosures and the system grounded conductor and GEC.

Main Bonding Jumper (MBJ) – The connection at the service equipment that ties together the EGC, the grounded conductor, and the GEC.

MASTHEAD

SERVICE PANEL

BONDING JUMPER

BUSBAR

MAIN AND EQUIPMENT BONDING JUMPERS

250-28 Main bonding jumper shall not be spliced.

250-28(a); 250-102(a) MBJ and EBJ shall be copper or other corrosion-resistant material.

250-28(b) If MBJ is a screw, it shall have a green finish and be visible.

250-28(c) MBJ and EBJ shall be attached per 250-8.

250-28(d); 250-102(c) EBJ on supply side of service and MBJ shall be sized per Table 250-66 but not less than 12½% of largest phase conductor. Where service-entrance conductors are paralleled in separate raceways as permitted in 310-4, the size of the bonding jumper for each raceway shall be based on size of conductors in each raceway. Size of the

bonding jumper connected to grounding bushing for each raceway shall be based on the size of the service conductors in each raceway.

250-50(a)(2) Supplemental electrode of type specified in 250-50 or 250-52 when underground metal water pipe is used as grounding electrode.

250-52(c) Pipe and rod electrodes to be a minimum of 8' long and top end of electrode to be flush or below ground level unless special protection from physical damage per 250-10.

250-68(b) To assure effective bonding of the metal piping system, bonding jumpers shall be provided around insulated joints and equipment that may need

to be repaired. When used, the bonding jumpers shall be long enough to allow removal of the equipment without removing the bonding jumper.

250-8 Grounding electrode connection shall be by exothermic welding or other listed means or fittings. Where used on buried electrodes, connectors shall be listed for direct burial applications. Connectors for more than one conductor shall be listed for multiple conductors.

250-70 Grounding electrode connection shall be by exothermic welding or other listed means. See (1 – 4).

Main Bonding Jumper (MBJ) – The connection between the grounded circuit conductor and the equipment grounding conductor (EGC) at the service.

Equipment Bonding Jumper (EBJ) – The connection between two or more portions of the equipment grounding conductor (EGC).

METER SOCKET

SERVICE EQUIPMENT ENCLOSURE

MBJ AND EBJ
• 250-28(a)
• 250-102(a)

SE CONDUCTORS PARALLELED IN TWO OR MORE RACEWAYS

GROUNDING BUSHING

BONDING JUMPER SIZE
• 250-102(d)

GEC

MBJ AND EBJ
• 250-28(a)
• 250-102(a)

BONDING JUMPER
• 250-68(b)

GEC

WATER METER

CLAMP
• 250-70

GROUND ROD
• 250-52(c)

PRIMARY SIDE
HIGH VOLTAGE

120 V   120 V

TWO EQUAL SECTIONS

240 V
SECONDARY SIDE

PRIMARY LINES 2400 V TO 13,800 V

SERVICE DROP

NEUTRAL WIRE

TYPICAL POLE TRANSFORMER

GROUNDED CONDUCTOR BROUGHT TO SERVICE EQUIPMENT — 250-24(b)

TRANSFORMERS FOR SECONDARY DISTRIBUTION SYSTEM

120/240 V, 3ϕ, 4-WIRE SECONDARY DISTRIBUTION SYSTEM

B PHASE HAS HIGHER VOLTAGE TO GROUND
• 384-3(f)

SECONDARY SYSTEM GROUNDED AT TRANSFORMER (COULD BE POLE OR PAD-MOUNTED)

GROUNDED CONDUCTOR
• 250-24(b)

SERVICE TO BUILDING FOR 240 V, 3ϕ POWER ONLY

SERVICE DISCONNECTING MEANS (240 V, 3ϕ)

EBJ

SERVICE FEEDS 240 V LOAD ONLY

GEC

3ϕ, 3-WIRE SERVICE NEUTRAL NOT REQUIRED

230-56 High-leg or the phase conductor with the higher voltage to ground shall be identified by orange color or other effective means of each connection point where neutral is present.

250-24(b) Where an AC system operating at 1000 V or less is grounded at any point, the grounded conductor shall be run to each service. This conductor shall be sized per 250-66 and, in addition, for service-phase conductors larger than 1100 kcmil Cu or 1750 kcmil Al, the grounded conductor shall not be smaller than 12½% of largest phase conductor except the grounded conductor need not be larger than the largest phase conductor.

384-3(e) High-leg of the phase conductors with the higher voltage to ground shall be orange in color or marked by other effective means.

384-3(f) High-leg to be B phase except that it may be same phase as metering equipment. On 3ϕ, 4-wire delta-connected systems, phase arrangement shall be such that B phase shall have the higher voltage to ground. Where additions to existing installations are made, other arrangements shall be permitted but shall be marked.

EXAMPLE

Each service phase consists of three 600,000 cm Cu conductors in parallel. Total area of phase conductors is $3 \times 600,000$ cm, or 1,800,000 cm. The minimum size of the grounded conductor run to the service equipment is $1,800,000 \times 12½\%$, or 225,000 cm Cu, which requires a 250 kcmil conductor.

The purpose of this requirement is to ensure a low impedance path for any line to ground fault currents that could develop in the premises served. Section 250-2(d) requires the path to ground shall have sufficiently low impedance to facilitate the operation of circuit protective devices in the circuit.

METER SOCKET

HIGH-LEG IS B PHASE
• 384-3(f)

HIGH-LEG IS ORANGE
• 230-56
• 384-3(e)

NEUTRAL

SERVICE EQUIPMENT

B PHASE

GEC

SERVICE FEEDS 120/240 V LOAD

TRANSFORMERS

CURRENT	
AIR CORE	OR
MAGNETIC CORE	
AUTO	OR

GROUNDING OF TWO OR MORE BUILDINGS SUPPLIED
FROM SINGLE SERVICE EQUIPMENT – 250-32

GROUNDED SYSTEM

UNGROUNDED SYSTEM

225-31 Disconnecting means required for all ungrounded conductors and located per 230-70.

225-36 Shall be suitable for service equipment.

225-36, Ex. A 1ϕ, 3-way or 4-way branch-circuit snap switch is acceptable for residential garage or outbuilding.

225-37 Where Building No. 2 is supplied by more than one service, branch circuit, feeder, or any combination of these, a plaque or directory shall be placed at each disconnect location denoting all other services, branch circuits, or feeders that supply the building.

250-32(a) Grounded systems shall have a grounding electrode which shall be connected to the system grounded circuit conductor (neutral) in each building.

250-32(a), Ex. Grounding electrode not required in Building No. 2 if only one branch circuit and no equipment in Building No. 2 requires grounding.

250-32(b)(1) When an EGC is run from Building No. 1 to Building No. 2 for the purpose of grounding all metal equipment and parts, the neutral is not connected to this conductor or to any metal equipment enclosures in Building No. 2.

250-32(c) Ungrounded systems shall have a grounding electrode connected to enclosure of the building or structure disconnecting means.

GROUNDED CONDUCTORS FOR AC SYSTEMS – 250-26

UTILITY SUPPLY SYSTEM

SERVICES TO BUILDING ON PREMISES

1φ, 2-WIRE SYSTEM

1φ, 3-WIRE SYSTEM

GROUND AT POLE OR PAD-MOUNTED TRANSFORMER

ONE CONDUCTOR GROUNDED
• 250-26(1)

NEUTRAL CONDUCTOR GROUNDED
• 250-26(2)

GEC ON PREMISES

WYE SYSTEM

GROUND AT POLE OR PAD-MOUNTED TRANSFORMER

COMMON CONDUCTOR GROUNDED
• 250-26(3)

GEC ON PREMISES

DELTA SYSTEM

GROUND AT POLE OR PAD-MOUNTED TRANSFORMER

ONE PHASE CONDUCTOR GROUNDED
• 250-26(4)

GEC ON PREMISES

DELTA SYSTEM FROM WHICH 1φ, 3-WIRE SYSTEM IS DERIVED

GROUND AT POLE OR PAD-MOUNTED TRANSFORMER

NEUTRAL CONDUCTOR GROUNDED
• 250-26(5)

GEC ON PREMISES

250-20(a) Alternating current circuits of less than 50 V shall be grounded as follows:

(1) Where supplied by a transformer whose primary exceeds 150 V to ground.

(2) Where supplied by a transformer whose primary is ungrounded.

(3) Where installed as overhead conductors outside of buildings.

250-20(b) Alternating current circuits of 50 V to 1000 V which supply premises wiring and premises wiring systems shall be grounded as follows:

(1) Where system can be grounded so that maximum voltage-to-ground does not exceed 150 V.

(2) Where the neutral is used as a circuit conductor on 3φ, 4-wire wye systems.

(3) Where the midpoint of one phase winding is used as a circuit conductor on 3φ, 4-wire delta systems.

250-26 For alternating current premises wiring systems, the conductor to be grounded shall be as follows:

(1) 1φ, 2-wire – one conductor

(2) 1φ, 3-wire – neutral conductor

(3) Multiphase systems with one wire common to all phases – common conductor

(4) Multiphase systems with one grounded phase – one phase conductor

(5) Multiphase system with 1φ, 3-wire system derived – neutral conductor

GROUNDING TWO TO SIX SERVICE SWITCHES FED FROM ONE SET OF SE CONDUCTORS

230-75 When the service disconnecting means does not disconnect the grounded conductor from premises wiring, other means shall be provided in the service equipment.

250-24(a)(1) Premises wiring supplied by an AC service that is grounded shall have a GEC connected to a grounding electrode at any accessible point from the service drop to the terminal or bus to which the grounded service conductor is connected at the service disconnecting means.

250-142(a) The grounded circuit conductor is permitted to ground noncurrent-carrying metal parts of equipment, raceways, and other enclosures at the supply side of:

(1) service disconnecting means

(2) main disconnecting means for separate buildings

(3) disconnecting means or overcurrent devices of separately derived systems

373-6(c) Insulating material with temperature rating not less than that of installed conductors or bushing required for No. 4 or larger conductors. See 300-4(f).

EGCs ON LOAD SIDE OF SERVICE – 250-122

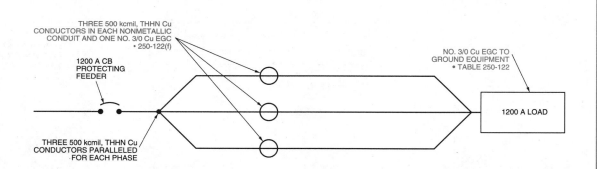

250-122(f) Where conductors are run in parallel, either in multiple raceways or cables, as permitted by 310-4, the EGC shall also be run in parallel. Each parallel EGC shall be sized in relation to size of overcurrent device protecting circuit conductor.

EXAMPLE

Each of the 3″ nonmetallic conduits contain three 50,000 cm, THHN Cu conductors. The three conductors are connected in parallel and are protected by a 1200 A CB. Table 250-122 requires a No. 3/0 Cu EGC and 250-122(f) requires that these three No. 3/0 conductors shall be installed in each 3″ conduit and connected in parallel. If all nine 500 kcmil conductors were installed in one raceway, one No. 3/0 Cu EGC would satisfy the requirements of 250-122.

OUTLET, DEVICE, PULL AND JUNCTION BOXES, CONDUIT BODIES AND FITTINGS — 370 . . .

WOODEN WALL

R

WIRING IS ACCESSIBLE

6" OF FREE CONDUCTOR

C

FLUSH BOX K

PLASTER RING E

GROUNDED BOX

D

L

$\frac{1}{8}$" GAP MAXIMUM

PLASTER OR DRYWALL

8'

FLOOR

A

4" x $2\frac{1}{8}$" OCTAGON BOX

6 WIRES BROUGHT OUT

2 WIRES STRAIGHT THROUGH

TOTAL OF 8 WIRES IN BOX

F

DEFINITIONS

100: Outlet – A point on the wiring system at which current is taken to supply utilization equipment.

100: Device – A unit of an electrical system that is intended to carry but not utilize electric energy.

Pull Box – A box used as a point to pull or feed electrical conductors into the raceway system.

Junction Box – A box in which splices, taps, or terminations are made.

Conduit Body – A conduit fitting that provides access to the raceway system through a removable cover at a junction or termination point.

Fitting – An electrical system accessory that performs a mechanical function.

BOX FILL

Wire Size	Maximum No. of Wires* Table 370-16(a)	Volume Per Wire** Table 370-16(b)
18	14	1.5
16	12	1.75
14	10	2
12	9	2.25
10	8	2.5
8	7	3
6	4	5

* for 4" x $2\frac{1}{8}$" round or octagonal box
** in cu in.

110-13(a) Wooden plugs driven into holes in masonry, concrete, plaster, etc. shall not be used to mount electric equipment.

A. 250-110(1) Metal boxes located within 8' vertically or 5' horizontally of ground or grounded metal objects and subject to contact by persons shall be grounded.

B. 250-148(a) When a metal box is required to be grounded, a grounding screw or other listed means (such as grounding clip) that serves no other purpose than to connect box to equipment ground wire shall be used.

C. 300-14 At least 6" of free conductor shall be left at each outlet and switch point, except for conductors pulled straight through box.

370-2 Round boxes are not permitted where conduits or connectors using locknuts or bushings are to be connected.

370-3 Nonmetallic boxes shall be used only with nonmetallic raceway, NMC, knob and tube wiring, ENT, or open wiring.

D. 370-4 Metal boxes shall be grounded per 250.

370-15(a) In damp or wet locations, boxes, conduit bodies, and fittings shall be positioned to prevent moisture from entering. Those installed in wet locations shall be listed for the purpose.

370-16 The volume of the box shall be calculated per 370-16(a). It shall not be less than the fill as calculated per 370-16(b). Boxes and conduit bodies containing conductors No. 4 or larger shall also comply with 370-28.

E. 370-16(a) The volume of the original box shall be increased by cu in. marked on extension ring, plaster ring, or domed cover.

F. 370-16(a)(1) The volume of standard boxes that are not marked with their cubic inch capacity shall be per Table 370-16(a).

370-16(a)(2) Boxes 100 cu in. or less, other than those in Table 370-16(a), and nonmetallic boxes shall be marked.

B

BONDING SCREW AND BOND WIRE

4" x $1\frac{1}{8}$" SQUARE BOX

EGCs

G

FIXTURE STUD

9 No. 14 WIRES

... OUTLET, DEVICE, PULL AND JUNCTION BOXES, CONDUIT BODIES AND FITTINGS — 370

G. **370-16(b)(1–5)** Where one or more fixture studs, cable clamps, or hickeys are contained in the box, the number of conductors shall be reduced by one with an additional reduction of two wires for each strap containing one or more devices, and a further deduction of one wire for one or more ground wires for second set of EGCs passing through a box per 250-74, Ex. 4.

H. **370-17(c)** Nonmetallic boxes shall be suitable for the lowest temperature rated conductor entering the box. Open wiring or concealed knob-and-tube conductors shall enter through individual holes. Conductors in flex tubing or Romex shall have the outer sheath extend into the box not less than ¼″.

I. **370-17(c), Ex.** Nonmetallic cables shall be secured to nonmetallic bodies so that cable cannot be pulled in or out of the box except for single gang boxes (not larger than 2¼″ × 4″) where cable is secured within 8″ of the box. Nonmetallic cable sheath shall extend at least ¼″ into box.

J. **370-18** Unused openings in boxes or fittings shall be closed. Metal plugs or plates shall not be used with nonmetallic fittings unless recessed ¼″.

370-19 Substantial support shall be provided for the device. Screws for supporting the box shall not be used for device attachment.

K. **370-20** Boxes and fittings, when installed in walls or ceilings constructed of noncombustible material, shall be permitted to be recessed no more than ¼″. Boxes in wall of combustible material must be flush with surface or project therefrom.

L. **370-21** Gaps in plaster, drywall, or plasterboard shall be repaired so that they do not exceed ⅛″ at the edge of the box.

370-22 Boxes and extension rings shall be permitted for surface extensions from a box of concealed wiring system. The box or ring shall be securely fastened to the outlet box and grounded per 250. See Exception.

M. **370-23(a)** Boxes shall be securely and rigidly fastened to the surface on which they are mounted.

N. **370-23(b)** Boxes shall be rigidly supported from a structural member either directly or by using an approved metallic or wood brace.

O. **370-23(b)(2)** Wood brace shall have a nominal cross section of 1″ × 2″.

P. **370-24** No box shall be less than ½″ deep. Where box is intended to contain flush device, it shall be at least ¹⁵⁄₁₆″ deep.

Q. **370-25(c)** Covers through which flexible cord pendants pass shall have smoothly rounded edges or bushings. Hard rubber or composition bushings are not permitted.

370-27(a) Boxes for lighting fixtures shall be designed or installed so fixture may be installed.

370-28 Boxes and conduit bodies used as pull or junction boxes shall comply with 370-28(a–d).

R. **370-29** Boxes and conduit bodies shall be installed so that wiring within them can be rendered accessible without damage to the building. Listed boxes shall be permitted to be covered by gravel, soil, and light aggregate under some conditions. See Exception.

370-40(a) Metal boxes, conduit bodies, and fittings shall be corrosion resistant or similarly treated.

S. **370-40(d)** A tapped hole, or equivalent, shall be provided in each metal box to connect an EGC.

Outlet boxes shall be securely fastened.

At least 6″ of conductor shall hang out of the box.

CIRCUIT BREAKERS

ITCB STANDARD AMPERE RATINGS – 240-6(a)

15, 20, 25, 30, 35, 40, 45, 50, 60, 70, 80, 90, 100, 110, 125, 150, 175, 200, 225, 250, 300, 350, 400, 450, 500, 600, 700, 800, 1000, 1200, 1600, 2000, 2500, 3000, 4000, 5000, and 6000 A.

DEFINITIONS

Circuit Breaker (CB) – A device which opens and closes a circuit by nonautomatic means and automatically opens on a predetermined overcurrent within its rating without damaging itself.

Adjustable Trip Circuit Breaker (ATCB) – A CB which can be set to trip at various values of current, time, or both.

Instantaneous Trip Circuit Breaker (ITB) – A CB with no delay purposely introduced in the tripping action.

Inverse Time Circuit Breaker (ITCB) – A CB with a delay purposely introduced in the tripping action.

480 V, 3φ
3-WIRE

STRAIGHT MARKING

120/208 V, 3φ
4-WIRE OR
277/480 V, 3φ
4-WIRE

SLASH MARKING

BQ1B020
I-T-E Circuit Breaker
INTERRUPTING RATING
MAX. RMS. SYM. AMPS
10,000   120/240 V.A.C.
TYPE BQ   40°C
UND. LAB. INC. LIST. ®
CIR. BRK. 1 POLE UNIT
ISSUE LM-9858
HACR TYPE
SIEMENS-ALLIS, INC.

RESET HANDLE
FLAG TRIP INDICATOR
BIMETALLIC STRIP THERMAL STRIP
ADJUSTING SCREW (AT FACTORY)
LOAD LUG
LATCH ARM
LINE SIDE CONNECTION
ARMATURE
CONTACTS (POINTS)
ARC BLOWOUT
MAGNETIC ELEMENT

Square D Company

110-9 Equipment shall have an interrupting rating sufficient for nominal voltage and current available at line terminals of equipment.

When checking interrupting rating of a CB, also check the voltage rating because generally the interrupting current (IC) is lower as the voltage rating is higher. For example:

at 240 V – 65,000 A IC

at 480 V – 25,000 A IC

at 600 V – 18,000 A IC

If the foregoing is not evaluated, this can result in a dangerous misapplication of the breaker insofar as interrupting rating is concerned.

110-10 CBs shall be selected and coordinated to clear a fault without damage to the electrical components of the circuit.

240-8 CBs shall not be connected in parallel. CBs which are factory-assembled in parallel and listed as a unit are permitted.

A. **240-81** CBs shall clearly indicate when open (OFF) or closed (ON). When CB handles operate vertically, the up position shall be ON.

240-82 CBs shall be designed so that any alteration of the trip point or time for operation shall require dismantling or breaking the seal.

B. **240-83(a)** Ampere ratings on CBs shall be durable and visible after installation. The removal of the trim or cover is permitted to see the markings.

C. **240-83(b)** Ampere ratings shall be marked into handles of CBs rated at 100 A or less and 600 V or less.

D. **240-83(c)** Interrupting ratings shall be marked into handles of CBs with interrupting ratings other than 5000 A. Interrupting rating marking is not required for CBs used for supplementary protection.

240-83(d) CBs used to switch 120 V or 277 V fluorescent fixtures shall be listed and marked SWD.

E. **240-83(e)** Voltage ratings shall be marked.

240-85 CBs with straight marking can be used in neutral grounded, grounded wye, and ungrounded or grounded delta systems. CBs with slash markings can be used only in grounded neutral systems.

380-2(b) CBs shall not disconnect the grounded conductor of a circuit.

380-2(b), Ex. CBs are permitted to disconnect the grounded conductor when all circuit conductors are disconnected at the same time. CBs are permitted to disconnect the grounded conductor when the device allows only the ungrounded conductors to be opened first.

380-3(a) CBs shall be externally operable and installed in an enclosure listed for the intended use.

380-8(a) CBs used as switches shall be readily accessible. The center of the grip of the operating handle shall not be over 6′-7″ from the floor or platform. See 380-8(a), Ex. 1, 2, and 3.

GROUND-FAULT CIRCUIT INTERRUPTERS (GFCIs)

210-8(a) All 125 V, 1φ, 15 A and 20 A receptacles installed in dwelling units as specified in 210-8(a)(1 – 7) shall have GFCI protection for personnel.

210-8(a)(1) Bathrooms.

210-8(a)(2) Garages and grade-level areas of storage or work buildings. See Exceptions.

210-8(a)(3) Outdoors. See Exception.

210-8(a)(4) Crawl spaces at or below grade.

210-8(a)(5) Unfinished basements. See Exceptions.

210-8(a)(6) Kitchens where the receptacles serve countertops.

210-8(a)(7) Wet bar sinks where receptacles serve countertops and are within 6′ of the outside edge of the sink.

Ground-Fault Circuit Interrupter (GFCI) – A device whose function is to de-energize a circuit within established time when current to ground exceeds a predetermined value which is less than that required to operate protective device of supply circuit.

FUSETRON BOX COVER UNITS

SOU	SOW	SOX	SOY	SSU	SSW	SSX	SSY	SOY-B

SRU	SRW	SRX	SRY	SSY-I	SSY-RL	SCY	STY	SKA

Covers fit standard outlet or switch boxes. Units are UL® listed and are made with Edison-base fuseholder. Units without switch for motor are ¾ HP or smaller. Units with switch for motors are ½ HP or smaller. Do not use on DC.

240-24(a) Overcurrent devices required to be readily accessible. See Applications (1 – 4).

240-50(d) There shall be no exposed live parts after fuses or fuses and adapters are installed.

240-52 Edison-base fuseholders shall be installed only where made to accept Type S fuses by use of adapters.

PLUG FUSES, FUSEHOLDERS, AND ADAPTERS – 240, PART E

A. 240-30(a)(b) In general, overcurrent devices shall be enclosed in cutout boxes or cabinets. See Exceptions.

240-32 Enclosures for overcurrent devices in damp or wet locations shall be mounted so there is a minimum of ¼″ air space between the enclosure and wall. Cabinets or cutout boxes installed in wet locations shall be weatherproof. See 373-2(a).

B. 240-33 Enclosures shall be mounted in vertical position except where impracticable and in compliance with 240-81.

240-50(a) Plug fuses shall not be used in circuits over 125 V between conductors except in grounded circuits at not more than 150 V to ground. See 240-50(a)(2).

C. 240-50(b) Each fuse, fuseholder, and adapter shall be marked with amp rating.

D. 240-50(c) Fuses 15 A or less shall have hexagonal top or cap.

240-50(e) The screw shell of plug-type fuseholder shall be connected to load side of circuit.

E. 240-51(a) Edison-base plug fuses shall not be classified over 125 V and 0 A – 30 A.

240-51(b) Plug fuses of Edison-base type to be used only as replacements where there is no evidence of overfusing or tampering.

F. 240-53(a) Type S fuses shall be plug type and classified not over 125 V and 0 – 15, 16 – 20, and 21 – 30 A.

240-53(b) Type S fuses shall not be interchangeable with lower amp classification and designed to be used only with Type S fuseholder or Type S adapter inserted.

G. 240-54(a) Type S adapters to fit Edison-base fuseholders.

240-54(b) Type S fuseholders and adapters are only for Type S fuses.

240-54(c) When once inserted, Type S adapters are to be nonremovable.

240-54(d) Type S fuseholders and adapters are designed to make tampering or shunting difficult.

240-54(e) Type S fuses, fuseholders, and adapters to be standardized by manufacturers to permit interchangeability.

H. 384-14(a) A panelboard is classified as a lighting and appliance panelboard if 10% of its circuits are 30 A or less and have provisions for neutral connections.

I. 384-15 Lighting and appliance branch-circuit panelboards shall not contain more than 42 overcurrent devices in any one cabinet or cutout box. A 2-pole device counts as 2; a 3-pole device counts as 3.

384-16(a) General rule: Each lighting and appliance

ORDINARY TYPE W, EDISON-BASE (NO TIME DELAY)

FUSETRON, DUAL-ELEMENT EDISON-BASE PLUG FUSE (TIME DELAY)

DUAL-ELEMENT TYPE S FUSE

ON OVERLOAD SOLDER IS HEATED AND OPENS CIRCUIT

DUAL-ELEMENT TYPE S FUSE

ADAPTER

FUSE LINK BLOWS AND OPENS CIRCUIT

LOCKING DEVICE

PLUG FUSES

TERMINAL BAR MOUNTING SCREWS

EGC TERMINAL BAR

EQUIPMENT GROUNDING TERMINAL BAR

0 – 30 A, TWIN PLUG

0 – 60 A, FERRULE PULL-OUT TYPE

70 – 600 A, KNIFE-BLADE, PULL-OUT TYPE

FUSEHOLDERS

CONNECTION BUS

PROVISIONS TO MOUNT TERMINAL BAR IN PANELBOARD

LIGHTING AND APPLIANCE PANELBOARD

Square D Company

OVERCURRENT PROTECTION – CARTRIDGE FUSES . . .

Fuse – An overcurrent protection device with a fusible link that melts and opens the circuit when an overload condition or short circuit occurs.

DUAL-ELEMENT, TIME-DELAY CARTRIDGE FUSE

The dual-element, time-delay fuse is designed to provide a time delay in the harmless low overload range and also to be fast-opening when exposed to dangerous short circuits. This fuse contains two fusible elements in series and the magnitude of a potentially damaging current flow determines which one of these elements opens. The thermal cutout element will open at currents of up to about 500% of the fuse rating (overload range), whereas the short-circuit element will open at 500% or more of the fuse rating.

DUAL-ELEMENT, TIME-DELAY CURRENT-LIMITING FUSE

Dual-element, current-limiting fuses function similar to regular dual-element, time-delay fuses except that they are faster operating in short-circuit range, thus are more current-limiting. The short-circuit element is normally made of silver, with a quartz sand arc quenching filter.

SINGLE-ELEMENT, CURRENT-LIMITING FUSE

Single-element, current-limiting fuses function similar to regular dual-element, time-delay low-overload range. This fuse has lowest let-through values in comparison with other fuses. It is used to protect circuit components having inadequate interrupting, bracing, or withstand rating, such as low IC breakers.

DUAL-ELEMENT TIME DELAY
10 SECONDS (MINIMUM) AT 50%
RATED CURRENT

AMPERE RATING 60 A

VOLTAGE RATING
600 VAC, 300 VDC

Cooper Industries Bussmann Division

Low-Peak® dual-element TDFs provide long time-delay for temporary motor startup.

110-9 Fuses shall have sufficient interrupting capacity for the voltage and the current that must be interrupted.

110-10 Fuses shall be selected and coordinated to clear fault without extensive damage to electrical components of the circuit.

240-1, FPN Overcurrent protection opens circuit when current reaches a value that can damage either equipment or conductors. Overcurrent devices are selected based on circuit impedance, short circuit, and withstand ratings. Equipment intended to interrupt current at fault levels shall have sufficient interrupting rating. See 110-9 and 110-10.

240-2 Equipment shall be protected against overcurrent.

240-3 Conductors shall be protected at all ampacities per Tables 310-16 – 310-19 except where ampacity of conductor and standard rating of fuse do not correspond, then next larger fuse may be used only up to 800 A per 240-3(a – g).

240-6(a) Standard ratings for ITCBs shall be: 15, 20, 25, 30, 35, 40, 45, 50, 60, 70, 80, 90, 100, 110, 125, 150, 175, 200, 225, 250, 300, 350, 400, 450, 500, 600, 700, 800, 1000, 1200, 1600, 2000, 2500, 3000, 4000, 5000, and 6000 A. Additional standard ratings for fuses are: 1, 3, 6, 10, and 601 A.

240-8 Fuses shall not be connected in parallel. Fuses in parallel are permitted if factory assembled and listed as a unit.

240-11 Current-limiting fuse when interrupting currents in its range will reduce current in faulted circuit to magnitude substantially less than obtainable in same circuit if replaced with solid wire if same impedance.

240-20 Fuses shall be connected in series with each ungrounded conductor.

240-21 Overcurrent device connected where conductor to be protected receives supply per 240-21(a – g).

240-22 In general, overcurrent devices shall not be placed in series with any grounded conductor. See Applications for use in (1) motor overload protection and (2) simultaneous disconnection.

240-24(a) Fuses shall be readily accessible except: (1) busways – 364-12; (2) supplemental – 240-10; (3) OCPD – 225-9(b); (4) utilization equipment.

240-30 Fuses shall be enclosed in cabinets or cutout boxes. See Ex. 1 – 3.

240-32 In damp or wet locations, fuses shall be in enclosures identified for such use and shall have ¼″ air space between enclosure and mounting surface.

240-33 Enclosures shall be mounted in vertical position except where impracticable and in compliance with 240-81.

240-40 Disconnect required on supply side of cartridge fuses where accessible to other than qualified person except: (1) service as per 230-82; (2) group operation of motors per 430-112, and fixed space heating per 424-22(c).

240-60(a) In general, cartridge fuses and fuseholders of the 300 V type are not permitted to be installed in circuits of over 300 V between conductors.

240-60(a)(1) Cartridge fuses and fuseholders are permitted in 1ϕ, line-to-neutral circuits supplied from a solidly grounded neutral system. However, fuses and fuseholders are permitted when the line-to-neutral voltage does not exceed 300 V.

240-60(b) Cartridge fuseholders (0 A – 6000 A) shall be so designed to make it difficult to put fuse of given class into fuseholder for current lower or voltage higher than intended. Current-limiting fuseholders shall permit insertion only of current-limiting fuses.

240-60(c) Fuses shall be marked to show: (1) amp rating; (2) volt rating; (3) interrupting rating if other than 10,000 A; (4) current-limiting where applicable except where used as supplementary protection; (5) trademark of manufacturer.

240-61 Classification of cartridge fuses and fuseholders shall be on a basis of amperage and voltage. Fuses of a higher rating are permitted to be used on lower rated circuits, but fuses of a lower rating are not permitted to be used with higher voltages. 600 V rated fuses could be used on a 277 V circuit, but 250 V rated fuses could not.

240-101(a) Fuse rating in continuous amps shall not exceed 3 times amp rating of conductor. For CBs, long-time trip setting or minimum-trip setting of an electronically-activated fuse shall not exceed 6 times ampacity of conductor.

240-101(b) Conductors tapped to a feeder can be protected by the feeder OCPD where the OCPD also protects the tap conductor.

... OVERCURRENT PROTECTION – CARTRIDGE FUSES

CLASS H FUSES

ONE-TIME FUSE   **RENEWABLE FUSE**

Class H fuses are probably the oldest and best known. In the past, referred to as "NEC" or "Code" fuses. Interrupting rating not marked on label. Tested by UL® on circuits of 10,000 A. Rated 0 A – 600 A in both 250 V and 500 V ranges.

CLASS J FUSES

LIMITRON FAST-ACTING FUSE   **HI-CAP, TIME-DELAY FUSE**

Class J fuses are 600 VAC, 0 A – 600 A, current-limiting and listed by UL® with interrupting rating of 200,000 symmetrical rms amps. Fuseholders are smaller than Class H, thus are noninterchangeable.

CLASS K FUSES

LIMITRON FAST-ACTING FUSE CLASS K1   **LOW-PEAK DUAL-ELEMENT FUSE**   **FUSETRON DUAL-ELEMENT FUSE**
                                                        CLASS K5

Class K fuses have ratings of 0 A – 600 A in both 250 V and 600 V. Dimensions same as Class H fuses. Grouped into three categories: K1, K5, and K9, and may have UL® listing of 50,000 A, 100,000 A, or 200,000 A symmetrical rms amps. UL® has assigned a maximum level of peak-let-through current (1p) and amp-squared seconds (I^2t) for each K rating. K rating determines degree of current limiting. K9 has least current-limiting ability and K1 the greatest. May also be listed by UL® as time delay; then are required to have minimum time delay of 10 sec at 500% of rated current.

CLASS L FUSES

LIMITRON FAST-ACTING FUSE   **HI-CAP, TIME-DELAY FUSE**

Class L is listed by UL® in sizes 601 A – 6000 A AC. They bolt into place, have a current-limiting characteristic and an interrupting rating of 200,000 symmetrical rms amps and are used in bolted pressure contact switches.

CLASS R FUSES

61 – 600 A KNIFE-BLADE TYPE

0 – 60 A FERRULE TYPE

Class R fuses are of the nonrenewable, cartridge type and are the most recently listed fuse in the standards of UL®. This fuse is the same size as Class H and K fuses except in 0 A – 60 A one end-cap ferrule will have an annular grooved ring and in 61 A – 600 A one knife blade will have a slot in its side. Fuseholders are designed to accept only Class R fuses, but rejection features in fuse would not prohibit the use of Class R fuse in a Class H or K fuseholder. Fuse rated to interrupt fault current up to 200,000 symmetrical rms amps. If Class R to have UL® label as time-delay, it would basically have a minimum of 10 sec delay at 5 times fuse rating.

CLASS G FUSES

BUSS TYPE SC FUSES

Class G are available in sizes 0 A – 60 A for circuits of 300 V or less to ground (i.e., 277/480 V systems). They are a small dimension cartridge fuse (commonly referred to as Type SC fuses) and basically Type S cartridge fuses because they are noninterchangeable in that a 15 A fuse will fit only into a 15 A fuseholder and a 20 A fuse will fit only into a 20 A fuseholder. They are available in sizes 15, 20, 30, and 60 A. They are listed by UL® at 100,000 symmetrical rms amps interrupting capacity.

Cooper Industries Bussmann Division

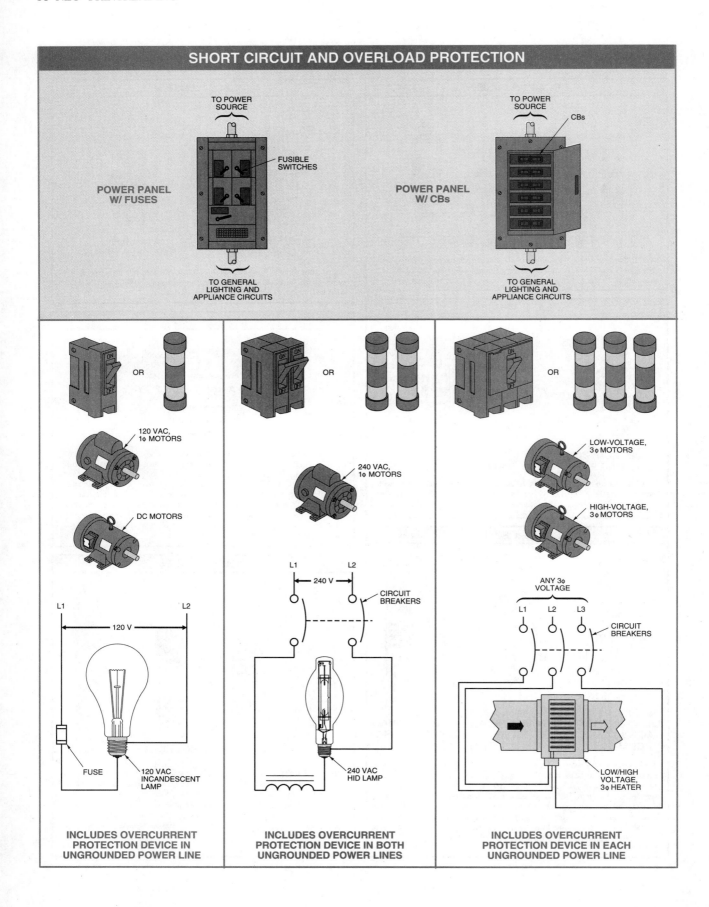

SHORT CIRCUIT AND OVERLOAD PROTECTION

TO POWER SOURCE

POWER PANEL W/ FUSES

FUSIBLE SWITCHES

TO GENERAL LIGHTING AND APPLIANCE CIRCUITS

TO POWER SOURCE

CBs

POWER PANEL W/ CBs

TO GENERAL LIGHTING AND APPLIANCE CIRCUITS

OR

120 VAC, 1φ MOTORS

DC MOTORS

L1   L2

120 V

FUSE

120 VAC INCANDESCENT LAMP

INCLUDES OVERCURRENT PROTECTION DEVICE IN UNGROUNDED POWER LINE

OR

240 VAC, 1φ MOTORS

L1   L2

240 V

CIRCUIT BREAKERS

240 VAC HID LAMP

INCLUDES OVERCURRENT PROTECTION DEVICE IN BOTH UNGROUNDED POWER LINES

OR

LOW-VOLTAGE, 3φ MOTORS

HIGH-VOLTAGE, 3φ MOTORS

ANY 3φ VOLTAGE

L1   L2   L3

CIRCUIT BREAKERS

LOW/HIGH VOLTAGE, 3φ HEATER

INCLUDES OVERCURRENT PROTECTION DEVICE IN EACH UNGROUNDED POWER LINE

PLUG FUSES

RATINGS

30 A AND BELOW

ROUND WINDOW FOR OVER 15 A

HEXAGONAL WINDOW FOR 15 A OR LESS

EDISON-BASE

RATINGS

0 A – 15 A
16 A – 20 A
21 A – 30 A

FUSE LINK
SPRING
SOLDER CUP
COIL

AMPERE RATING
LOCKING DEVICE

FUSEHOLDER OR ADAPTER

TYPE S

CLASS T FUSES

The 300 A, 300 V T-Tron fuse is no larger than a nickel. Its interrupting rating is 200,000 A.

This 200 A Load Center with pull-out disconnect has Class T fuses bolted into the panelboard.

JJN (300 V) T-Tron fuses (all case sizes).

JJS (600 V) T-Tron fuses (all case sizes).

Cooper Industries Bussmann Division

Class T fuses are listed by UL® in sizes 1 A – 600 A in both 300 V and 600 V AC range. They are fast-acting with a high degree of current limitation and a high interrupting rating (200,000 A). They provide overload and high-performance short-circuit protection of main, feeder, and branch circuits, as well as general-purpose equipment. Class T fuses are the smallest and most compact of all UL®-listed power distribution fuse devices. They comply with 110-9 "Interrupting Rating," which requires service equipment to be suitable for the short-circuit current available at its supply terminals. In addition, Class T fuses, when properly selected, can limit the short-circuit current for the protection of low-interrupting capacity (IC) CBs. (In a 10,000 A, IC branch-circuit, the CBs are protected by Class T fuses in the mains.)

ARMORED CABLE (AC) – 333

333-1: Armored Cable (AC) – A fabricated assembly of insulated conductors in a flexible metallic enclosure.

USES PERMITTED – 333-3

- Branch circuits and feeders in both exposed and concealed work
- In cable trays where identified for such use

USES NOT PERMITTED – 333-4

(1) In theatres and places of public assembly

(2) In motion picture studios

(3) In hazardous locations –
501-4(b), Ex. – Class I, Div. 2 locations with wiring in nonincendive circuits
502-4(b), Ex. – Class II, Div. 2 locations with wiring in nonincendive circuits
504-20 – Intrinsically safe apparatus

(4) In areas with corrosive fumes or vapors

(5) On cranes or hoists

(6) In storage battery rooms

(7) In hoistways or on elevators

(8) In commercial garages where prohibited per 511

A. 333-7 Spacing between supports is 4½′ and 12″ within each box and fitting.

333-7(b) Type AC cable permitted to be unsupported:

(1) Where cable is fished
(2) Up to 2′ without support where flexibility is required to motors, etc.
(3) Not more than 6′ for lighting fixture whips

B. 333-8 Minimum radius for bends is 5x cable diameter.

C. 333-9 Fittings and bushings shall be provided at boxes unless the box design provides protection.

D. 333-10 Shall comply with 300-4 when installed through or parallel to joists, studs, or rafters.

333-11 Exposed work shall follow the surface of the building finish.

E. 333-12(a) Guard strips are required for cable run across the top of floor joists or within 7′ of floor joists on faces of rafters or studs.

F. 333-12(b) Guard strips are not required when cable is parallel to sides of rafters, studs, or floor joists.

333-22 Marking shall be per 310-11.

METAL-CLAD CABLE (MC) — 334

CONDUCTORS — G

SMOOTH METALLIC SHEATH — H

CORRUGATED METALLIC SHEATH

INTERLOCKED METALLIC SHEATH

334-1: Metal-Clad Cable (MC) – A factory assembly of one or more conductors with or without fiber optic members.

A. 334-10(a) MC cable shall be supported every 6′.

B. 334-10(a) MC cable with four or fewer No. 10 or smaller conductors shall be secured within 12″ of every box, cabinet, etc.

C. 334-10(f) MC cables run through or parallel to framing members shall comply with 300-4 for protection from physical damage.

D. 334-11(a) Bending radius for smooth sheath:
(1) not over ¾″ D: radius 10x cable diameter
(2) over ¾″ D – 1½ D: radius 12x cable diameter
(3) over 1½″ D: radius 15x cable diameter

E. 334-11(b) Bending radius for interlocked or corrugated sheath:

radius = 7x cable diameter

F. 334-12 Fittings shall be listed and identified for such use.

G. 334-20 Conductors shall be Cu, Al, Cu-clad Al, solid or stranded. The minimum conductor size shall be No. 18 Cu and No. 12 Al or Cu-clad Al.

H. 334-22 The metallic sheath shall be smooth, corrugated, or interlocking. It shall be smooth and continuous.

334-24 Marking shall be per 310-11.

USES PERMITTED — 334-3

Unless otherwise specified and where not subject to physical damage
(1) Services, feeders, and branch circuits
(2) Power, lighting, control, and signal circuits
(3) Indoors or outdoors
(4) Exposed or concealed
(5) Direct burial when identified for such use
(6) In cable trays
(7) In raceways
(8) As open runs
(9) As aerial cable on a messenger
(10) In hazardous locations per 501, 502, 503, 504, and 505
(11) In dry locations, embedded in plaster finish or masonry except in damp or wet locations
(12) In wet locations (with conditions)

USES NOT PERMITTED — 334-4

- Exposed to destructive corrosive conditions
- Direct burial in earth or concrete
- Exposed to cinder fills, strong chlorines, caustic alkalis, vapors of chlorine, or hydrochloric acids with conditions

INTERLOCKED OR CORRUGATED METALLIC SHEATH

CABLE DIAMETER

7X CABLE DIAMETER

BENDING RADIUS

FOUR NO. 10 Cu CONDUCTORS

JUNCTION BOX

12″

6′

FRAMING MEMBERS

SUPPORT

NONMETALLIC-SHEATHED CABLE (NM) – 336

336-2: Nonmetallic-Sheathed Cable (NM) – A factory assembly of two or more insulated conductors having an outer sheath of moisture-resistant, flame-retardant, nonmetallic material.

A. 336-4 NM, NMC, and NMS cable can be used in one- and two-family dwellings, multifamily dwellings, and other structures.

B. 336-6(a) Exposed work shall follow surface of building finish.

C. 336-6(b) Exposed work shall be protected from physical damage.

D. 336-6(c) Exposed work not smaller than two No. 6 or three No. 8 conductors in unfinished basements may be secured to lower edge of joists.

E. 336-6(d) Exposed work in attics shall be protected by guard strips if on top or within 7′ of floor joists per 333-12(a).

F. 336-6(d) Exposed work in attic does not require guard strips if installed parallel to rafters, studs, or floor joists per 333-12(b).

G. 336-9 Shall comply with 300-4 when installed through or parallel to joists, studs, or rafters.

H. 336-16 Bending radius shall not be less than 5x cable diameter.

I. 336-18 NMC shall be supported within 12″ of boxes, etc. and at intervals not over 4½′ apart.

USES PERMITTED – 336-4

(1) In one- and two-family dwellings

(2) In multifamily dwellings and other structures

(3) In identified cable trays

USES NOT PERMITTED – 336-5

NM, NMC, and NMS shall not be used in any building over 3 floors above grade, as service-entrance cable, or embedded in concrete. See 336-5(a) (1 – 9).

CONSTRUCTION SPECS – 336-30(a)

(1) **NM** – Overall covering shall be flame retardant and moisture resistant.

(2) **NMC** – Overall covering shall be flame retardant, moisture resistant, fungus resistant, and corrosion resistant.

(3) **NMS** – Sheath shall be moisture resistant and flame retardant.

SERVICE ENTRANCE CABLE (SE AND USE) – 338

A. 230-24(a) Conductors shall have a vertical clearance of 8′ above roof surfaces. See Exceptions.

B. 230-24(b) The minimum clearance of service-entrance conductors that are not over 600 V is from 10′ – 18′ above final grade.

C. 230-51(a) Service cables shall be supported within 12″ of equipment and at intervals not exceeding 30″.

D. Table 300-5 Minimum cover requirements for USE cable are per Table 300-5.

E. 338-1(b) USE is listed for direct burial.

F. 338-4(a) SE is permitted for use in interior wiring systems.

G. 338-6 Bending radius shall not be less than 5x cable diameter.

338-5 Marking shall be per 310-11.

338-1: Service-Entrance Cable – A single or multiconductor assembly with or without an overall covering.

SE – Service-entrance cable with a flame-retardant, moisture-resistant covering.

USE – Underground service-entrance cable with a moisture-resistant covering that is not flame-resistant.

8′ MINIMUM

CABLE DIAMETER

5X CABLE DIAMETER

BENDING RADIUS

DWELLING

24″

ELECTRIC RANGE

DRIP LOOPS

12″

30″

10′

12′ OR 15′

18′

SERVICE-DROP CONDUCTORS CLEARANCE FROM FINAL GRADE
• 230-24(b)

• 10′ – SUBJECT TO PEDESTRIAN TRAFFIC ONLY AND 150 V OR LESS TO GROUND

• 12′ – NOT SUBJECT TO TRUCK TRAFFIC AND 300 V OR LESS TO GROUND

• 15′ – NOT SUBJECT TO TRUCK TRAFFIC AND OVER 300 V TO GROUND

• 18′ – SUBJECT TO TRUCK TRAFFIC AND NOT OVER 600 V TO GROUND

FINISHED GRADE OR SIDEWALK

DRIVEWAY – NO TRUCK TRAFFIC

ROAD, STREET, ALLEY, COMMERCIAL DRIVEWAY
• 230-24(b)

USES PERMITTED – 338-2

- SE cable used for service-entrance conductor per 230
- USE cable used for service laterals may emerge above ground outside at terminations in meter bases per 300-5(d)

RIGID METAL CONDUIT (RMC) — 346

Rigid Metal Conduit (RMC) – A conduit made of metal.

SIZES • 346-6, FPN	
• 10′ LENGTHS	
Inches	**Millimeters**
$\frac{1}{2}$	16
$\frac{3}{4}$	21
1	27
$1\frac{1}{4}$	35
$1\frac{1}{2}$	41
2	53
$2\frac{1}{2}$	63
3	78
$3\frac{1}{2}$	91
4	103

A. 346-3(a) RMC permitted in all atmospheric conditions.

B. 346-3(c) RMC in or under cinder fill shall be encased in 2″ concrete or be 18″ below cinder fill.

C. 346-5 Supports, straps, screws, etc. shall be corrosion resistant.

D. 346-6(a)(b) RMC is available in 10′ lengths from ½″ to 6″ diameters.

346-7 The number of conductors in a single conduit shall not exceed the percentages of fill in Ch 9, Table 1.

E. 346-8 All cut ends shall be reamed to remove rough edges.

346-15 Where RMC enters boxes, fittings, or other enclosures, bushings shall be provided to protect the wire from abrasion.

F. 346-9(b) Running threads prohibited.

G. 346-11 Bends shall not total more than 360° in one run.

H. 346-12 RMC shall be supported within 3′ of boxes and at intervals of not more than 10′.

I. 346-16(a) RMC is shipped in 10′ lengths and shall be identified.

RIGID NONMETALLIC CONDUIT (RNMC) – 347

FIELD BENDING

① HEATING

② FORMING

③ COOLING

SMALL DIAMETERS
(½" – 1½")

LARGE DIAMETERS
(2" – 6")

Carlon, an Indian Head Co.

HEATING 1. Rotate the conduit to provide even heating. Time varies with conduit size. In large sizes, insert bending plugs as shown before heating conduit.

FORMING 2. Use bend guide for small diameters, or a jig, or a line drawn on the work surface. Large sizes can be formed using a corner for an extra set of hands. Plugs must remain tight during handling.

COOLING 3. The bend will set when cooled. Use a wet rag or sponge to speed the cooling process. When set, remove plugs. The bend is now ready to install.

Rigid Nonmetallic Conduit (RNMC) – A type of conduit of suitable nonmetallic material that is resistant to moisture and chemical atmospheres.

| SUPPORT – TABLE 347-8 ||
Diameter*	Maximum Spacing**
$\frac{1}{2}-1$	3
$1\frac{1}{4}-2$	5
$2\frac{1}{2}-3$	6
$3\frac{1}{2}-4$	7
6	8

* in in.
** in feet

PVC BELL ENDS

EXPANSION JOINT

45°   90°
PVC ELBOWS

PVC PLUG

COUPLING

FEMALE ADAPTER

TERMINAL ADAPTER

USES PERMITTED – 347-2

(a) Concealed in ceilings, walls, and floors

(b) In corrosive areas per 300-6 and where exposed to chemicals

(c) In cinder fill

(d) In wet locations where walls are washed, such as dairies or canneries

(e) In dry or damp areas where not prohibited by 347-3

(f) For exposed work if identified for such use and not exposed to physical damage

(g) Underground, per 300-5 and 710-4(b)

(h) Support of conduit bodies

USES NOT PERMITTED – 347-3

(a) In hazardous locations except as permitted in 503-3(a); 504-20; 514-8; 515-5; and 501-4(b), Ex.

(b) To support fixtures and equipment. See Exceptions for nonmetallic conduit bodies.

(c) Where exposed to physical damage unless identified for such use

(d) Where ambient temperatures exceed those of conduit

(e) Where conductor insulation temperature limitations exceed conduit insulation

(f) In theatres, except as permitted in 520

4" RADIUS

½" DIA

BENDING RADIUS

ENCLOSURE

ON

OFF

3' MAXIMUM

RNMC

A B C D E F

A. **347-1** For above ground use, RNMC shall be flame retardant and resistant to crushing, distortion, low temperatures, and sunlight effects. For underground use, RNMC shall be moisture resistant and withstand crushing during installation. It is permitted to be installed underground in continuous lengths from a reel.

B. **347-5** All cut ends shall be trimmed inside and outside.

C. **347-8** RNMC shall be installed as a complete system per 300-18. It shall be securely fastened within 3' of each box, conduit body, or other termination. Spacing between supports is based on diameter. See Table 347-8.

347-9 Expansion fittings shall be used to compensate for thermal expansion and contraction.

347-10 The minimum size is ½" diameter. The maximum size is 6" diameter.

347-11 The maximum number of conductors shall be per Chapter 9, Table 1.

D. **347-12** Bushings are required where RNMC enters a box, fitting, or other enclosure.

E. **347-13** The radius of bends (on the inner edge) is based on diameter per Table 346-10, Ex.

F. **347-14** The maximum number of quarter bends shall not exceed four (360°) between pull points.

347-17 Each length of RNMC shall be marked within every 10'.

GROUNDING ELECTRODE SYSTEM – 250-50

COLD WATER PIPE

GROUNDING ELECTRODE CONDUCTOR

Ⓐ

LISTED GROUND CLAMP

250-50 Interior metal waterpiping more than 5' from point of entrance to building shall not be used as GEC or as a conductor to interconnect electrodes that form GES.

If available at building or structure, the following grounding electrodes shall be bonded together and used as the "Grounding Electrode System":

250-50(a) Metal water pipe with at least 10' in contact with earth (including well casing if bonded to pipe). Water pipe to be supplemented by additional electrode per 250-81 or 250-83.

250-50(b) Effectively grounded metal frame of building.

250-50(c) Electrode consisting of rebar or rod at least ½" diameter and 20' bare copper wire not smaller than No. 4 and encased in at least 2" of concrete at bottom of footing.

250-50(d) A ground ring encircling building, consisting of at least 20' of bare copper wire not smaller than No. 2 and in direct contact with earth at least 2½' below surface.

250-64(a) Al or Cu-clad aluminum GECs not permitted in direct contact with masonry, earth, or where subject to corrosive conditions.

Metallic enclosures for grounding conductors shall be continuous from cabinet to grounding electrode. Enclosures which are not continuous from cabinet to electrode can be bonded at each end to grounding conductor.

Al or Cu-clad Al conductors shall not be used within 18" of earth when used outside.

A. **250-68(a)(b)** Connection of GEC to grounding electrode shall be permanent, effective, and accessible. The connections to concrete encased or buried electrodes need not be accessible. Where water pipe is used, bonding around water meter, valves, unions, etc., is required.

250-70 Ground clamps suitable for conductor and electrode materials, and suitable for direct soil burial where buried. One conductor only unless listed for multiple conductors.

CONDUCTORS – GENERAL RULES

3-WIRE CIRCUIT
{
BLACK
WHITE
RED
}

200-2 Grounded conductor, on premises, shall be identified per 200-6. Grounded conductor, when insulated, shall have suitable insulation as ungrounded conductor.

200-7 Conductors having a white or natural gray covering shall be used only for grounded conductors.

200-7(c)(1) Where necessary to use these conductors in circuits of 50 V or more, they may be reidentified to indicate use.

200-7, (c)(2) Conductors that are part of a cable shall be reidentified and they shall be used as a supply, but not as a return, to single-pole, 3-way, or 4-way switch loops.

210-6(a) Voltage between conductors that supply terminals of lighting fixtures or cord-and-plug connected loads less than 1440 VA or ¼ HP shall not exceed 120 V.

210-19(a), FPN 4 Voltage drop should not exceed 3% to the farthest outlet on branch circuit. Maximum voltage drop for feeders plus branch circuits should not exceed 5%. See 215-2, FPN 2.

310-2(a), Ex. Conductors shall be insulated except where covered or bare conductors are specifically permitted.

- For covered service-drop conductors, see 230-22, Ex.
- For bare service-entrance grounded conductors, see 230-41, Ex.
- For bare or covered underground service grounded conductors, see 230-30, Ex..

310-12(c) Ungrounded conductors shall be clearly distinguishable from grounded and grounding conductors by colors or combination of colors plus distinguishing marking, and in either case the color shall be other than white, natural gray, or green.

384-16(d) Continuous loads shall not exceed 80% of the branch-circuit rating where the load shall continue for 3 hr.

384-16(d), Ex. Circuits supplied by assembly are listed for continuous operation at 100% of their rating.

CONDUCTORS AND CIRCUITS

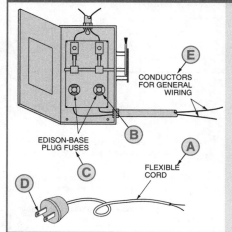

Ⓔ

CONDUCTORS FOR GENERAL WIRING

Ⓑ

EDISON-BASE PLUG FUSES

Ⓒ

Ⓐ

FLEXIBLE CORD

Ⓓ

A. **240-4(a)** Flexible cord approved for specific appliances is considered protected by branch-circuit overcurrent device per 210 as follows:

20 A circuit – Tinsel cord No. 18 or larger

30 A circuit – No. 16 or larger cord

40 A circuit – 20 A or more rated cord

50 A circuit – 20 A or more rated cord

B. **240-50(a)** Plug fuses and fuseholders shall not be used in circuits exceeding 125 V between conductors except in systems having a grounded neutral and of which no conductor operates at more than 150 V to ground.

C. **240-51(b)** Edison-base plug fuses shall be used only as a replacement item in existing installations where there is no evidence of overfusing or tampering.

D. **250-114, Exceptions** Some cord-and-plug connected equipment is not required to be grounded.

310-4 No. 1/0 and larger conductors may be run in parallel if of same length, same conductor material, same size, same insulation type, and terminated in the same manner. If in separate raceways or cables, these must have the same physical characteristics. See Exceptions.

E. **310-5** Conductors (solid or stranded) for general wiring shall not be smaller than No. 14 Cu or No. 12 Al or Cu-clad Al. See Exceptions.

WIRING IN ATTIC OR ROOF SPACES

A. 324-3 Special permission is required for installation of concealed knob-and tube-wiring unless it is used for extensions of existing installations only.

B. 324-11(b); 333-12; 336-6(d); 339-3(a)(4); 350-4 In general, conductors and cables shall be run through or on sides of joists, studs, and rafters.

C. 324-11(b), Ex. In unfinished attics and roof spaces not accessible by stairs or permanent ladder and head room at all points is less than 3′, knob-and-tube wiring may be run on face of timber.

D. 333-12; 336-6(d); 339-3(a)(4); 350-4 NM, NMC, AC, UF, and Greenfield (FMC) run across face of roof rafters or studding within 7′ of floor or floor joists

shall be protected by proper guard strips if attic is accessible by stairs or ladder. 339-3(a)(4) refers to 333-12.

E. 333-12; 336-6(d); 339-3(a)(4); 350-4 In attics accessible by stairs or permanent ladder, BX, NM, NMC, UF, and Greenfield shall not be run across top of floor joists, unless protected by guard strips at least as high as the cable.

F. 333-12(a) If attic not accessible by stairs or permanent ladder, the cable need be protected only within 6′ of scuttle hole.

G. 333-12(a) Guard strips shall be substantial and at least as high as the cable.

H. 333-12(b); 324-11(b); 336-6(d); 339-3(a)(4); 336; 350-4 No running boards or guard strips necessary for protection on sides of joists, studs, or rafters.

WIRING RUN AT ANGLES TO JOISTS IN UNFINISHED BASEMENTS

A. 336-6(c); 336-9; 339-3(a)(4) Assemblies of NM, NMC, or UF equal to two No. 6, three No. 8, or larger may be run on underside of joists.

B. 336-9; 339-3(a)(4) Smaller size of cable shall be passed through holes or fastened to running boards.

C. 336-9 Cables through studs, joists, or rafters shall be installed per 300-4.

CABLE FISHED INTO MASONRY WALLS

A. 333-3; 336-4(a)(b)(c) Types AC, NM, NMC, or NMS may be fished in masonry walls if no moisture is present and above grade line.

333-3; 336-4(b); 339-3(a)(4) If moisture is present, Types NMC, UF, or AC shall be used.

B. 336-4(c) NMS is permitted for both exposed and concealed installations in dry locations. NMS cable shall be installed per 780.

336-5(a)(8) NM, NMC, and NMS are not permitted to be embedded in poured cement, concrete, or aggregate.

336-5(b)(2) NM and NMS are not permitted to be embedded in masonry, concrete, fill, or plaster, nor run in a shallow chase in masonry or concrete and covered with plaster or similar finish.

LIGHTING FIXTURES, LAMPHOLDERS, LAMPS, AND RECEPTACLES – 410 . . .

NONFERROUS METAL .04″ THICK

COMPOSITION .10″ THICK

P

RECEPTACLE

FLUORESCENT FIXTURE
• 410-8(b)(2)
• 410-8(d)(2)

RECESSED FIXTURE
• 410-8(b)(1)(2)
• 410-8(d)(3)(4)

INCANDESCENT FIXTURE
• 410-8(b)(1)
• 410-8(d)(1)

6″

12″

SHELF WIDTH OR 12″ WHICHEVER IS GREATER

SHELF

ROD

STORAGE AREA

H

24 ″ WIDE TO 6 ′-0″ ABOVE FLOOR OR BOTTOM OF ROD WHICHEVER IS GREATER

CLOSET

CEILING

METAL CANOPY

FIXTURE CHAIN

E

I

CEILING

METAL CANOPY

SWITCH

FIXTURE CHAIN

D

C

L

OUTDOOR RECEPTACLE

HINGED COVER

F

Q

R

WHITE OR NATURAL GRAY FINISH

SCREW SHELL

A **B**

J

CAST OUTLET BOX

N

SOCKET SUPPORT
• 225-13

WEATHERPROOF PIN SOCKET

WEATHERPROOF PIGTAIL SOCKET

K

INSULATED BUSHING

METAL LAMPHOLDER

M

410-1, FPN: Luminaire – A complete lighting unit containing a lamp or lamps and associated equipment for distributing or positioning the light. *Note:* Luminaire is the international term for a lighting fixture.

A. 200-10(c) For devices with screw shells, the grounded conductor terminal shall be connected to the screw shell.

B. 200-10(d) For devices with attached leads, the conductor attached to the screw shell shall have white or natural gray finish.

C. 250-110(1) The metal fixture canopy or socket is required to be grounded if within 8′ vertically or 5′ horizontally of ground or grounded objects and subject to contact.

D. 402-8 Grounded conductor to be white or natural gray or per 400-22(a – e).

E. 410-3 No live parts of fixtures, lampholders, lamps, and receptacles shall be normally exposed to contact. Exposed terminals shall be installed in metal fixture canopies or in open bases of table and floor lamps.

F. 410-4(a) Fixtures installed in wet or damp locations shall be marked.

G. 410-4(d) Cord-connected fixtures, lighting track, pendants, and ceiling fans shall not be installed in a 3′ horizontal and 8′ vertical zone from the bathtub rim.

H. 410-8(a) Storage is the space above the shelf and below the rod. Storage space above the rod is shelf width or 12″, whichever is greater. Storage space below the rod is 24″ to the rod or 6′ above the floor, whichever is greater.

410-11 Fixtures shall be constructed and installed so that temperatures in outlet boxes shall not exceed conductor rating. Wires are not permitted through outlet boxes unless identified for such use.

410-13 Combustible wall or ceiling finish exposed between the edge of a fixture pan or canopy and an outlet box shall be covered with noncombustible material.

410-16(a) Fixtures over 50 lb shall be supported independently of the box unless listed for the weight to be supported.

I. 410-22 Fixture wiring shall be neatly arranged and not exposed to physical damage.

J. 410-23; 410-47 The grounded conductor shall be connected to the screw shell of a screw-shell lampholder.

K. 410-24(b) Fixture wires shall be suitable for all conditions to which they will be subjected.

L. 410-28(e) Stranded conductors are required on fixture chains.

M. 410-30(a) Metal lampholders which are attached to flexible cords shall be equipped with insulating bushings.

N. 410-49 Weatherproof lampholders are required in wet or damp locations.

410-56(a) Receptacles for portable cords shall be rated at not less than 15 A at 125 V and 15 A at 250 V. They shall be suitable for use as lampholders.

O. 410-56(c) Isolated ground (IG) receptacles, identified by an orange triangle on the receptacle face, are only permitted to be installed when an isolated equipment grounding conductor is provided.

. . . LIGHTING FIXTURES, LAMPHOLDERS, LAMPS, AND RECEPTACLES — 410

P. **410-56(d)** Faceplates of nonferrous metal shall be not less than .04″ thick, ferrous .03″ thick, and composition .10″ thick, unless formed or reinforced to provide adequate mechanical strength. Metal faceplates shall be grounded.

410-56(e) Faceplates shall be installed to completely cover wall openings and seat against the surface.

Q. **410-57(a)** Receptacles in damp location shall be installed in an enclosure that is weatherproof only when the plug cap is not inserted.

R. **410-57(b)** Receptacles in wet locations shall maintain weatherproof integrity when the plug cap is inserted.

S. **410-57(c)** Receptacles shall not be installed in bathroom or shower space.

410-57(f) Outdoor receptacles shall be installed so that water will not accumulate on the outlet cover or plate.

T. **410-66(a)** Unless identified for direct contact, recessed parts of fixture enclosures shall maintain at least ½″ spacing from combustible materials.

U. **410-66(b)** Thermal insulation not permitted within 3″ of fixture and not above fixture so as to entrap heat and prevent free air circulation except when identified for the purpose.

410-67(a) Conductor insulation shall be suitable for the temperature.

V. **410-67(b)** Branch-circuit wires permitted to fixture if insulation is suitable for fixture temperature.

W. **410-67(c)** Tap wires with insulation suitable for the fixture temperature are permitted from fixture terminal to the outlet box if in a suitable raceway. The tap conductors shall be at least 18″ long but not longer than 6′.

PLAN VIEW

ELEVATION VIEW

BATHROOMS

ISOLATED GROUND RECEPTACLE

RECESSED FIXTURES

WALL-MOUNTED OVEN AND COUNTERTOP RANGES

A. 210-19(c), Ex. 1 Where tap conductors supply a wall-mounted oven or countertop from 50 A circuit, they shall have not less than 20 A rating and shall not be longer than necessary for servicing.

Table 220-19, Note 3 In lieu of the values given in Column A, the load for ranges over 1¾ kW but not over 8¾ kW may be taken as the sum of nameplate ratings multiplied by demand factors of Columns B and C.

Table 220-19, Note 4 The branch-circuit load for one wall-mounted oven or countertop unit shall be taken as the nameplate rating. Where countertop and not more than two wall-mounted ovens are in the same room, the nameplate ratings shall be added and the load treated as a single range.

B. 410-4(c) No fixtures are permitted in nonresidential hoods unless identified for the purpose.

DISCONNECTING MEANS FOR APPLIANCES AND FREE-STANDING RANGES

A. 210-50(c) Outlets for special appliances, such as laundry equipment, shall be within 6' of the appliance.

B. 422-30 Each appliance shall be provided with a means for disconnecting all underground conductors. Where supplied by more than one source, all disconnecting means shall be grouped and identified.

C. 422-32(a) For cord-and-plug connected appliances, including free-standing household ranges and clothes dryers, a separable connector or attachment plug and receptacle may serve as the disconnecting means where accessible. See 422-31.

422-32(b) For residential ranges, plug and receptacle if accessible by removal of a drawer is acceptable.

DEMAND FACTORS

For household range from 12 kW to 27 kW, calculate per Table 220-19, Note 1.

For household ranges from 8¾ kW to 27 kW of unequal ratings, calculate per Table 220-19, Note 2.

DISCONNECTING MEANS FOR APPLIANCES

Appliance – Utilization equipment installed or connected as a unit to perform one or more functions such as food mixing, clothes washing, etc.

A. 422-31(b) For permanently-connected appliances over 300 VA or ⅛ HP, the branch-circuit switch or CB may serve as the disconnecting means if it is readily accessible to the user of the appliance and can be locked open.

B. 422-33(c) In a one-family dwelling, the service switch is acceptable as the single disconnecting means if the unit switch on the appliance has an OFF position and disconnects all ungrounded conductors.

422-35 Disconnecting means for motor-driven appliances over ⅛ HP shall be within sight of motor controller and comply with 430, Part J.

422-35, Ex. Disconnect may be out of sight if unit switch on appliance has OFF position and disconnects all ungrounded conductors.

Name _____

Date _____

Trade Competency Test 1

NEC®	Answer	

NEC®   **Answer**

_____   T   F   **1.** The term "service-entrance conductors overhead system" includes the term "service drops."

_____   _____   **2.** The general rule with respect to the service disconnecting means states that a grounding connection shall not be made to any grounded circuit on the _____ side.
     A. supply   C. either A or B
     B. load   D. both A and B

_____   _____   **3.** The rating of any one cord-and-plug connected utilization equipment shall not exceed _____% of the branch circuit ampere rating.

_____   T   F   **4.** The smallest permissible size of hard-drawn copper wire for a service drop is No. 10.

_____   T   F   **5.** Service drops shall clear the ground by not less than 10′.

_____   _____   **6.** The GEC for an AC ungrounded system need not be larger than No. _____ Cu when it is the sole connection to a made electrode.

_____   T   F   **7.** The disconnecting means for a service shall disconnect all ungrounded conductors simultaneously.

_____   _____   **8.** Nonmetallic boxes may have nonmetallic or _____ covers.

_____   _____   **9.** An EBJ is permitted outside a raceway when it is not over _____′ long.
     A. 6   C. 10
     B. 8   D. neither A, B, nor C

_____   _____   **10.** The branch-circuit rating shall not be less than the noncontinuous load plus _____% of the continuous load.

_____   _____   **11.** In general, the free length of conductor at a switch outlet shall be not less than _____″.

_____   _____   **12.** The service disconnecting means need not be greater than 30 A where there are not more than _____ 2-wire branch circuits.

_____   _____   **13.** The clearance of service drops above a flat roof shall be not less than _____′.

_____   T   F   **14.** The service disconnecting means shall consist of not more than eight switches or circuit breakers.

_____   _____   **15.** The clearance of service drops from the side or bottom of windows that are designed to open shall not be less than _____′.

_____   _____   **16.** The minimum clearance of service drops above a public driveway on other than residential property is _____′.
     A. 10   C. 16
     B. 12   D. 18

17. Not more than six switches at one service location shall be used as the service disconnecting means if they are each _____ marked.

18. In residential occupancies, the NEC® requires one 15 A lighting circuit for every _____ sq ft.

19. The service disconnect rating for a one-family dwelling shall be not less than _____ A, 1φ, 3-wire.

  A. 30   C. 100
  B. 60   D. 200

20. One service lateral can consist of six underground sets of multiple conductors sized at No. _____ or larger.

21. In determining VA/sq ft area, the _____ dimensions of the building shall be used.

22. A bare copper neutral service-entrance conductor is permitted only if the voltage is not over _____ V to ground.

  A. 150   C. 600
  B. 300   D. A, B, and C

23. Under the optional calculation for a one-family dwelling, the air conditioning load is assessed at _____%.

24. Conduit shall not be connected to the _____ of round boxes.

25. Unless otherwise marked, a CB with a 60°C terminal shall be loaded per the _____°C Column of Table 310-16.

26. A metal fixture shall be grounded if located within _____′ vertically or 5′ horizontally of a kitchen sink.

27. Unused openings in boxes and fittings shall be effectively _____.

28. Disconnecting means for motor-driven appliances over _____ HP shall be located within sight of the motor controller.

29. Metal enclosures for GECs shall be _____.

  A. rigid conduit   C. electrically continuous
  B. isolated   D. neither A, B, nor C

30. A new receptacle outlet installed on a 15 A or 20 A branch circuit shall be of the _____ type.

31. The front edge of an outlet box in a noncombustible ceiling may be _____″ back from the surface.

32. For the purpose of determining the maximum number of conductors in boxes, two conductors shall be counted for each yoke or _____.

33. A receptacle outlet, in a dwelling, within 6′ of a wet bar sink shall be _____-protected.

34. Receptacles in living areas are spaced so that no point along the available floor line perimeter shall be more than _____′ from a receptacle outlet.

35. In general, the minimum ampacity of the ungrounded service-entrance conductors shall not be less than the minimum _____ of the service disconnecting means.

36. Where necessary, a shallow outlet box not less than _____″ deep may be used.

37. In general, conductors other than service conductors shall not be installed in the same service _____ or service cable.

Name _____

Date _____

Trade Competency Test 2

NEC® _____ **Answer** _____

_____ _____ 1. The unit lighting load for a dwelling unit, expressed in VA/sq ft, shall be _____ VA.

    A. 1½   C. 3

    B. 2   D. 5

_____ _____ 2. The cubic inch rating of a No. 18 fixture wire is _____ cu in.

_____ T   F 3. An insulated copper GEC used with a single service disconnecting means may be spliced by exothermic welding.

_____ _____ 4. Rooftop receptacle outlets used to service HACR equipment in other than dwelling units shall be _____-protected.

_____ T   F 5. The sheet metal straps ground clamp, constructed solely of strong sheet metal, is not permitted for connecting the GEC to the grounding electrode.

_____ _____ 6. A receptacle outlet is required for a hallway that has a length of _____′ or more.

    A. 4   C. 8

    B. 6   D. 10

_____ _____ 7. Water heaters rated over _____ VA shall have a disconnecting means located within 50′ and within sight.

_____ T   F 8. Additional plug receptacles in dwelling bedrooms, above the number required by the NEC®, shall be assessed at ½ A each.

_____ _____ 9. Limiting the number of _____ in an auxiliary gutter will minimize the effects from a short-circuit or ground fault in one circuit.

_____ _____ 10. The minimum clearance between the lug and wall of an enclosure for a No. 4/0 conductor being terminated as an S or Z bend is _____″.

_____ _____ 11. _____ means "connected to earth or to some conducting body that serves in place of earth."

_____ _____ 12. For other than temporary wiring, _____ shall not be used for the support of overhead conductor spans.

_____ _____ 13. When using the optional calculation for a dwelling service, all other loads above the initial 10 kVA are to be assessed at _____%.

    A. 40   C. 60

    B. 50   D. 75

_____ _____ 14. The branch-circuit switch or CB may serve as the disconnecting means for a permanently connected appliance of over 300 VA or ⅛ HP, where _____ and within sight.

_____ T   F 15. A No. 6 GEC that is secured to the building surface and is exposed to moderate physical damage requires metal covering or protection.

_____ _____ **16.** The rating of a cord-and-plug connected appliance used on a 20 A branch circuit having two or more outlets shall not exceed _____ A.

    A. 12   C. 16

    B. 15   D. 18

_____ _____ **17.** In general, the voltage between conductors on a branch circuit supplying 15 A receptacles shall be not greater than _____ V.

_____ T   F **18.** In a bedroom, if a point on any available wall is 7′ from the nearest plug receptacle, an additional receptacle is needed.

_____ _____ **19.** For outside branch circuits and feeders, the 8′ vertical clearance above the roof level shall be maintained for _____ ′ in all directions from the edge of the roof.

_____ _____ **20.** A service drop conductor shall not be permitted to exceed _____ ′ of length with 18″ of clearance above the overhang portion of the roof.

_____ _____ **21.** The maximum rating of the OCPD on a 17 A noncontinuous, single non-motor appliance circuit is _____ A.

_____ T   F **22.** Track lighting installed in dwelling units shall be calculated at 180 VA for each 2′.

_____ _____ **23.** The minimum clearance between the lug and wall of an enclosure for a No. 4/0 conductor being terminated as a L bend is _____ ″.

_____ _____ **24.** For the kitchen small appliance load in dwellings, the NEC® requires not less than _____ A circuit(s).

    A. one 15   C. two 15

    B. one 20   D. two 20

_____ _____ **25.** A single gang FS box can house five No. _____ conductors.

_____ _____ **26.** NMS cable shall be installed so no bend will have a radius less than _____ times the diameter of the cable.

_____ _____ **27.** The number of No. 12 conductors allowed in a 4″ × 1¼″ octagon box is _____.

_____ T   F **28.** A plug receptacle in the hallway outside of a kitchen shall not be connected to the small appliance circuit.

_____ _____ **29.** The cubic inch rating of a No. 10 conductor is _____ cu in.

_____ _____ **30.** A total of _____ No. 14 conductors is permitted in a 3″ × 2″ × 2″ device box.

_____ T   F **31.** A receptacle in the middle of the dining room floor is not counted toward fulfilling minimum NEC® requirements.

_____ _____ **32.** In dwelling occupancies, the voltage between conductors supplying screw shell lampholders shall not exceed _____ V.

_____ _____ **33.** Where raceways containing ungrounded conductors of No. _____ or larger enter a cabinet, box enclosure, or raceway, the conductors shall be protected by a substantial fitting providing a smoothly rounded insulating surface, unless the conductors are separated from the fitting or raceway by substantial insulating material securely fastened in place.

_____ _____ **34.** In general, where installed in raceways, conductors of size No. _____ and larger shall be stranded.

Trade Competency Test **3**

NEC®	Answer	

1. Tap conductors for recessed lighting fixtures shall be in a suitable raceway or cable at least _____″ in length.

T F 2. A general-purpose branch circuit supplies outlets for both lighting and appliances.

3. If a 1½″ × 4″ octagonal box contains a fixture stud and a cable clamp, the maximum number of No. 14 conductors shall be reduced to _____.

    A. 4   C. 7

    B. 5   D. 8

4. A ground ring conductor shall be installed at least _____′ below the earth's surface.

T F 5. Junction boxes shall be readily accessible.

6. An approved, listed stainless steel rod shall be not smaller than _____″ in diameter.

7. Stationary motors rated _____ HP or less can have their disconnecting means located in the service panel.

8. Hanging lighting fixtures are not permitted within _____′ vertically from the rim of a bathtub.

T F 9. The screw shell of a plug-type fuseholder shall be connected to the load side of the circuit.

10. The demand load for a 16 kVA electric range is assessed at _____ VA.

    A. 8200   C. 12,600

    B. 9600   D. 14,400

11. To be considered as a GEC, a metal water pipe shall be in contact with the earth for at least _____′.

12. NMS cable (Romex) shall be protected to a height of at least _____″ above the floor.

T F 13. Service equipment shall be identified as suitable for use.

14. In combustible walls or ceilings, the front edge of an outlet box or fitting may set _____″ back of the finished surface.

    A. ⅛   C. ½

    B. ¼   D. neither A, B, nor C

15. EMT shall be supported within _____′ of outlet boxes.

16. LTFNMC shall not be used for systems having conductors operating at over _____ V.

T F 17. Wooden supports for boxes in concealed work shall not be less than nominal 1″ × 2″.

_____ _____ **18.** Cartridge fuses and fuseholders shall be classified according to _____ and amperage ranges.

_____ _____ **19.** Plug fuses are never permitted in circuits that exceed _____.

    A. 125 V between conductors   C. 150 V between conductors
    B. 125 V to ground   D. 150 V to ground

_____ _____ **20.** Receptacle outlets for special appliances such as laundry equipment shall be located within _____′ of the appliances.

_____ _____ **21.** Recessed lighting fixtures shall have at least _____″ clearance from combustible material.

_____ _____ **22.** The GEC can be spliced by means of the _____ welding process.

_____ _____ **23.** Securely installed boxes and fittings that are threaded into two or more conduits from two or more sides and that do not support fixtures or contain devices shall not exceed _____ cu in.

    A. 100   C. 150
    B. 125   D. 200

_____ _____ **24.** Surface-mounted fluorescent lighting fixtures shall be located at least _____″ from the storage space in a clothes closet.

_____ _____ **25.** NMS cable (Romex) shall not be used to wire multifamily dwellings exceeding _____ floors above grade.

_____ _____ **26.** The minimum size conductor permitted to be used as an encased electrode is No. _____ Cu.

_____   T   F **27.** When two switches are mounted on the same strap, the number of conductors allowed in the box shall be reduced by one.

_____ _____ **28.** The maximum number of No. 14 conductors permitted in a $1\frac{1}{2}″ \times 4″$ octagonal outlet box is _____.

    A. 5   C. 8
    B. 7   D. 9

_____ _____ **29.** The internal depth of outlet boxes intended to enclose flush devices shall be at least _____″.

    A. $\frac{1}{2}$ C. $\frac{15}{16}$
    B. $\frac{7}{8}$ D. $1\frac{1}{2}$

_____ _____ **30.** Aluminum conductors shall not be installed within _____″ of the earth when used outside as a GEC.

_____ _____ **31.** Hanging lighting fixtures are not permitted within a zone measured _____′ horizontally from the rim of a bathtub.

_____ _____ **32.** A metal underground gas pipe shall not be used as a grounding _____.

_____ _____ **33.** Driven rods made of _____ shall not be used as electrodes.

_____ _____ **34.** The minimum size rebar permitted to be installed in concrete at the bottom of a footing is _____″ in diameter and 20′ long.

_____ _____ **35.** The smallest service drop permitted by the NEC® is No. _____ Cu or No. 6 Al.

_____ _____ **36.** The grounded conductor shall be connected to a wire from the _____ of a lighting socket.

Trade Competency Test 4

NEC®	Answer	

_____ T F **1.** All 125 V, 20 A, 1ϕ receptacles installed in bathrooms of dwelling units shall have ground-fault circuit-interrupter protection for personnel.

_____ _____ **2.** Conductor _____ protection shall not be required where the interruption of the circuit would create a hazard, such as in a fire pump circuit.

_____ _____ **3.** Conductors installed inside AC cable (BX) shall have ampacity values rated at _____°C per Table 310-16.

_____ T F **4.** Receptacle outlets located above 5½″ from the finished floor can be counted as a required outlet.

_____ _____ **5.** Where several cables are routed through the same hole at a length of more than _____″ without spacing, derating factors shall be applied.

_____ T F **6.** Parallel conductors in a conduit run shall have an equipment ground pulled through each conduit.

_____ _____ **7.** A single electrode consisting of a rod, pipe, or plate that does not have a resistance to ground of _____ Ω or less shall be augmented by one additional electrode.

A. 25   C. 75
B. 50   D. 100

_____ _____ **8.** In general, at least _____″ of free conductor shall be left at each outlet.

_____ T F **9.** Where one or more fittings are within a box, the count toward the fill is based on the largest conductor entering the box.

_____ _____ **10.** The _____ conductor of all 3-wire, DC systems supplying premises wiring shall be grounded.

_____ _____ **11.** Conductors installed inside nonmetallic-sheathed cable shall have a rating of at least _____°C.

_____ _____ **12.** When 100′ of No. 16 fixture wire is connected to a standard 120 V circuit, the branch-circuit overcurrent protection device shall not exceed _____ A.

A. 15   C. 30
B. 20   D. 40

_____ _____ **13.** Fittings such as capped elbows and service entrance elbows can be classified as _____.

A. conduit bodies   C. junction boxes
B. switch boxes   D. gutter boxes

_____ T F **14.** CBs that are not marked with an IC rating are rated at 5000 A.

_____ _____ **15.** Conductors or cables are not required to be connected in a nonmetallic box where secured within _____″ of the box.

_____ _____   **16.** A 4″ × 1¼″ round box normally has _____ cu in. of fill space area.

_____ _____   **17.** FMC routed over 6′ in length shall be provided with a(n) _____.

_____ _____   **18.** NM cable shall be permitted for both exposed and concealed work in normally _____ locations.

_____ _____   **19.** The metal raceway enclosing tap conductors for recessed fixtures shall not be less than _____′ nor more than _____′ long.

       A. 1½; 6   C. 3; 5
       B. 2; 6   D. 3; 6

_____ _____   **20.** A fuse which does not have an IC rating marked on its surface shall be considered as having an IC rating of _____ A.

_____ _____   **21.** Conductors connected in parallel in general wiring systems shall be at least No. _____ or larger.

       A. 1/0   C. 2
       B. 1   D. 3

_____ _____   **22.** The conductors of NM cable shall be rated at 90°C, but the ampacity of the conductors shall be _____°C.

       A. 40   C. 75
       B. 60   D. 105

_____ T   F   **23.** Where not permanently accessible, nonmetallic-sheathed cable within 5′ of a scuttle hole does not require protection.

_____ _____   **24.** When it is impractical to install the service drop below the weatherhead, the point of attachment can be located _____″ from the weatherhead.

_____ _____   **25.** Only service drop conductors should be attached to a service _____.

_____ T   F   **26.** Lighting switches mounted in or on the wall can be located above 6′- 7″ from the finished floor.

_____ _____   **27.** Service drop conductors shall have a clearance of not less than _____′ from a window that opens and closes.

_____ _____   **28.** The NEC® requires that recessed portions of fixture enclosures shall be spaced at least _____″ from combustible material.

       A. ⅜   C. ⅝
       B. ½   D. ¾

_____ _____   **29.** AC cable (BX) does not have to be supported where used in _____′ lengths for flexibility.

_____ _____   **30.** The ampacity of UF cable shall be that of the _____°C conductors in accordance with 310-15.

_____ _____   **31.** All receptacle outlets installed within _____′ of a wet bar sink in a dwelling unit shall be GFCI-protected.

_____ _____   **32.** A Class I circuit operates at _____ VA and 30 V or less.

_____ T   F   **33.** The metal boxes listed in Table 370-16(a) are permitted to have conductors contained within rated at sizes No. 4 and larger.

_____ _____   **34.** In general, raceways on exterior surfaces of buildings shall be _____ and arranged to drain.

MULTIFAMILY DWELLINGS 3

A *multifamily dwelling* is a building containing three or more dwelling units. Common service sizes are 120/240 V, 1φ; 120/208 V, 1φ; and 120/208 V, 3φ.

CONDUCTORS

Table 310-16 lists current-carrying values for copper and aluminum wires when not more than three conductors are grouped in raceway, cable, or buried in the ground. Table 310-17 lists values for copper and aluminum conductors in free air. Tables 310-16 through 310-19 list ampacities of conductor insulation for specialized applications.

Table 310-16

The upper portion of this table deals with ampacities of conductors that are used at a nominal room temperature of 30° Celsius (C), or 86° Fahrenheit (F). The lower part of Tables 310-16 through 310-19 supplies correction factors for limiting ampacities of conductors at room temperatures above 30°C or 86°F.

Column 1 in the main part of the table lists AWG sizes of conductors from No. 18 to 2000 kcmil inclusive. Succeeding columns show ampacities for each size of conductor according to the type of insulation it has. The columns relating to 60°C, 75°C, and 90°C insulation in Tables 310-16 and 310-17 are most commonly used for inside wiring.

With size No. 8, for example, Column 2 shows that with 60°C insulation, the conductor carries 40 A without subjecting the insulation to damaging heat. If the No. 8 conductor has 75°C insulation (Column 3), it may be loaded to 50 A without harmful result. This size conductor may carry 55 A with the insulations in Column 4.

Suppose that the room temperature for this particular application is 50°C (122°F) instead of 30°C (86°F). Referring to the lower part of Table 310-16, Ampacity Correction Factors, a multiplying factor of .58 shall be used to determine the ampacity of a conductor having any of the Column 2 insulations.

Use multipliers of .75 for Column 3 insulations and .82 for Column 4 insulations. The absence of a multiplying factor in some columns indicates that the insulations shall not be used for the listed ambient temperatures.

Using the indicated factors, the ampacity of No. 8 Type TW conductor used at a room temperature of 50°C (122°F) equals 40 A × .58, or 23 A. No. 8 Type THWN conductor is equal to 50 A × .75, or 37.5 A.

The table shows that Column 2 insulations are not to be used at room temperatures greater than 140°F and Column 4 insulations at temperatures greater than 194°F. Per 240-3(d), the rating of No. 14 Cu is 15 A, No. 12 Cu is 20 A, No. 10 Cu is 30 A, No. 12 Al is 15 A, and No. 10 Al is 25 A.

Table 310-17

This table is similar to Table 310-16 except that it deals with the ampacities of single insulated conductors rated 0 V through 2000 V in free air. Table 310-17 is based on an ambient air temperature of 30°C (86°F). Ampacities of conductors in free air are greater than those of conductors in raceways, cables, or earth.

Tables 310-18 and 310-19

These tables are similar in construction. They deal with allowable ampacities of conductors rated 150°C to 250°C. The correction factor tables are also similar, but they also relate to higher temperatures.

Notes to the Tables

In addition to the tables, 310-15(b) contains ampacity adjustment factors which should be carefully evaluated. Although the facts are clearly stated, notice (b)(3), which states that bare conductors have the same ampacity as the insulated conductors with which they are used, and (b)(2) a., which lists derating factors where more than three current-carrying conductors are installed in a raceway or cable.

Table 310-13 – Conductor Application and Insulations

This table lists conductor applications and the physical characteristics of the various types of insulated conductors. It should be noted that 310-13, Conductor Constructions and Applications, states that the conductors listed in this table may be installed for any of the wiring methods recognized by Chapter 3 as specified in the tables. It should also be noted that 310-13 states that insulated conductors used for 600 V or less shall comply with Table 310-13.

There are several types of conductors that are given two maximum operating temperatures depending upon how they are used, such as Type THW. Type THW has a 75°C rating when used as a general purpose conductor and installed in a wet or dry location, but when it is installed within 3″ of a ballast it is given a 90°C rating as required in 410-33.

Table 310-13 also lists several types of conductors for specialized applications as noted, such as MI cable, which has an outer metal sheath, Types TBS and SIS, which are limited to switchboard wiring only, and Type TFE, which is limited to leads within apparatus in a dry location. Type THNN is a 90°C conductor that can be used in dry and damp locations. The ampacity is used with 60°C or 75°C based on equipment terminals.

Table 402-3 – Fixture Wire

The operating temperature of fixture wires shall not exceed the temperatures listed in Table 402-3.

Table 402-5 – Ampacity of Fixture Wires

Table 402-5 lists the allowable ampacities of fixture wires in sizes No. 18 through No. 10 AWG. The ampacities of the fixture wires listed in this table are based on an ambient temperature of 30°C (86°F).

Flexible Cords

Table 400-5(A) shows ampacities of flexible cords where not more than three conductors are present. If there are more than three conductors in the group, ampacities given here are to be reduced to 80% of listed values. It may be noted that ordinary portable cord size No. 18 will carry 7 A and size No. 16 will carry 10 A. Notes 1 through 12 of Table 400-4 should be read carefully.

Table 400-4 – Flexible Cords and Cables

This table sets forth construction features of the various cords and their conditions of use and application. Flexible cords and cables shall conform to the description provided in Table 400-4. If flexible cords and cables not included in Table 400-4 are to be used, they must undergo a special investigation as to their suitability for the intended use.

APARTMENTS – FLOOR PLAN

This print shows a six-unit, one-story multifamily dwelling (Apartments A – F). Five of the apartments (Apartments A – E) are identical, having living room (LR), kitchen (KIT.), bedroom (BR), and bath (B). The lighting panel, a flush type, is placed in the entrance storeroom. The kitchen has a wall-mounted oven (8), a counter-mounted cooktop (7), a dishwasher (10), a garbage disposal (9), and an exhaust fan (4).

The caretaker's apartment, Apartment F, has a living room (LR), a kitchen (KIT.), and a bath (B). It has no separate bedroom, a wall bed being used in the living room. Kitchen equipment is similar to that in the larger apartments.

There is a washroom (WR), a storeroom (SR), and a combination heating equipment and switchboard room (H/SR). There is a house receptacle circuit, an electric water heater circuit, and a pump motor. A house lighting circuit supplies the three workrooms and three night-light standards. All these circuits originate in a special house panel, which is adjacent to service equipment in the switchboard room.

Arrows pointing in the direction of the lighting panel indicate circuit runs. The number of cross marks on the arrow shows the number of wires in the particular home run. Where no cross marks are found, it is understood that the circuit consists of two wires. Conductors are assumed to be No. 14 unless referred to in the specifications or marked adjacent to the arrow.

Alternate long and short dashed lines denote branch-circuit raceways concealed in the floor.

Solid black lines mark branch-circuit conduit concealed in the wall or ceiling. Various types of outlets, including lights, plug receptacles, recessed units, special outlets, telephones, and fans, are shown with standard symbols. Feeders are shown by arrows originating at the various panels and pointing in the direction of the switchboard.

It should be observed that a note taken from the electrical specifications states that electrical metallic tubing is to be used in concrete floor slabs. Flexible conduit is to be used in walls and ceilings.

Load Calculations

The unit load specified for dwelling occupancies in Table 220-3(a) is 3 VA per square foot. The allowance for two small appliance or utility circuits per 220-16(a) is 3000 VA. The load required for two cooking units per Column C of Table 220-19 is 65% of the sum of the nameplate ratings. Based on these NEC® sections, the load for each of the five apartments, A to E inclusive, is determined as follows:

LOAD CALCULATIONS FOR APARTMENTS A – E

Lighting and Appliance			
710 sq ft × 3 VA	2130 VA		
Two appliance circuits	3000 VA		
Total	5130 VA		
Feeder Demand Factors			
1st 3000 VA at 100%	3000 VA		
Remainder			
2130 VA at 35%	745 VA		
Total	3745 VA		**3745 VA**
Oven and Cooktop			
Oven rated	4800 VA		
Cooktop rated	6700 VA		
Total	11,500 VA		
11,500 VA × .65	7475 VA		**7475 VA**
Dishwasher			**1500 VA**
Garbage Disposal			**600 VA**
Exhaust Fan			**350 VA**
		Total	**13,670 VA**

FEEDER SIZE (120/240 V service)

13,670 VA ÷ 240 V = **56.9 A**

Table 310-16 shows that No. 6 THW Cu conductor is needed. Table C1, Appd C shows that a 1″ EMT is required for three No. 6 THW Cu conductors.

LOAD CALCULATIONS FOR APARTMENT F

Lighting and Appliance

448 sq ft × 3 VA	1344 VA	
Two appliance circuits	3000 VA	
Total	4344 VA	

Feeder Demand Factors

1st 3000 VA at 100%	3000 VA	
Remainder 1344 VA at 35%	470 VA	
Total	3470 VA	**3470 VA**
Oven and Cooktop		**7475 VA**
Dishwasher		**1500 VA**
Garbage Disposal		**600 VA**
Exhaust Fan		**350 VA**
Total		**13,395 VA**

FEEDER SIZE (120/240 V service)

13,395 VA ÷ 240 V = **55.8 A**

Table 310-16 shows that No. 6 THW Cu conductor is needed. Table C1, Appd C shows that a 1″ EMT is required for three No. 6 THW Cu conductors.

In some cases, the neutral feeder may be smaller than the outer conductors. NEC 220-22 allows a 70% demand factor on the portion of the neutral feeder load that supplies an electric range or equivalent cooking appliance. Although the specifications and prints state that three No. 6 THW Cu conductors shall be installed to each panel, it is useful to perform calculations to see if a smaller neutral conductor might have been otherwise admissible. Neutral calculations for Apartment F are determined as follows:

ADDITIONAL LOAD CALCULATIONS FOR APARTMENT F

Oven and Cooktop = 7475 VA

7475 VA × .7 = 5233 VA		**5233 VA**
Assumed Feeder Load		
Lighting and Appliance		**3470 VA**
Dishwasher		**1500 VA**
Garbage Disposal		**600 VA**
Exhaust Fan		**350 VA**
Total		**11,153 VA**

NEUTRAL FEEDER SIZE

11,153 VA ÷ 240 V = **46.5 A**

Per Table 310-16, No. 8 THW Cu conductor is acceptable for the neutral. Maximum unbalance of 50 A per 220-22.

Another factor that is sometimes important when the load appears close to the current-carrying capacity of a particular size conductor is that of balancing circuits. In Apartment A, for example,

a 3-wire lighting circuit, a 3-wire utility circuit, and a 3-wire cooking circuit are balanced across the ungrounded conductors and the neutral wire. But the 1500 VA, 120 V dishwasher, which is connected between one ungrounded conductor and the neutral, does not balance the 950 VA garbage disposal and exhaust fan load, which is connected between the other ungrounded conductor and the neutral.

In such a case, an imaginary (artificial) load may be assumed in parallel with the smaller load in order to gain balance, its value here being equal to 1500 VA minus 950 VA, or 550 VA. The listing is then:

LOAD CALCULATION ON APARTMENT A WITH ARTIFICIAL LOAD

Lighting and Appliance	**3745 VA**
Oven and Cooktop	**7475 VA**
Dishwasher	**1500 VA**
Garbage Disposal	**600 VA**
Exhaust Fan	**350 VA**
Artificial Load	**550 VA**
Total	**14,220 VA**

ADJUSTED LOAD

14,220 VA ÷ 240 V = **59 A**

Per Table 310-16, No. 6 THW Cu conductors may be used.

Although the calculation has not altered the result here, the method is worth noting. Some inspection authorities insist upon balancing loads, others do not.

Service Calculations

NEC 220-3(b)(9) provides that a load of 180 VA per outlet shall be allowed for outlets other than for general illumination. The two plug receptacles in the work area come within such classification. NEC 220-13 permits this type of load to be included with the general lighting load in applying demand factors. These plug receptacles and the three 200 VA night-light standards come within scope of the section.

NEC 220-17 states that where four or more fixed appliances other than ranges, clothes dryers, space heating units, or A/C equipment are connected to the same feeder or service in a multifamily dwelling, a demand factor of 75% may be applied to the load. The six dishwashers, six garbage disposals, six exhaust fans, and electric water heater qualify here. The service conductor sizes are governed by 230-

42, which refers to 220, so the service conductor sizes are calculated as follows:

SERVICE CALCULATIONS

Lighting and Appliance

4220 sq ft × 3 VA =	12,660 VA

Six Appliance Circuits

6 × 3000 VA =	18,000 VA
Laundry Circuit	1500 VA

Three Night-light Standards

3 × 200 VA =	600 VA

Two Receptacle Outlets
in Work Area

2 × 180 VA =	360 VA
Total	33,120 VA

1st 3000 VA at 100%	3000 VA
Remainder	
30,120 VA at 35%	10,542 VA
Total	13,542 VA	**13,542 VA**

Ovens and Cooktops

6 × 4800 VA =	28,800 VA
6 × 6700 VA =	40,200 VA
Total	69,000 VA

Service Demand Factor

32% for 12 ranges 69,000 VA × .32 =   **22,080 VA**

Other Loads

Dishwashers	6 × 1500 =	9000 VA
Disposals	6 × 600 =	3600 VA
Exhaust fans	6 × 350 =	2100 VA
Water heater	1 × 5000 =	5000 VA
Total		19,700 VA

19,700 VA × .75 =   **14,775 VA**

Motor Load

¾ HP, 1φ, 230 V × 6.9 A = 1600 VA

1600 VA × 125% =   **2000 VA**

Total   **52,397 VA**

SERVICE ENTRANCE CONDUCTORS

52,397 VA ÷ 240 V = **218 A**

Table 310-16 shows that No. 4/0 THW Cu conductors are acceptable.

NEUTRAL CONDUCTOR

Lighting and Appliance	**13,542 VA**

Cooking Units

22,080 VA × .7 =	**15,456 VA**
Fixed appliances	**14,775 VA**
Total	**43,773 VA**

43,773 VA ÷ 240 V = **182 A**

Table 310-16 shows that No. 3/0 THW Cu conductor is acceptable as the neutral.

SERVICE SIZE

Two No. 4/0 and one No. 3/0 THW Cu conductors, all in 2″ RMC, are acceptable. *Note:* The Optional Calculation per 220-30 may be applied.

APARTMENTS – DETAILS AND ELEVATIONS

The larger print shows one of the five similar apartments (A – E). The plan view drawn to a larger scale brings out details to better advantage than is possible on the general print, so that the position of equipment may be found with greater accuracy.

The conduit in the floor slab is indicated by a dashed line with alternating long and short dashes. The conduit in a wall or ceiling is indicated by solid black lines. All wires are No. 14 except the 3-wire No. 8 circuit to the cooking units and the 3-wire No. 12 circuit to the kitchen appliance receptacles.

Circuits may be traced to learn the reason for three or four wires, as the case may be, in various conduits. Certain runs that might otherwise have been combined are kept separate to avoid the penalty of derating where more than three wires are installed in one raceway.

One point that should not be overlooked is the arrangement for connecting bedroom plug receptacles. They are supplied by conduit laid in the floor slab. This run continues up to the switch outlet box in the wall at the side of the door and then to the ceiling outlet. A small circle and the word *UP* indicate this procedure. It should also be noted that one half of the symbol designating the duplex receptacle outlet in the living room is a solid color. This indicates that one half of this receptacle outlet is controlled by a wall switch.

Triangles *A, B,* and *C* marked in a semicircle on the kitchen floor represent one scheme for directing attention to elevations seen at the right of the drawing. From these elevations, it is possible to determine exact locations of outlets such as those for plug receptacles and for cooking units. The location of the garbage disposal and its method of connection are shown in elevation A.

APARTMENT A

28'

25'-4"

KIT.

BR

B

LR

3" R-1 CONC SLAB

3 No. 8
3/4" C

₵ 7-8

₵ 9-10

₵ 4-5

3 No. 12
₵ 2-3

₵ 1

₵ 6

₵ 6

PANEL A

FEEDER A
3 No. 6THW-1" C

UP

PS

PS

PS

GFCI

DW

ELEV A

ELEV B

ELEV C

Note 1: Cross lines on conduit runs indicate number of wires in raceway; where no cross lines are shown, two conductors are present

Note 2: "UP" notation in bedroom partition shows conduit riser from plug receptacle outlet box to switch box

Note 3: "(O)" notation on switches and outlets of same circuit indicates which switches control the outlets

Note 4: Notation "₵" sometimes used as a symbol for "circuit"

FIXTURE LIST					
DESCR	QUANT	RATING*	MFGR	CAT.	TYPE
r	5	100	MORBRITE	612	RECESSED
s	6	50	ALLRAY	1117	BRACKET
t	12	150	ALLRAY	1129	DRUM
u	5	100	MORBRITE	700	DRUM
v	12	75	ALLRAY	1119	CEILING
j	6	75	MORBRITE	804	BRACKET
l	12	50	TITAN	322	PORC PC SOC
n	3	50	TITAN	323	PORC SOC
p	3	200	FRANKLIN	1791	WP FLOOD

* in W

SERVICE FOR MULTIPLE OCCUPANCY BUILDINGS

225-36 In general, when more than one building is on the same property, each building or structure shall have a separate disconnecting means. Disconnecting means shall be suitable for use as service equipment except for 3-way and 4-way snap switches for garages and out buildings on residential property. See Exceptions.

A. **230-2(a)** A building or other structure shall be supplied by only one service. See 230-2(a – d).

230-2(e) Where exceptions permit more than one service, a permanent plaque or directory is required at each service and branch-circuit disconnect location which indicates all other services and area served by each.

B. **230-40** In general, each service drop or lateral shall supply only one set of service-entrance conductors.

C. **230-40, Ex. 1** A multiple occupancy building is permitted to be supplied by two separate sets of service-entrance conductors.

D. **230-40, Ex. 2** Separate sets of service-entrance conductors are permitted for 2 to 6 service disconnects.

230-40, Ex. 3 A one-family dwelling can also supply a separate structure.

230-40, Ex. 4 A set of service-entrance conductors can supply a common panel per 210-25.

230-40, Ex. 5 A set of service-entrance conductors can supply a fire pump or life safety per 230-82 4.

SERVICE — LIGHTING AND APPLIANCE PANELBOARD

110-14(a) One conductor per terminal is permitted in the equipment grounding terminal bar, unless the terminal is identified for two conductors.

230-43 Wiring methods for service conductors are limited to: (1) open wiring on insulators; (2) IGS cable; (3) RMC; (4) IMC; (5) EMT; (6) ENT; (7) SE cable; (8) wireways; (9) busways; (10) auxiliary gutters; (11) RNMC; (12) cablebus; (13) MC cable; (14) MI cable; (15) FMC or LTFMC, 6′ maximum length, with EBJ; and (16) LTFNMC.

373-5(c) Where cable is used, each cable shall be connected to the cabinet. Two cables from one connector are not permitted unless the connector is approved for two cables.

SERVICES AND FEEDERS

DISCONNECT

GROUNDED CONDUCTOR

GROUNDED CONDUCTOR

E

D

A

3φ, 4-WIRE CIRCUIT

B

C

A. 110-10 Impedance and other circuit characteristics shall be considered so that the selected overcurrent device will clear the fault before serious damage results.

B. 220-22 The neutral conductor shall be large enough to carry the maximum unbalanced load.

C. 230-42(c) The grounded (neutral) conductors shall not be smaller than required by 250-24(b)(1).

D. 240-20 CBs are permitted as overcurrent devices and shall open all ungrounded conductors. Individual single-pole CBs may be used with handle ties

in grounded systems in: (1) 1φ circuits; (2) 3-wire DC circuits; (3) lighting and appliance 3φ, 4-wire or 2φ, 5-wire branch circuits supplied from a system with a grounded neutral and not operating at over 150 V to ground.

250-24(b)(1) The service grounded conductor shall not be smaller than the GEC per Table 250-66 and if phase conductors are larger than 1100 kcmil Cu or 1750 kcmil Al, the grounded conductor shall not be smaller than 12½% of the area of the largest phase conductor except it need never be larger than the largest service ungrounded conductor.

E. 230-90(b) No overcurrent device shall be inserted in a grounded service conductor except a CB which simultaneously opens all conductors.

DISTRIBUTION PANELBOARDS

FEEDER TO OTHER PANELBOARDS

FEED-THROUGH PANEL

NEUTRAL TERMINAL BAR

SIDE WIRING SPACE

TERMINAL FOR EQUIPMENT GROUND ONLY

C

E

WIRE BENDING SPACE

B

A

D

FEEDER TO PANELBOARD

A. 250-142(b) A grounded circuit conductor shall not be used for grounding noncurrent-carrying metal parts of equipment on the load side of the service disconnecting means except for separate buildings per 250-32(b) and separately derived systems per 250-30(a)(1).

B. 373-6(b) Conductors entering and leaving the cabinet shall have wire bending space at the terminals per Table 373-6(a) and Table 373-6(b).

373-6(c) Ungrounded conductors No. 4 and larger shall be protected by a substantial insulating material, securely fastened in place. See 300-4(f).

C. 373-11(c) Cabinets and cutout boxes which contain devices or apparatus connected to more than 8 conductors, not including the supply circuit or a continuation thereof, shall have back or side-wiring spaces.

384-16(a) Each branch-circuit panelboard shall be individually protected on the supply side by not more than two CBs with a combined rating that does not exceed that of the panelboard.

D. 384-16(a), Ex. 1 Individual protection is not required if the panelboard feeder has overcurrent protection not greater than the panelboard's rating.

E. 384-20 An identified terminal bar shall be provided inside cabinet and secured to cabinet for connection of all feeder and branch-circuit EGCs. The terminal bar for EGCs shall be secured to cabinet.

MADE AND OTHER ELECTRODES — 250-52

250-52 If a metal underground water pipe, metal frame of the building, concrete-enclosed electrode, or a ground ring is not available, other local metal underground systems or structures, rod and pipe electrodes, or plate electrodes shall be used.

Made electrodes shall be below moisture level and shall not be painted. If more than one electrode is used, each electrode of a grounding system shall be at least 6′ from any electrode of another grounding system. Two or more electrodes which are bonded together shall be considered a single system.

A. 250-52(a) Underground metal gas piping shall not be used as a grounding electrode.

B. 250-52(b) Other local metal underground systems which are permitted include piping systems and underground tanks.

C. 250-52(c) Rod and pipe electrodes shall be at least 8′ in length.

D. 250-52(c)(1) Conduit or pipe electrodes shall be at least ¾″ in diameter and shall be corrosion-resistant.

E. 250-52(c)(2) Iron or steel rod electrodes shall be at least ⅝″ in diameter. Nonferrous electrodes shall be at least ½″ in diameter.

250-52(c)(3) At least 8′ of the electrode shall be in contact with the soil. If rock is encountered, it shall be driven at up to a 45° angle from vertical or be buried in a trench at least 2′ deep.

250-52(d) Plate electrodes shall be at least 2 sq ft. Iron or steel plates shall be at least ¼″ thick. Nonferrous plates shall be at least .06″ thick.

250-56 If resistance of made electrodes exceeds 25 Ω, an additional electrode is required and shall be at least 6″ away.

F. The GEC shall be accessible and installed in a manner that makes connection both permanent and effective. Bonding around insulated joints like water meters helps to ensure an effective ground path.

GROUNDING MULTIPHASE SYSTEMS

3φ NETWORK SYSTEM

A. 250-26(3) The common conductor shall be grounded for multiphase systems having a conductor common to all phases.

B. 250-66 The GEC is sized per Table 250-66.

TERMINAL CONNECTIONS

A. 110-14(a) Connection of conductor to terminals shall be made with pressure connectors, solder lugs, or splices to flexible leads. Terminals for more than one conductor and terminals to connect Al conductors shall be identified for such use. See 110-14(c) for 60°C and 75°C terminal ratings.
110-14 Conductors of dissimilar metals not to be intermixed unless terminal, lug, or connector identified for purpose.

B. 110-14(a) Connection with wire bending screws or studs and nuts with upturned lugs is permitted for No. 10 or smaller conductors.

SWITCH ENCLOSURES

A. 110-12 Equipment shall be installed in a neat manner.

B. 110-13(a) Electrical equipment to be securely fastened to mounting surface. Wooden plugs driven into masonry or concrete are not acceptable for support.
110-3(b) Evidence of suitability of equipment may be by listing or labeling.

C. 373-8 Switch enclosures are not to be used as auxiliary gutters, raceways, or junction boxes unless adequate space is provided. These conductors permitted if not more than 40% of cross-sectional area is used. For splices and taps, not over 75% of area may be used.

D. 380-8(a) Center of grip of operating handle when in highest position shall not be more than 6′-7″ from floor or platform.

SWITCHES

A. 380-14(a) AC general-use snap switch, suitable for use on AC only, may be used to control resistive and inductive loads not to exceed the ampere rating of the switch or the voltage of the circuit, or tungsten-filament loads of rating of switch at 120 V, or motor loads not to exceed 80% of rating of switch of rated voltage. Snap switches rated 20 A or less shall be listed and marked CO/ALR when used with aluminum wire. See 380-14(c).

B. 380-8(a), Ex. 1, 2, 3 Switches and CBs shall be readily accessible and not over 6'-7" from floor or platform except: (1) busway switches permitted at same level as busway; (2) switches located adjacent to motors, appliances, or other equipment; (3) hookstick operated isolating switches.

380-8(b) Voltage between adjacent snap switches in same box shall not exceed 300 V. If voltage between adjacent switches exceeds 300 V, then permanent barriers shall be installed.

380-14(b)(1, 2, 3) AC/DC general-use snap switches suitable for AC or DC may be used to control resistive loads not exceeding the ampere rating of the switch or the voltage of the circuit, or inductive loads not exceeding 50% of the ampere rating of the switch at applied voltage, when "T" rated. See 430-83, 430-109, 430-110 for motors, and 600-6(b) for signs.

RECEPTACLE GROUNDING

A. 210-7(a) Receptacles installed on 15 A and 20 A circuits shall be of grounding type.

210-7(b)(c) Receptacles having grounded contacts shall be rounded. Branch circuits shall include a grounding conductor unless the raceway provides grounding.

210-7(c), FPN 2 For extension to non-grounding existing circuits, an adjacent water pipe may be used for grounding. See 250-130(c).

B. 210-7(d) Grounding-type receptacles to be used as replacements for existing nongrounding types except where receptacle box does not have grounding means, a GFCI or nongrounding-type receptacle is permitted.

210-7(f) Attachment plugs shall not be interchangeable on receptacles connected to circuits of different voltages, frequencies, or types of current.

C. 250-146(b) Contact device or yoke, if designed and listed for purpose, may be used to bond receptacle to flush-type boxes.

410-56(a) Receptacles for portable cords shall be rated not less than 15 A for 125 V or 250 V. See Ex. for 10 A, 250 V nonresidential. Also see 210-21(b).

"HOT" WIRE SCREWS (BRASS)

EQUIPMENT GROUNDING SCREW (GREEN AND HEXAGONAL)

GROUNDING-TYPE RECEPTACLE

NEUTRAL WIRE SCREWS (SILVER)

NONGROUNDING-TYPE RECEPTACLE

ELECTRIC-DISCHARGE LIGHTING

SIGN TRANSFORMER

NEON TUBING

RESIDENTIAL COVE OR SPECIAL EFFECT LIGHTING

LOW-VOLTAGE LEADS

HIGH-VOLTAGE TERMINALS

FLUORESCENT FIXTURE

A. 410-73(a) Equipment used with electric discharge lighting and designed for open-circuit voltage of 1000 V or less shall be of a type intended for such use.

B. 410-73(e)(1) In general, all fluorescent fixtures installed indoors shall have ballasts with integral thermal protection. See Special Applications for (2) simple reactance ballasts; (3) exit fixtures; and (4) exit lighting.

410-75 Equipment having an open-circuit voltage of more than 300 V shall not be installed in dwelling occupancies. Equipment having an open-circuit voltage of more than 300 V shall not be installed in or on dwelling occupancies unless so designed that no live parts are exposed when lamps are being handled.

410-76(b) Fluorescent fixtures mounted on combustible low-density cellulose fiberboard are required to have 1½" air space or listed for direct mounting.

410-76(b), FPN Combustible low-density cellulose fiberboard includes sheets, panels, and tiles formed of bonded plant fiber and has a density of 20 lb or less per cubic foot. Does not include solid or laminated wood or material integrally treated with fire-retardant chemicals to limit flame spread to not over 25'.

GROUNDING – METAL ENCLOSURES

A. 250-86 Metal enclosures for other than service conductors shall be grounded.

250-86, Ex. 1 Additions to existing installations, metal enclosures and raceways for open wiring, knob-and-tube, and NMC without an EGC need not be grounded if less than 25′ and suitably isolated or guarded against contact by persons.

B. 250-86, Ex. 2 Short sections of metal conduit used to protect cables from physical damage need not be grounded.

Table 250-122 EGC for 15 A, 20 A, and 30 A circuits shall have ampacity at least equal to overcurrent protection of circuit. For circuits over 30 A, use Table 250-122.

RACEWAY GROUND CLAMPS

GROUNDING CORD-AND-PLUG CONNECTED EQUIPMENT

A. 250-114 3. In residential occupancies, the following cord-and-plug connected equipment shall be grounded: clothes washers and dryers, refrigerators, freezers, air conditioners, dishwashers, electrical aquarium equipment, sump pumps, portable hand-held power tools and motor-operated appliances such as saws, sanders, drills, lawn mowers, wet scrubbers, snow blowers, portable headlamps, and hedge clippers.

B. 250-114, Ex. Listed tools and appliances protected by a double insulation system and so marked need not be grounded.

250-138 Noncurrent-carrying metal parts of cord-and-plug connected equipment which shall be grounded may be grounded: (a) by an EGC which is part of the cable or flexible cord, and which is either bare or has a green or green with one or more yellow stripes on the outer covering; or (b) by a separate flexible wire or strap. See Exception.

250-122(e) Fixture wires and cords are permitted to contain a No. 18 or larger Cu EGC, provided it is not smaller than circuit conductors per 240-4.

410-58(c) A grounding terminal or grounding device shall be used only for grounding purpose.

PORTABLE DEVICE RECEPTACLE

DOUBLE-INSULATED TOOL DOES NOT REQUIRE GROUNDING CORD WITH GROUNDING CONDUCTOR

SWITCHING CONNECTIONS

A. 380-2(a) All 3-way and 4-way switches shall be wired so that the ungrounded conductor is switched.

HOT WIRE

GROUNDED CONDUCTOR RETURN WIRE

TRAVELERS 3-WAY SWITCH SWITCH LEG LIGHTING FIXTURE

3-WAY SWITCH CONDUCTORS IN SAME RACEWAY NEUTRAL

NEUTRAL SPLICED IN EACH SWITCH BOX

3-WAY SWITCH LIGHTING FIXTURE

CONDUCTORS IN SAME RACEWAY 4-WAY SWITCH 3-WAY SWITCH

SWITCHING ON CIRCUITS

GROUNDED NEUTRAL CONDUCTOR LIGHT IN HOUSE LIGHT ON PORCH YARD LIGHT LIGHT AT GARAGE LIGHT IN GARAGE

SWITCH LEG

UNGROUNDED HOT CONDUCTOR • 380-2 4-WAY SWITCH 3-WAY SWITCHES

NOT AFFECTED BY OTHER SWITCHES

SWITCH IN GARAGE

RECEPTACLE IN GARAGE HOT AT ALL TIMES

THREE SWITCHES TO CONTROL FOUR LIGHTS

3-WAY SWITCH 4-WAY SWITCH 3-WAY SWITCH LIGHT GROUNDED NEUTRAL CONDUCTOR

THREE SWITCHES TO CONTROL ONE LIGHT

GROUNDED NEUTRAL CONDUCTOR

FLOODLIGHT AT HOUSE

LIGHT AT GARAGE LIGHT AT GARAGE

3-WAY SWITCH 3-WAY SWITCH

CIRCUIT IN GARAGE HOT AT ALL TIMES

RECEPTACLE

SWITCH IN GARAGE

HOT CIRCUIT PROVIDED IN GARAGE

NONMETALLIC EXTENSIONS – 342

NONMETALLIC SURFACE EXTENSIONS

Hubbell Incorporated (Delaware), Wiring Device-Kellems

Nonmetallic surface extensions may be mounted directly on walls or ceilings.

> **342-1: Nonmetallic Extensions** – Assemblies of two insulated conductors in a nonmetallic jacket or an extruded thermoplastic covering. Nonmetallic extensions may be either surface extensions or aerial cable.

AERIAL CABLE

USES PERMITTED – 342-3

(a) When from an existing outlet on a 15 A or 20 A branch circuit per 210

(b) When exposed and in a dry location

(c) For nonmetallic surface extensions in residential or office buildings. See 336-5(a)(1)

(c1) For aerial cable in industrial buildings

USES NOT PERMITTED – 342-4

(a) As aerial cable to substitute for a general wiring method

(b) In unfinished basements, attics, or roof spaces

(c) For nonmetallic surface extensions where voltage between conductors exceeds 150 V. For aerial cable where voltage between conductors exceeds 300 V

(d) Where subject to corrosive vapors

(e) When run through a floor or partition

A. 342-5 Surface extensions shall be continuous without splices. Approved fittings are permitted for taps.

B. 342-7(a)(1) One or more surface extensions can run in any direction from an existing outlet. They shall not be run on the floor or within 2″ of the floor.

C. 342-7(a)(2) Surface extensions shall be supported within every 8″.

D. 342-7(a)(2) If connection is an attachment plug, the first support shall be within 12″.

E. 342-7(b)(1) Aerial cable shall be attached to turnbuckles and clamps. It shall be supported every 20′ and not be within 2″ of building steel or conductive material.

F. 342-7(b)(2) Clearances above floors shall be 10′ for pedestrian traffic and 14′ for vehicular traffic.

G. 342-7(b)(3) Clearances above work benches shall be 8′.

H. 342-7(b)(4) Aerial cables are permitted to support lighting fixtures if load does not exceed the intended load.

342-7(b)(5) Identified messenger cables may serve as EGC when installed per 250. They shall not serve as branch-circuit conductors.

FIXED ELECTRIC SPACE-HEATING EQUIPMENT – 424 . . .

A. 424-3(a) Fixed electric space-heating equipment may be supplied by any size branch circuit. For two or more outlets for fixed electric space-heating equipment, branch circuits shall be 15 A, 20 A, or 30 A.

424-3(a), Ex. In non-residential occupancies, fixed infrared heating equipment may be supplied from branch circuits not exceeding 50 A.

424-3(b) Branch-circuit conductor's ampacity and the rating of OCPDs for electric space-heating equipment with resistance elements shall not be less than 125% of the total heating load.

424-9 Factory-installed receptacles or outlets which are part of a separate listed assembly are permitted to be installed in place of those in 210-50(b).

B. 424-11 Fixed electric space heaters with supply conductors of over 60°C shall be clearly and permanently marked.

C. 424-12(a) Fixed electric space heaters shall be adequately protected in an approved manner where exposed to physical damage.

D. 424-12(b) Fixed electric space heaters installed in damp or wet locations shall be approved for such use.

424-13 Unless approved for direct contact with combustible material, fixed electric space heaters shall have required spacing.

E. 424-20(a) Thermostatic devices are permitted as both controllers and switching devices provided they:

(1) have a marked OFF position

(2) directly open all ungrounded conductors when in the OFF position

(3) are so designed that they will not energize automatically after being placed in the OFF position

(4) are located per 424-19

424-20(b) Thermostats that do not open all ungrounded conductors and thermostats that operate remote-control circuits are not required to meet 424-20(a) and shall not be used as disconnecting means.

F. 424-34 Heating cables shall have factory-assembled nonheating leads at least 7′ in length.

G. 424-35 Nonheating leads of heating cable shall be marked with the identifying name or symbol, catalog number, and ratings in volts and watts or volts and amps. The following color identification shall be used to mark nominal voltage on each nonheating lead within 3″ of the terminal end:

120 V – yellow

208 V – blue

240 V – red

277 V – brown

480 V – orange

H. 424-38(a) Heating cables shall not extend beyond the room or area where they originate.

424-38(b) Heating cables shall not be installed in closets, over walls or partitions that extend to the ceiling, or over cabinets without proper clearance from the ceiling.

424-38(c) Heating cables may be used to control relative humidity in closets when shelves or other permanent fixtures do not obstruct the space.

I. 424-39 Heating cables shall be separated at least 8″ from outlet and junction boxes, surface fixtures, and at least 2″ from any part of recessed fixtures, ventilating, or other room openings. Surface fixtures shall not be mounted under heating cables.

424-40 Embedded heating cables may be spliced by approved means where necessary. The length of the cable shall not be altered.

424-41(a) In general, heating cables are not permitted in walls.

424-41(a), Ex. Single runs of heating cable are permitted to run down a vertical surface to reach a dropped ceiling.

J. 424-41(b) Adjacent runs of heating cable not exceeding 2¾ W per foot shall be installed at least 1½″ OC.

K. 424-41(c) Heating cables shall be applied only to gypsum board, plaster lath, or other nonconducting surfaces. A layer of plaster shall be applied to metal lath or conducting surfaces before applying heating cables.

424-41(e) The entire ceiling surface shall have a finish of thermally noninsulating material at least ½″ in nominal thickness or noninsulating material suitable for the use and installed per specifications.

L. 424-41(f) Heating cables shall be secured at intervals not exceeding 16″ by means of staples, tape, plaster, or other approved means. Metal fasteners shall not be used with metal lath or other conducting surfaces.

424-41(h) Cables shall not contact metal or other conducting surface.

M. 424-43(a) Free nonheating leads from a junction box to a location within the ceiling shall be installed with approved wiring methods. Conductors in raceways, Type UF, NMC, MI cable, or other approved conductors may be used.

N. 424-43(b) At least 6″ of free nonheating lead shall be within the junction box. Markings of the lead shall be visible.

424-43(c) Excess leads shall not be cut. They shall be embedded in plaster or other approved material.

O. 424-44(a) Constant wattage heating cables in concrete or poured masonry floors shall not exceed 16½ W per linear foot.

P. 424-44(b) The minimum spacing between adjacent runs of heating cable in concrete or poured masonry floors is 1″ OC.

Q. 424-44(c) Heating cables in concrete shall be held in place by nonmetallic frames while concrete is poured.

424-44(d) Spacing shall be maintained between heating cable and metal in concrete floors unless the cable is grounded MC cable.

R. 424-44(e) Leads leaving the concrete floor shall be protected by RMC, IMC, PVC, EMT, or other approved means.

S. 424-44(f) Bushings or approved fittings shall be used in the floor slab where the leads emerge.

424-45 Cable installations shall be inspected and approved before cables are covered or concealed.

424-58 Heaters installed in air ducts shall be identified as suitable for such use.

424-65 Controller equipment for duct heaters shall be accessible with the disconnect at or within sight of the controller or as permitted by 424-19(a).

424-91(a): Heating Panel – A complete assembly with a junction box or a length of flexible conduit for connection.

424-91(b): Heating Panel Set – A rigid or nonrigid assembly with nonheating leads or a terminal junction suitable for connection.

T. 424-92(a) Panels to be marked in a location readily visible before finish is applied.

424-92(b) Heating panels shall be suitable for the installation.

U. 424-92(c) Each heating panel shall be marked with identifying name or symbol, catalog number, volts and watts, or volts and amps.

424-93(a)(2) Heating panels and heating panel sets shall not be:

(a) installed in or behind surfaces where subject to physical damage

(b) run through or above walls and ceilings

(c) run in or through thermal insulation

V. 424-93(a)(3) Panels shall be separated at least 8″ from outlet and junction boxes for surface fixtures and at least 2″ from any part of recessed fixtures, ventilating, or other room openings. Surface fixtures shall not be mounted under heating panels.

424-93(b)(4) Heating panel sets shall be installed complete unless identified for field cutting.

W. 424-94 Wiring installed above a heated ceiling shall be considered as operating at an ambient temperature of 50°C and shall be derated accordingly. Wiring shall be spaced at least 2″ above the heated ceiling.

X. 424-94 Wiring located above thermal insulation which is at least 2″ thick need not be derated.

Y. 424-95(a) Wiring methods are per 300 and 310-10.

...FIXED ELECTRIC SPACE-HEATING EQUIPMENT — 424

DAMP LOCATION **D**

PROTECTION FROM PHYSICAL DAMAGE **C**

A

E

B

MARKING

OFF LOW HIGH

THERMOSTATIC DEVICE

FIXED ELECTRIC SPACE HEATER

COIL OF HEATING CABLE

F 7′ NONHEATING LEAD

COLORED MARKINGS FOR VOLTAGE **G**

JUNCTION BOX

I

6″

8″ MINIMUM

N

M

CEILING

K GYPSUM BOARD

J

$2\frac{3}{4}$ W/FT

$1\frac{1}{2}$″ OC MINIMUM

2″ MINIMUM

VENTILATOR

H

X 2″ THERMAL INSULATION

W WIRING 2″ ABOVE THERMAL INSULATION

HEATING PANEL

T

U

HEATING PANEL

2″ MINIMUM

8″ MINIMUM

V

Y

HEATING CABLES

16″ MAXIMUM

FASTENERS

L

WALL

WALL

VENTILATOR OPENING

R

S

Q NONMETALLIC FRAMES

METALLIC REINFORCING SCREEN

RMC

BUSHING

SUPPORT

1″ OC MINIMUM **P**

HEATING CABLES

CHAIRS

CONCRETE SLAB

CONSTANT WATTAGE CABLE — MAXIMUM $16\frac{1}{2}$ W/FT **O**

SPACE-HEATING CABLES

FIXED OUTDOOR ELECTRIC DEICING AND SNOW-MELTING EQUIPMENT – 426

Heating System – A complete system of components such as heating elements, fasteners, nonheating circuit wiring, leads, temperature controllers, safety signs, junction boxes, raceways, and fittings.

Resistance Heating Element – A specific element to generate heat that is embedded in or fastened to the surface.

Impedance Heating System – A system in which heat is generated by pipe(s) and rod(s) by an electrical source.

426-4 Branch circuit conductors for fixed electric deicing and snow-melting equipment shall have an ampacity of at least 125% of total heater load.

426-10(1) Equipment for outdoor electric deicing and snow melting shall be suitable for the chemical, thermal, and physical environment.

426-10(2) Equipment for outdoor electric deicing and snow melting shall be installed per the manufacturer's plans and instructions.

426-20(a) Panels shall not exceed 120 W/sq ft of heated area.

A. **426-20(b)** Spacing between adjacent cables shall be at least 1″ OC.

B. **426-20(c)** Units, panels, or cables shall be installed:

(1) On 2″ of asphalt or masonry base with at least 1½″ of covering

(2) On other approved bases at least 3½″ thick and 1½″ minimum top covering

(3) Equipment for other forms of installation shall be thoroughly investigated.

426-20(e) Cables and panels shall not bridge expansion joints without provision.

426-22(a) Nonheating leads with a grounding sheath or braid may be embedded in asphalt or masonry without other protection.

426-22(b) All but 1″ to 6″ of nonheating leads not having a grounded conductor shall be enclosed in metal raceways within asphalt or masonry.

C. **426-23(a)** Not less than 6″ of nonheating leads shall be in junction box.

D. **426-28** GFCIs shall be provided for fixed outdoor electric deicing and snow-melting equipment unless the cable is Type MI in noncombustible medium.

OVERCURRENT PROTECTION OF FIXED ELECTRIC SPACE-HEATING EQUIPMENT – 424-22

A. **220-15** Fixed electric space-heating loads are computed at 100% of the total connected load. The feeder load current rating shall not be smaller than the rating of the largest branch circuit supplied.

220-15, Ex. The AHJ may grant permission for feeder conductors to have an ampacity less than 100% when units operate on duty cycle, intermittently, or not at one time.

B. **424-3(b)** The branch-circuit conductors and overcurrent protective devices which supply fixed electric space-heating equipment shall be rated not less than 125% of the total load of the motors and heaters.

424-22(a) Fixed electric space-heating equipment, except motor-operated equipment per 430 and 440, is permitted to be protected against overcurrent when supplied by a branch circuit per 210.

424-22(b) When resistance-type heating elements of over 48 A are used in space-heating equipment, they shall be subdivided into smaller circuits of not over 48 A and protected at not more than 60 A. The rating of the overcurrent device shall be per 424-3(b) where the subdivided load is less than 48 A.

INFRARED LAMP INDUSTRIAL HEATING APPLIANCES – 422-48

A. **422-48(a)** Infrared heating lamps rated at 300 VA or less may be used with lampholders of the medium-base unswitched type or other types identified for the purpose.

422-48(b) Infrared lamps over 300 VA shall not use screw-shell lampholders. See Exception.

LOADS FOR ADDITION TO EXISTING BUILDINGS

A. **220-3(c)(1)** When calculating loads in existing dwellings for previously unwired portions or new additions, either of which exceeds 500 sq ft, the VA-per-square-foot basis shall be used. For new circuits or extended circuits, the VA-per-square-foot or the actual load served per 220-3(c) shall be used.

220-3(c)(2) In other than dwellings, adding new or extending existing circuits shall be calculated per 220-3(b)(c).

OVERCURRENT PROTECTION OF APPLIANCES – 422-11

A. 422-11(a) If appliance is marked with protective device rating, the branch-circuit OCPD cannot exceed the appliance marked rating.

800 VA HEATER — PLUG RECEPTACLE — LIGHTING OUTLET — 15 A OR 20 A CIRCUIT

DISCONNECTING MEANS FOR MOTOR-DRIVEN APPLIANCES – 422-31

422-31(a) The disconnecting means for a motor-drive appliance of more than ⅛ HP shall be located within sight of the motor controller and shall comply with 430, Part I.

422-35, Ex. Where the appliance unit switch disconnects all ungrounded conductors, and switch is marked with OFF position, then switch or CBs permitted to be out of sight, only if conforming to 422-33 (a)(b)(c)(d).

DISCONNECT IN CIRCUIT-BREAKER PANEL ONLY ACCEPTABLE IF IN SIGHT OF MOTOR CONTROLLER

CONTROL SWITCH FOR FOOD DISPOSER

FOOD DISPOSER

IDENTIFIED CONDUCTOR USED AS SWITCH LOOP

A. 200-7(c)(2) A cable containing an identified conductor may be used for single-pole, three-way, or four-way switch loops if the unidentified wire is made the return conductor from the switch to the outlet and is permanently reidentified to indicate its use by painting or other effective means.

TWO-WIRE CABLE — BLACK WIRE — IDENTIFIED (WHITE) WIRE — PLUG RECEPTACLE — BLACK SUPPLY WIRE IS SPLICED TO WHITE SWITCH-LOOP CONDUCTOR — WHITE WIRE PERMANENTLY REIDENTIFIED — UNGROUNDED (BLACK) WIRE — SWITCH BOX

CORD-CONNECTED APPLIANCES

APPLIANCE — KNOT IN CORD, WINDING WITH TAPE, OR FITTING DESIGNED FOR PURPOSE — FLEXIBLE CORD — TERMINAL BLOCK

A. 210-23(a) One cord-and-plug appliance on a 15 A or 20 A circuit shall not exceed 80% of circuit rating. Total fastened in place appliances not to exceed 50% of circuit rating if lights and other portable equipment or both are supplied.

B. 400-10 Flexible cords shall be so connected to devices and fittings that tension will not be transmitted to joints or terminal screws.

C. 422-16; 400-7(a)(6) Flexible cord is permitted for connection to appliances to facilitate frequent interchange.

FIXED APPLIANCE AND NONCOINCIDENT LOADS – DWELLINGS

SERVICE — WATER HEATER — FOOD DISPOSER — DISHWASHER — GARBAGE COMPACTOR — NONCOINCIDENT LOADS — 75% DEMAND FACTOR MAY BE APPLIED

A. 220-17 When four or more fixed appliances, other than electric ranges, clothes dryers, air conditioning equipment, or space-heating equipment are connected to the same feeder in a one-family, two-family, or multifamily dwelling, a demand factor of 75% may be applied to the fixed appliance load.

B. 220-21 In making load calculations, the smaller of two noncoincident loads may be omitted from the total when it is unlikely that both loads will be served simultaneously.

ELECTRIC RANGES

RECEPTACLE OUTLETS

SINGLE RECEPTACLE OUTLET	
DUPLEX RECEPTACLE OUTLET	
TRIPLEX RECEPTACLE OUTLET	
SPLIT-WIRED DUPLEX RECEPTACLE OUTLET	
SPLIT-WIRED TRIPLEX RECEPTACLE OUTLET	
SINGLE SPECIAL-PURPOSE RECEPTACLE OUTLET	
DUPLEX SPECIAL-PURPOSE RECEPTACLE OUTLET	
RANGE OUTLET	R

A. 210-19(c) Conductors for ranges 8¾ kW and larger shall not be rated less than 40 A.

B. 210-19(c), Ex. 2 The neutral conductor for a household electric range may have an ampacity not less than 70% of the branch-circuit rating and shall not be smaller than No. 10.

C. 220-19 Where single-phase ranges are supplied by a 3φ, 4-wire feeder, the load shall be computed per Table 220-19 on a demand basis of twice the maximum number between any two phases.

D. 250-140 For new branch-circuit installations, frames of electric ranges, cooktops, ovens, dryers, and their associated outlet or junction boxes shall be grounded the same as other electrical equipment. See 250-134 and 250-138. For existing branch-circuit installations, the grounded circuit conductor (neutral) may be used as the EGC, if all following conditions are met: (a) circuit is 120/240 V, 1φ, 3-wire or 120/208 V derived from a 3φ, 4-wire system; (b) grounded conductor is No. 10 Cu or No. 8 Al; (c) grounded conductor is insulated or is uninsulated and part of an SE cable and is fed from service equipment; or (d) grounded contacts of receptacles are bonded to equipment.

E. 400-10 Cord shall be connected so tension not transmitted to terminals.

PHASE	NO. OF RANGES
A	10
B	10
C	10

(TWICE THE NUMBER OF RANGES PER PHASE = 20)

120/208 V, 3φ, 4-WIRE

220-19: 10 12 kW RANGES /φ x 2 = 20 RANGES
TABLE 220-19: 20 RANGES = 35 kW
35 kW ÷ 2 = 17,500 VA x 3 = 52,500 VA
52,500 VA ÷ 360 = 145.83 A = **146 A**

3φ VOLTAGE x √3

for 208 V x 1.732, use 360
for 230 V x 1.732, use 398
for 240 V x 1.732, use 416
for 440 V x 1.732, use 762
for 460 V x 1.732, use 797
for 480 V x 1.732, use 831
for 2400 V x 1.732, use 4157
for 4160 V x 1.732, use 7205

AIR CONDITIONING EQUIPMENT

Room air conditioners are placed in the window or wall of the room being cooled.

MODEL #: RAKA-037JA2   MFD: XXX
VOLTS: 208/230   PHASE: 1   HERTZ: 60
COMPRESSOR RLA: 16.5/16.5   L.R.A.: 95
OUTDOOR FAN MOTOR: F.L.A. 13   HP: 1/5
MINIMUM SUPPLY CIRCUIT AMPACITY: 22/22A
MAXIMUM FUSE OR CB SIZE: 35/35A
MINIMUM FUSE OR CB SIZE: 30/30A
DESIGN PRESSURE HIGH: 300 PSIG
DESIGN PRESSURE LOW: 150 PSIG
OUTDOOR UNIT FACTORY CHARGE: 79 oz - R22
HACR TYPE BREAKER REQUIRED FOR USA

Carrier Corporation

INSTALLATION AND OPERATING INSTRUCTIONS LIST FUSES FOR OVERCURRENT PROTECTION

FUSED DISCONNECT ON UNIT

NAMEPLATE

CENTRAL AIR CONDITIONER LISTED FOR OUTDOOR USE

WHERE PROPER FUSE IS USED IN PANEL, THEN NON-FUSED DISCONNECT OR CB CAN BE USED. WHEN NAMEPLATE STATES THAT UNIT IS TO BE PROTECTED BY "OVERCURRENT DEVICE" OR "CB OR FUSE", THEN A HACR-LISTED BREAKER IS PERMITTED

ROOM AIR CONDITIONER

ROOM AC CORDS

Volts	Length*
120	10
208	6
240	6

* in feet

A. 110-3(b) Equipment that is listed or labeled shall be installed or used per the instructions included in the listing or labeling.

250-110 Exposed noncurrent-carrying metal parts of fixed equipment shall be grounded if: (1) within 8' vertically or 5' horizontally of ground or grounded objects and possible contact by persons; (2) in damp or wet location and not isolated; (3) in electrical contact with metal; (4) in hazardous location; (5) if fed by metal enclosed wiring (except for 25' of isolated raceway per 250-33; or (6) see Exceptions if over 150 V to ground.

B. 250-114 Exposed metal noncurrent-carrying parts of cord-and-plug connected equipment shall be grounded if: (1) in hazardous location; (2) see Exceptions if over 150 V to ground; (3) in residential occupancies: refrigerators, freezers, air conditioners, clothes washers, clothes dryers, dishwashers, sump pumps, and portable tools unless double insulated. See 250-114 (4) for grounding of cord-and-plug equipment in other than residential occupancies.

440-12(a) Disconnect selected on basis of nameplate rated-load current or branch-circuit selection current (whichever is greater), and locked-rotor current: (a)(1) ampere rating at least 115% of FLA where non-horsepower disconnect (such as CB) is used; (a)(2) to determine size of HP rated disconnect, as required by 430-109, the HP corresponding to the FLA in Tables 430-148, 430-149, 430-150; the HP corresponding to the LRA in Table 430-151, where different HP ratings are obtained, the largest HP shall be used.

C. 440-14 Disconnect within sight and readily accessible for air conditioning or refrigerating equipment except for appliances connected by cord and plug. Permitted on or within equipment. See Ex. 2 for cord-and-plug connected appliances.

D. 440-60 Generally, a room air conditioner must be in the window or wall or room being cooled, incorporate a hermetic-refrigerant motor-compressor, and operate at not over 250 V, 1φ. A 3φ unit rated over 250 V shall be directly wired with a recognized wiring method.

E. 440-61 Grounding or room air conditioning units shall be in accordance with 250-110, 250-112, and 250-114.

F. 440-62(a) For determining branch-circuit requirements, a room air conditioner shall be considered a single motor when it is (1) cord-and-plug connected; (2) rating is less than 40 A at 250 V, 1φ; (3) total rated-load current is provided on nameplate; and (4) branch-circuit, short-circuit, and ground-fault protective device is less than ampacity of branch-circuit conductors or receptacle rating.

G. 440-62(b) The rating of a cord-and-plug connected room air conditioner shall not exceed 80% of its individual branch-circuit rating where no other loads are supplied.

440-62(c) The total load of cord-and-plug room air conditioning equipment shall not exceed 50% of the rating of a branch circuit which also supplies lighting units or appliances.

H. 440-63 A plug and receptacle may serve as disconnecting means for a 1φ air conditioner rated 250 V or less when the manual controls are readily accessible and within 6' of floor, or a manually operable switch is readily accessible and within sight of unit.

I. 440-64 Supply cords for room air conditioners shall not exceed 10' at 120 V or 6' at 208 V or 240 V.

VOLTAGE DROP . . .

VOLTAGE DROP

Voltage drop is the voltage that is lost due to the resistance of conductors. The longer the conductor for a given size, the larger the voltage drop. To compensate for voltage drop, a larger conductor is selected, or the voltage may be increased (if available). Per 215-2, FPN 2, the maximum recommended voltage drop for conductors for feeders is 3%. The maximum recommended voltage drop for conductors for feeders and branch circuits is 5%. The voltage drop is figured to the last outlet.

Chapter 9, Table 8, Conductor Properties, lists wire sizes from No. 18 AWG through No. 2000 kcmil. The circular mil (CM) area is given for No. 18 AWG through No. 4/0 AWG sizes. For example, No. 4 AWG has a cross-sectional area of 41,740 CM. The CMs for No. 250 kcmil through No. 2000 kcmil can be determined by adding three zeroes to the kcmil size. For example, a No. 500 kcmil conductor has a cross-sectional area of 500,000 CM.

Resistance of conductors is given in the last three columns of Ch 9, Table 8. These resistances are used to find exact K and in solving voltage drop problems.

VOLTAGE DROP VARIABLES
V = Voltage (in V)
VD = Voltage drop (in V)
$\%VD$ = Percent voltage drop (in %)
V_l = Voltage loss (in V)
V_s = Supply voltage (in V)
K = Resistivity of conductor (in Ω)
R = Resistance of conductor (in Ω/kft)
CM = Circular mils (in area)
I = Current (in A)
L = Length of conductor (in ft)
1000 = 1000′ or less of conductor
$.866 = \dfrac{\sqrt{3}}{2}$
Note: x2 for 1φ   x2 × .866 for 3φ

VOLTAGE DROP FORMULAS
Note: Add .866 for 3φ.
$VD = \dfrac{2 \times R \times L \times I}{1000}$
$\%VD = \dfrac{V_l}{V_s} \times 100$
$CM = \dfrac{2 \times K \times L \times I}{VD}$
$I = \dfrac{VD \times CM}{2 \times K \times L}$
$L = \dfrac{VD \times CM}{2 \times K \times I}$
$K = \dfrac{R \times CM}{1000}$ (Exact K Method)
K = 10.7 Cu; 17.4 Al (Approximate K Method)

The *K factor (constant)* is the resistance of a circular-mil-foot of wire at a set temperature. A *circular-mil-foot* of wire is a wire 1′ long with a cross-sectional area of one kcmil. Two methods are used to find K: one for approximate K and one for exact K. Either K may be used in the voltage drop formula. The approximate K is most commonly used.

The approximate K is 10.7 Ω for Cu and 17.4 Ω for Al for all size conductors. The exact K is found by selecting the resistance for the conductor from Ch 9, Table 8 and moving the decimal point three places to the right and applying the formula:

Ch 9, Table 8: No. 10 = 10,380 CM

$K = \dfrac{R \times CM}{1000}$

For example, the exact K for a No. 10 solid, uncoated Cu conductor is 12.56 Ω.

$K = \dfrac{R \times CM}{1000}$

$K = \dfrac{1.21 \times 10,380}{1000}$

$K = \mathbf{12.56 \ \Omega}$

AC VOLTAGE MEASUREMENT

. . . VOLTAGE DROP . . .

CIRCUIT A

120 V, 1φ, 15 A
BRANCH CIRCUIT

85'

L

NO. 10 SOLID, UNCOATED
Cu CONDUCTOR

Ch 9, Table 8:
R: No. 10 Cu = 1.21Ω /kft
CM: No. 10 Cu = 10,380 CM

Approximate K method

Cu = 10.7 Ω

1. Find the voltage drop.

$$VD = \frac{2 \times R \times L \times I}{1000}$$

$$VD = \frac{2 \times 1.21 \times 85 \times 15}{1000}$$

$$VD = \frac{3085.5}{1000}$$

$$VD = 3.0855 = \textbf{3.09 V}$$

2. Find the % of voltage drop.

$$\%VD = \frac{V_l}{V_s} \times 100$$

$$\%VD = \frac{3.09}{120} \times 100$$

$$\%VD = .02575 \times 100$$

$$\%VD = \textbf{2.575\%}$$

3. Find the conductor size.

$$CM = \frac{2 \times K \times L \times I}{VD}$$

$$CM = \frac{2 \times 10.7 \times 85 \times 15}{3.09}$$

$$CM = \frac{27,285}{3.09}$$

$$CM = 8830.09 = 8830.1$$

Ch 9, Table 8:
8830.1 < 10,380
10,380 CM = No. 10
Conductor = **No. 10**

4. Find the current.

$$I = \frac{VD \times CM}{2 \times K \times L}$$

$$I = \frac{3.09 \times 8830.1}{2 \times 10.7 \times 85}$$

$$I = \frac{27,285.01}{1819}$$

$$I = \textbf{15 A}$$

5. Find the length.

$$L = \frac{VD \times CM}{2 \times K \times I}$$

$$L = \frac{3.09 \times 8830.1}{2 \times 10.7 \times 15}$$

$$L = \frac{27,285.01}{321}$$

$$L = \textbf{85'}$$

Exact K Method

Ch 9, Table 8:
R: No. 10 Cu = 1.21 Ω/kft
CM: No. 10 Cu = 10,380 CM

$$K = \frac{R \times CM}{1000}$$

$$K = \frac{1.21 \times 10,380}{1000}$$

$$K = \frac{12,559.8}{1000}$$

$$K = 12.559 = \textbf{12.56 Ω}$$

1. Find the voltage drop.

$$VD = \frac{2 \times R \times L \times I}{1000}$$

$$VD = \frac{2 \times 1.21 \times 85 \times 15}{1000}$$

$$VD = \frac{3085.5}{1000}$$

$$VD = 3.0855 = \textbf{3.09 V}$$

2. Find the % of voltage drop.

$$\%VD = \frac{V_l}{V_s} \times 100$$

$$\%VD = \frac{3.09}{120} \times 100$$

$$\%VD = .02575 \times 100$$

$$\%VD = \textbf{2.575\%}$$

3. Find the conductor size.

$$CM = \frac{2 \times K \times L \times I}{VD}$$

$$CM = \frac{2 \times 12.56 \times 85 \times 15}{3.09}$$

$$CM = \frac{32,028}{3.09}$$

$$CM = 10,365$$

Ch 9, Table 8:
10,365 < 10,380
10,380 CM = No. 10
Conductor = **No. 10**

4. Find the current.

$$I = \frac{VD \times CM}{2 \times K \times L}$$

$$I = \frac{3.09 \times 10,365}{2 \times 12.56 \times 85}$$

$$I = \frac{32,027.85}{2135.2}$$

$$I = 14.99 = \textbf{15 A}$$

5. Find the length.

$$L = \frac{VD \times CM}{2 \times K \times I}$$

$$L = \frac{3.09 \times 10,365}{2 \times 12.56 \times 15}$$

$$L = \frac{32,027.85}{376.8}$$

$$L = 84.99 = \textbf{85'}$$

. . . VOLTAGE DROP

CIRCUIT B

208 V, 3Φ, 50 A BRANCH CIRCUIT

80'

NO. 8 STRANDED, UNCOATED Cu CONDUCTOR

Ch 9, Table 8:
  R: No. 8 Cu = .778 Ω/kft
  CM: No. 8 Cu = 16,510 CM

Approximate K method

Cu = 10.7 Ω

1. Find the voltage drop.

$$VD = \frac{2 \times R \times L \times I \times .866}{1000}$$

$$VD = \frac{2 \times .778 \times 80 \times 50 \times .866}{1000}$$

$$VD = \frac{5389.98}{1000}$$

$$VD = 5.38998 = \textbf{5.39 V}$$

2. Find the % of voltage drop.

$$\%VD = \frac{V_l}{V_s} \times 100$$

$$\%VD = \frac{5.39}{208} \times 100$$

$$\%VD = .02591 \times 100$$

$$\%VD = \textbf{2.591\%}$$

Note: Other 3Φ voltage drop problems are solved using the formulas for 1Φ problems with the addition of .866. The square root of 3 is 1.732 and $\frac{1.732}{2} = .866$.

BRANCH CIRCUITS

FOUR-WIRE CIRCUIT

BLACK
RED
BLUE
WHITE OR NATURAL GRAY
GREEN

UNGROUNDED CONDUCTORS — F

GROUNDED CONDUCTOR — A   D

EQUIPMENT GROUNDING CONDUCTOR — B

E

BRANCH CIRCUIT — C
LOAD

BRANCH CIRCUIT — C

FLOOR OUTLET

RECEPTACLE

CONDUCTOR IDENTIFICATION

210-4(d) Specific color requirements for ungrounded conductors are not provided, but if there is more than one nominal voltage system in a building, each system's ungrounded conductor color shall be permanently posted at each panelboard.

A. 210-5(a) The grounded conductor of a branch circuit shall be white, natural gray, or have three continuous white stripes its entire length. Each other system in the same raceway or enclosure shall have the grounded conductor (if required) identified by outer covering of white with colored stripe (not green). See 200-6.

B. 250-134(b) The EGC of a branch circuit shall be green, green with one or more yellow stripes, or bare.

210-21(b)(1) A single receptacle on an individual branch circuit shall have an ampere rating not less than that of the branch circuit.

C. 210-23 An individual branch circuit may supply any load for which it is rated. Branch circuits having two or more outlets may supply only loads as follows: (a) 15 A and 20 A circuits may supply lighting units and/or appliances, the rating of any cord-and-plug connected appliance not exceeding 80% of circuit, and total rating of fixed appliances not over 50%; (b) 30 A circuits may supply heavy-duty lampholders in other than dwellings, or appliances anywhere the rating of a cord-and-plug connected appliance does not exceed 80% of circuit; (c) 40 A or 50 A branch circuits may supply fixed cooking appliances in any occupancy or fixed lighting units with heavy-duty lampholders or

LOADS

infrared heating units or other utilization equipment in other than dwellings; (d) circuits over 50 A shall supply only non-lighting outlet loads.

D. 310-12(a) Grounded conductors shall be identified per 200-6.

E. 310-12(b) EGCs shall be identified per 250-134(a)(b).

F. 310-12(c) Ungrounded conductors shall be identified by colors or a combination of colors plus a distinguishing marking, other than white, natural gray, or green. These colors shall clearly distinguish the ungrounded conductor from the grounded or grounding conductor and shall not conflict with markings required by 310-11(b)(1).

Name _____

Date _____

Trade Competency Test 1

NEC®	Answer

_____ _____ **1.** The minimum size of a copper EGC required for equipment connected to a 40 A circuit is No. _____.
A. 8   C. 12
B. 10   D. 14

_____ _____ **2.** Splices that will be soldered shall first be made mechanically and _____ secure.

_____ T   F **3.** If the resistance of a driven ground rod exceeds 25 Ω to ground, an additional rod shall be added.

_____ _____ **4.** The smallest permissible diameter of a bronze ground rod is _____″.

_____ _____ **5.** For a service run of rigid metal raceway interrupted by an expansion joint, the sections of rigid conduit shall be bonded by a(n) _____ jumper or other means.

_____ _____ **6.** The minimum branch circuit rating to a single electric range of over 8¾ kVA is _____ A.

_____ T   F **7.** Stranded wires No. 10 and smaller may be connected to equipment by means of screws used with terminal plates having upturned lugs.

_____ _____ **8.** Double lockouts are required where rigid metal raceways in other than service runs enter a box or cabinet, if the raceway contains any wire of more than _____ V to ground.

_____ T   F **9.** The grounding conductor in a cable assembly for cord- and plug-connected equipment need not be insulated.

_____ _____ **10.** The three wires of a 1φ, 120/240 V service are designated as *A, N,* and *B.* A 100 A, 240 V load is connected across *A* and *B.* A 50 A, 120 V load is connected across *A* and *N.* The neutral conductor shall be No. _____ TW Cu.
A. 4   C. 8
B. 6   D. 10

_____ T   F **11.** It is recommended that metal raceways of an electrical system should be at least 6′ from lightning protection conductors.

_____ _____ **12.** The neutral feeder conductor shall be capable of carrying the maximum _____ load.

_____ _____ **13.** The minimum area of exposed surface offered by a plate electrode shall be _____ sq ft.
A. 1   C. 3
B. 2   D. 4

_____ _____ **14.** Wiring run above heated ceilings and within thermal insulation shall be derated on the basis of _____°C ambient temperature.

15. Receptacles connected to a 30 A branch circuit supplying two or more outlets shall be rated not less than _____ A.

16. Noncurrent-carrying metal parts of cord-and-plug connected equipment may be grounded by a(n) _____.

T   F   **17.** Iron wire shall not be used as an equipment bonding jumper.

18. A 15 A or 20 A receptacle can be installed on a _____ A general purpose circuit.

A. 20   C. 30
B. 25   D. neither A, B, nor C

19. The maximum standard ampere rating of a fuse or inverse time circuit breaker shall be considered to be _____ A.

T   F   **20.** In general, switch enclosures shall not be used for junction boxes.

21. If a change occurs in the size of the _____ conductor, a similar change may be made in the size of the grounded conductor.

22. Leads of 240 V heating cables are marked with _____.

23. The maximum permissible open-circuit voltage of electric discharge lighting equipment used in a dwelling is _____ V.

A. 500   C. 1500
B. 1000   D. 2000

24. _____ are permitted around service conduits entering concentric or eccentric knockouts.

T   F   **25.** A single 15 A receptacle may be installed on an individual 20 A branch circuit.

26. A metal elbow for a service raceway installed at least _____″ underground does not have to be grounded.

27. The _____ shall be grounded in a 1ϕ, 3-wire, AC system.

28. Ungrounded conductors shall be clearly distinguishable from _____ and grounding conductors by color or other marking.

29. Ground rods are permitted to be buried if the trench is at least _____′ deep where rock bottom is encountered.

30. The minimum diameter of a steel ground rod shall be _____″.

A. ⅜   C. ⅝
B. ½   D. ¾

31. Flexible cord shall not be used as a substitute for _____ wiring.

32. Unless special precautions are taken, exposed live parts operating at _____ V or more shall be guarded against accidental contact.

33. In dwelling units, equipment in which live parts are exposed while changing lamps shall have an open-circuit voltage not exceeding _____ V.

34. A single grounding electrode is permitted when the resistance to ground does not exceed _____ Ω.

A. 5   C. 15
B. 10   D. 25

35. Heating cables on ceilings shall be kept free from contact with metal or other _____ surfaces.

Name _____

Date _____

Trade Competency Test 2

NEC®   **Answer**

_____   T   F   **1.** The metal well casing enclosing a submersible pump shall be bonded to the EGC of the pump circuits.

_____   _____   **2.** Fittings and connectors shall be designed and _____ for the specific wiring method with which they are used.

_____   _____   **3.** Heating panels shall be separated from outlet boxes to be used for mounting surface lighting fixtures by not less than _____ ″.
A. 6   C. 10
B. 8   D. 12

_____   T   F   **4.** Not less than 10′ of No. 4 Cu bare conductor encased under the concrete footing in contact with the earth is an approved made electrode.

_____   _____   **5.** The service disconnecting means for a one-family dwelling shall have a rating of not less than _____ A.

_____   T   F   **6.** NMS cable shall be permitted to be used in storage battery rooms.

_____   _____   **7.** The earth shall not be used as the _____ equipment grounding means.

_____   _____   **8.** The MBJ in a panelboard may be a screw, a bus, or a(n) _____.

_____   _____   **9.** It is recommended that metal raceways, enclosures, frames, and other non-current-carrying metal parts of electric equipment be kept at least _____ ′ away from lightning rod conductors, unless they are bonded to such conductors.
A. 5   C. 8
B. 6   D. 10

_____   _____   **10.** Wiring located above electric heating panels shall be spaced not less than _____ ″ above the heated ceiling.
A. 2   C. 4
B. 3   D. 5

_____   T   F   **11.** Time switches need not be of the externally operable type.

_____   _____   **12.** The minimum size grounded conductor required for a service with 600 kcmil per phase is No. _____ Cu.

_____   _____   **13.** A 15 A or 20 A branch circuit shall be permitted to supply _____ loads.
A. lighting   C. both A and B
B. utilization equipment   D. neither A nor B

_____   _____   **14.** Equipment with open-circuit voltage exceeding _____ V shall not be installed in dwelling units.

_____   _____   **15.** In general, flexible cords other than tinsel cords shall not be smaller than No. _____.

_____   _____   **16.** A 30 A receptacle outlet supplying power to a cord-and-plug connected load shall not be loaded to more than _____ A.

_____ _____ 17. If more than one nominal voltage system exists in a building, each ungrounded conductor of a multiwire branch circuit in each phase and system shall be _____.
A. black   C. the same
B. orange   D. identified

_____ _____ 18. Wiring that supplies equipment, located behind panels designed to allow _____, shall be adequately supported and kept off the ceiling tiles.

_____ _____ 19. The general rule requires a(n) _____ to be installed at an auxiliary building.

_____ _____ 20. The minimum burial depth of RMC installed under a street or parking lot is _____″.

_____ _____ 21. A single receptacle installed on an individual branch circuit shall have a(n) _____ rating not less than that of the branch circuit.

_____ _____ 22. In general, copper conductors of size No. _____ and larger comprising each phase, neutral, or grounded circuit conductor shall be permitted to be connected in parallel.
A. 1/0   C. 3/0
B. 2/0   D. 4/0

_____ _____ 23. Ground-fault circuit protection for personnel is required for all 1ϕ, 125 V, 15 A and 20 A receptacles installed in the _____ of a dwelling unit.
A. attic   C. laundry
B. bathroom   D. utility room

_____ _____ 24. _____ boxes shall not be used where conduits or connectors requiring the use of locknuts or bushings are to be connected to the side of the box.

_____ _____ 25. The minimum burial depth for listed PVC conduit routed beneath a building is _____″.

_____ _____ 26. An insulated grounded conductor smaller than No. 6 shall be identified by a white or natural _____ outer finish.

_____ _____ 27. _____ conductors shall be used with chain fixtures.

_____ _____ 28. Disconnecting means for circuits to feeders and branch circuits shall be properly _____ to indicate the purpose.

_____ T   F 29. Connections that depend solely on solder are not permissible for grounding equipment.

_____ _____ 30. Heavy-duty lampholders shall have a rating of not less than _____ W.

_____ _____ 31. Side- or back-wiring spaces are required in cabinets when, in addition to the supply circuit or a continuation thereof, there are devices connected to more than _____ conductors.
A. 6   C. 10
B. 8   D. 12

_____ _____ 32. Screw-shell lampholders shall not be used with infrared lamps of over _____ W except when identified as suitable for such use.

_____ _____ 33. Cables routed up a pole shall be protected at least _____′ above grade.

_____ _____ 34. The smaller of two noncoincident loads may be omitted in calculating the load on a feeder if it is unlikely that both loads will be connected _____.

_____ T   F 35. Permanently installed 15 A and 20 A, 125 V receptacles that are part of the building wiring and used for construction personnel shall be GFCI-protected.

Name _____

Date _____

Trade Competency Test 3

NEC®	Answer	
_____	_____	**1.** No. _____ conductor shall be attached to any terminal or lead so as to reverse designated polarity.
_____	T F	**2.** The center grip of the operating handle for a switch enclosure when in the highest position shall not be more than 6′-7″ from the floor or platform.
_____	_____	**3.** The feeder load for one 10 kVA, one 16 kVA, and one 17 kVA household electric range is _____ kVA. A. 14.5 C. 15.6 B. 15 D. 16.1
_____	_____	**4.** The grounded conductor shall be _____ when SE cable is used in a feeder to supply a subpanel located in the same building.
_____	_____	**5.** Conduits, including fittings, entering a switchgear shall not rise more than _____ ″ above the bottom of the enclosure.
_____	T F	**6.** Branch-circuit conductors for fixed resistance space heaters shall be rated at 125% of the total load of heaters and motors.
_____	_____	**7.** The rating of any one cord-and-plug connected appliance used on a 30 A branch circuit shall not exceed _____ A. A. 21 C. 27 B. 24 D. 30
_____	_____	**8.** The maximum radius of a conduit bend enclosing THHN Cu conductors is _____ ″ for 2″ RMC.
_____	_____	**9.** A 20 A circuit in a dwelling unit supplies only fixed resistance space heaters. This circuit may be loaded to a maximum value of _____ A. A. 12 C. 18 B. 16 D. 20
_____	T F	**10.** Heating panels shall not be extended beyond partitions.
_____	_____	**11.** A feeder supplying a 5 kVA wall-mounted oven and a 7 kVA counter-mounted cooktop shall have the capacity for _____ kVA. A. 7 C. 9.2 B. 8 D. 12
_____	_____	**12.** The general rule requires that raceways be installed _____ between junction boxes prior to the installation of conductors.
_____	_____	**13.** _____ circuits in dwelling units shall supply only loads within that dwelling unit or loads associated only with that dwelling unit.
_____	_____	**14.** A 3PCB counts as _____ device(s) in determining the count toward a lighting and appliance panelboard.
_____	T F	**15.** Heating cable shall not be used with metal lath ceilings.
_____	_____	**16.** _____ shall be formed on individual service drop conductors to prevent the entrance of moisture.

_____ _____ **17.** No. 8 THWN conductors supply a central air conditioner, consisting of a sealed (hermetic) motor-compressor. The branch circuit may be loaded to a maximum value of _____ A.

A. 30   C. 36
B. 32   D. 50

_____ _____ **18.** In general, in all dwelling units, at least _____ wall switched-control lighting outlet(s) shall be installed in every habitable room.

_____ _____ **19.** Conduits with No. 4 and larger conductors shall be provided with _____ for protection of conductors.

_____ _____ **20.** Type AC cable _____ in walls do not require support.

_____ _____ **21.** Conductors normally used to carry current shall be _____ unless otherwise provided in the NEC®.

_____ _____ **22.** The total rating of fastened-in-place appliances connected to a 20 A branch circuit that also supplies lighting units shall not exceed _____ A.

A. 10   C. 16
B. 12   D. 20

_____ _____ **23.** Interior metal water piping, in multifamily dwelling units, located more than _____ ' from the point of entrance to the building shall not be used as part of the GES or as a conductor to interconnect electrodes that are part of the GES.

_____ T   F **24.** Heating cables in a concrete floor shall be placed on at least 2' centers.

_____ _____ **25.** A 4″ RMC installed with threaded couplings in a straight run shall be supported every _____ '.

_____ _____ **26.** A 6″ RNMC shall be supported at _____ ' intervals.

_____ _____ **27.** In general, MC cables shall be supported at _____ ' intervals.

_____ T   F **28.** Embedded heating cables are permitted to be spliced only where necessary and if done by approved means.

_____ _____ **29.** The minimum radius of the curve of the inner edge of any bend in AC cable shall not be less than _____ times the diameter of the cable.

_____ _____ **30.** Six current-carrying conductors routed through the same raceway shall be derated _____ %.

_____ _____ **31.** The total load on overcurrent protection devices supplying continuous duty loads shall not exceed more than _____ % of their rating.

_____ _____ **32.** A No. _____ is the smallest THWN neutral feeder conductor permitted to supply two 240 V, 12 kVA household ranges.

A. 4   C. 8
B. 6   D. 10

_____ _____ **33.** In walls or ceilings of concrete, tile, or other noncombustible material, boxes shall be installed so that the front edge of the box will not be set back of the finished surface more than _____ ″.

_____ T   F **34.** The nonheating leads of heating cables installed in concrete may be protected by RNMC.

_____ _____ **35.** The grounded circuit conductor can be used as an equipment grounding conductor and circuit conductor on the _____ side of the service equipment.

_____ T   F **36.** A separable connector will meet NEC® requirements for a disconnecting means for a wall-mounted oven.

NEC®	Answer	

_____   T   F   **1.** In a 1000 sq ft addition to a dwelling unit, the "VA-per-square-foot" method is not used.

_____   _____   **2.** Heating elements that exceed _____ A in a heating unit shall be subdivided.

_____   _____   **3.** The neutral conductor of a 3-wire branch circuit to a household electric range shall not be smaller than No. _____ THW.
  A. 4   C. 8
  B. 6   D. 10

_____   _____   **4.** Unbroken lengths of EMT up to _____ ′ are not required to be supported where structural members do not allow fastening.

_____   T   F   **5.** The overcurrent protection device for a branch circuit supplying power to a 120 gal. fixed water heater can be rated less than 125% of the unit's nameplate rating.

_____   _____   **6.** The grounded conductor in a subpanel shall be insulated from the metal enclosure unless the terminal bar is _____ for the purpose.

_____   _____   **7.** Frames of electric clothes dryers may be grounded by connection to the grounded circuit conductor only when this copper conductor is not smaller than _____ and it is an existing branch circuit.
  A. No. 8   C. No. 12
  B. No. 10   D. the other conductors

_____   T   F   **8.** Unit switches can be used as a disconnecting means for a fixed appliance where all conductors are disconnected simultaneously.

_____   _____   **9.** Lighting track shall be calculated at _____ VA for each 2′ or fraction thereof.

_____   T   F   **10.** Infrared lamps rated over 300 W shall not be used as screw-shell lampholders unless the lampholders are identified for the use.

_____   _____   **11.** The nameplate on heating equipment shall be located to be _____ after installation.

_____   _____   **12.** Flexible cords identified for hard or extra-hard usage shall be permitted to be used for _____ wiring on branch circuits.

_____   _____   **13.** A heavy-duty, medium lampholder shall have a rating of not less than _____ W.
  A. 150   C. 600
  B. 300   D. 660

_____   _____   **14.** In general, all conductors, and when used the EGC, shall be routed in the _____ cable, raceway, trench, or conduit system.

_____ _____ 15. A clearance of not less than _____″ shall be maintained between the conductors of a concealed knob-and-tube wiring system.

_____ T F 16. In multifamily dwellings, the disconnecting means for a permanently connected appliance rated at not over 300 VA, is permitted to be the branch circuit OCPD in the service equipment for the occupancy.

_____ _____ 17. Heating elements of cables shall have a separation of at least _____″ from the edge of outlet boxes used for mounting surface lighting fixtures.

_____ _____ 18. It is recommended that the voltage drop on a feeder circuit be held to _____%.

    A. 2   C. 5
    B. 3   D. 7

_____ T F 19. A No. 4 grounded conductor in a 277/480 V system can be identified white or natural gray.

_____ _____ 20. A(n) _____ A or less OCPD with neutral loads shall be counted to determine if a panel is a lighting and appliance panelboard.

_____ _____ 21. A circuit operating for _____ hours or more is considered a continuous duty load.

_____ _____ 22. Receptacle outlets shall be installed above a kitchen countertop no more than _____″ apart.

    A. 18   C. 36
    B. 24   D. 48

_____ _____ 23. The exposed blades of knife switches shall be de-energized when in the _____ position.

_____ T F 24. A 20 A branch circuit supplying air conditioning equipment shall not supply lighting outlets.

_____ _____ 25. Heavy-duty lighting track is identified for use with systems exceeding _____ A.

_____ T F 26. A lockable branch-circuit overcurrent protection device may serve as the disconnecting means for a motor-driven, permanently-connected appliance rated over ⅛ HP.

_____ _____ 27. Angle connectors used on Greenfield shall not be _____.

_____ _____ 28. Neon systems having an open circuit voltage exceeding _____ V shall not be installed in dwelling units.

_____ _____ 29. A disposal installed in a dwelling unit shall be equipped with a cord that does not exceed _____″ in length.

_____ T F 30. The total load of four or more fixed appliances in a dwelling unit can have an 80% demand factor applied.

_____ _____ 31. Ceiling fans, mounted to a listed outlet box identified for the purpose, shall not weigh more than _____ lb.

_____ _____ 32. At least one receptacle outlet shall be installed above a show window for each _____′ or major fraction thereof.

    A. 9   C. 18
    B. 12   D. 24

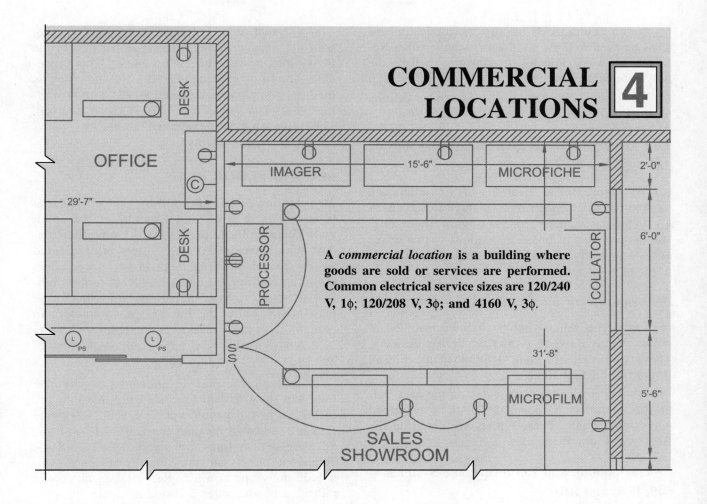

DESK

OFFICE

IMAGER

15'-6"

MICROFICHE

2'-0"

29'-7"

C

DESK

PROCESSOR

A *commercial location* is a building where goods are sold or services are performed. Common electrical service sizes are 120/240 V, 1φ; 120/208 V, 3φ; and 4160 V, 3φ.

COLLATOR

6'-0"

L PS

L PS

S S

31'-8"

MICROFILM

5'-6"

SALES SHOWROOM

READING PRINTS

In preparing to handle a large installation, the electrical supervisor glances through the complete set of plans to obtain an overview of the entire project. The electrical plans are surveyed to determine the extent of the work, kinds of circuits and equipment, and locations of main elements. The electrical specifications are read, and the plans are referred to from time to time as necessary.

The electrical supervisor must refer to the plot, architectural, structural, framing, mechanical, and electrical plans to observe all the details essential to the work. Following are excerpts from electrical specifications for a store building, along with notes that might be taken from the architectural plans.

EXCERPTS FROM SPECIFICATIONS

Note: The NEC® references indicate where the main topic is found in the National Electrical Code®. Always follow manufacturer's directions.

29-1 Main Service – 230

(a) The main service shall consist of a 4-wire, 120/208 V, 3φ secondary system from utility company utility hole on Sheppard Avenue to the main switchboard.

(b) All underground conduit outside the building shall be buried at least 18″ below grade and shall be painted with bitumastic paint.

(c) All service conductors shall be insulated, including the neutral conductor.

29-4 Conductors – 310

(a) Conductors shall be THW unless otherwise specified or required by applicable provisions of the NEC®. All conductors are to be copper unless otherwise noted.

(b) No conductors for lighting, plug receptacles, or power shall be smaller than No. 12 AWG Cu.

(c) No block and tackle or other mechanical means shall be used to draw conductors smaller than No. 2 into raceways. Powdered soapstone shall be the only lubricant permitted.

(d) The feeder voltage drop from the service location to any distribution panel shall not exceed 3%.

29-5 Wiring Methods and Materials – 300 to 384

(c) The approximate locations of all outlets are marked on the drawings. The contractor shall check all such measurements and, in case of doubt, consult the architect as to exact locations so that boxes will be properly centered with acoustical tile, interior trim, paneling, etc.

29-6 Equipment for General Use – 400 to 480

(b) Lock-type wall switches shall be installed at points where a switch is marked SK on drawing. These wall switches control lights in the Fitting Room on the first floor.

(m) Two special plug receptacles on the east wall of the Alteration Department shall be equipped with signal lights. These outlets are noted on the Basement Floor Plan.

29-7 Telephone and Loud Speaker Systems – 640, 800, and 810

(a) Telephone outlets shall be located approximately as shown.

29-8 Miscellaneous Equipment – 422

(j) Toilet exhaust fans shall be Nuway, 130 VA, Type C2. See Basement Floor Plan.

ELECTRICAL PLANS

Before undertaking a detailed analysis of the store building installation, notice that four drawings are included. The Basement, First Floor, and Second Floor floor plans are included with a special drawing that shows a Feeder Riser Diagram, Switchboard Elevation, Electrical Details, and Fixture Symbols. Discussion could be started with the Basement plan, then carried on to the First Floor and Second Floor plans in turn. Or, the First Floor might be dealt with, then Second Floor, and finally Basement.

A different order is followed, however, starting with the Second Floor, working downward to the First Floor, and lastly the Basement. This procedure simplifies explanation. The Second Floor is the least complicated area and makes a good starting point. The First Floor plan is more involved. The Basement plan is examined last because it covers service requirements.

SECOND FLOOR PLAN

Two lighting panels are used for this floor, Panel E at the north wall, and Panel D at the south wall. Each of these panels has 13 circuits. Conduit runs between lighting outlets are indicated by solid lines showing them to be concealed in the hung ceiling. All conduit is RMC unless otherwise noted. Arrows indicating homeruns to the panelboards show the number of conductors in conduit with hash marks. For example, Circuits 1, 2, and 3 to Panel E have four conductors in 3/4″ EMT.

Plug receptacle runs are indicated by alternate long and short dashed lines showing that they are to be run in the poured concrete slab. Telephone runs are indicated by lines marked — T — T —. Speaker system lines are marked — S — S —. Telephone outlets are indicated by standard symbols in the shape of a triangle. Speaker outlets are indicated by the letter S inside a square.

Conduit riser indications are shown in the south wall. At the left or west side, the first such riser is marked DN (DOWN) to show that a conduit run from the stair outlet leads to the floor below. The next two risers are marked UP, and since no conduit runs lead to them from the second floor, it is apparent that they originate below and continue upward toward the roof.

The final riser is marked DN. It is run of conduit leading downward from a lighting outlet at the head of the stairs. Two risers in the east wall of the passenger elevator shaft are both shown as DN. They are run in conduit leading downward from a lighting outlet at the head of the stairs. One of these risers is marked T to indicate a telephone run, the other S to indicate a speaker run. The PBX telephone board and the sound amplifier are in the small room to the right of the elevator shaft.

In determining the load on the VA-per-square-foot basis, 220-3(a) requires that the floor area be computed from outside dimensions of the building. The measurements are 72′ × 96′, which equals 6912 sq ft (72 × 96 = 6912 sq ft). From this amount, it is permissible to deduct 520 sq ft, which is the area of the office space and the area of the two elevators (6912 − 520 = 6392 sq ft). The elevator shaft areas and the office space area do not come under the heading of general lighting in Table 220-3(a).

Table 220-3(a) lists the unit load for a store as 3 VA per sq ft. NEC 210-20(a) states that this value shall be increased by 25% where the load on a branch circuit is continuous (load is expected to continue for 3 hours or more). The unit load in the present case is 1.25 × 3 VA, or 3.75 VA per sq ft. The minimum provision for the lighting load is 6392 sq ft × 3.75 VA, or 23,970 VA. The lighting load may also be found by multiplying the sq ft area by 3 VA per Table 220-3(a) and then multiplying by 125% per 210-20(a). For example, 6392 sq ft × 3 VA = 19.176 VA × 125% = 23,970 VA. Panels D and E must have a sufficient number of lighting circuits to furnish this amount of general illumination.

The Specifications provide that all lighting circuits shall be run with No. 12 THW Cu wire, protected by 20 A CBs. NEC 210-3 states that branch circuits shall be classified according to the rating of the overcurrent protection devices. Because the loads are continuous, these 20 A circuits, each of which can supply 2400 VA of power (20 A × 120 V = 2400 VA), can be loaded to only 80% of this value, or 1920 VA (2400 VA × .80 = 1920 VA) per 384-16(d). A *continuous load* is a load that, in normal operation, lasts for three hours or more.

Calculations for Panel E

The loads on the circuit are:

Circuit	Load*	Circuit	Load*
1	1400	8	1536
2	1536	9	972
3	972	10	900
4	1400	11	720
5	1536	12	540
6	972	13	1080
7	1400		

* in VA

The fixture list, fixture specifications, and the print show the load on Circuit 1 as seven 200 VA incandescent units, totaling 1400 VA. Continuing to Circuit 2, 220-4(b) requires that the load on a lighting fixture that has a ballast shall be calculated according to the ampere rating of the ballast and not the VA of the lamps. The current used in the two ballasts that are part of each fixture on the circuit is 3.2 A or 384 VA (3.2 A × 120 V = 384 VA).

The total VA for Circuit 2, which supplies four such units, is 1536 VA (384 VA × 4 = 1536 VA). Circuit 3 has two of these units and a third with a ballast rating of 1.7 A. The load is 972 VA for all units in Circuit 3 (384 VA × 2 = 768 VA; 1.7 A × 120 V = 204 VA; 768 VA + 204 VA = 972 VA). The load for this group of three circuits is 3908 VA (1400 VA + 1536 VA + 972 VA = 3908 VA).

Loads on the two groups 4, 5, 6 and 7, 8, 9 are the same as the loads for groups 1, 2, and 3, so the total lighting power to be supplied by Panel E amounts to 11,724 VA (3 × 3908 VA = 11,724 VA). Table 220-11 lists a demand factor of 100% for lighting loads of this nature.

In addition to the lighting outlets, a number of wall receptacles and floor outlets are supplied from the panel. The Specifications require that No. 12 THW Cu wire be used for these circuits. NEC 220-3(b)(9)(11) designates a load of 180 VA or 1½ A (180 VA ÷ 120 V = 1.5 A) for each of such other outlets. The eighteen plug receptacles connected to this panel represent a current input of 27 A × 120 V, or 3240 VA. *Note:* 18 × 1.5 A = 27 A.

The load imposed on the feeder is 11,724 VA + 3240 VA = 14,964 VA. At this point, determine if the feeder will carry its required proportion of minimum lighting load. The floor plan shows that circuits from Panel E provide general illumination for an area of 3456 sq ft (36′ × 96′ = 3456 sq ft).

Since the unit load cannot be less than 3.75 VA per sq ft, the amount of general lighting burden allotted to Panel E cannot be less than 12,960 VA (3456 sq ft × 3.75 VA = 12,960 VA). This value is used in the feeder calculations because it is larger than the actual connected load of lighting circuits.

The minimum value of lighting VA does not have to be used, but must be present in circuit and feeder capacity. The corrected load on the feeder then is 16,200 VA (12,960 VA + 3240 VA = 16,200 VA).

With a balanced, 4-wire, 120/208 V, 3φ system, the current per conductor may be determined by dividing VA by 360 (1.73 × 208 V = 360). The current in the feeder to Panel E is 45 A (16,200 VA ÷ 360 V = 45 A). Table 310-16 lists the nearest size of THW Cu conductor to carry 45 A as No. 8, which is rated at 50 A if there are not more than three current-carrying conductors in a raceway.

Note 8(a) of Notes to Ampacity Tables of 0 to 2000 Volts states that where the number of conductors in a raceway exceeds three, the allowable ampacity of each conductor shall be limited to 80% of the values given in the table. Since there are four conductors, it seems that the capacity of the No. 8 wire is reduced to 80% of 50 A, or 40 A. Note 10, however, states that a neutral conductor that carries only the unbalanced current from other conductors shall not be counted in determining ampacities where there are more than three conductors in a raceway.

Such a neutral conductor is included in the feeder under consideration, and the carrying capacity of the No. 8 wire will not be reduced in value. Nevertheless, the rule may be altered later on because the neutral conductor of a 4-wire, 3ϕ wye system that supplies circuits for electric discharge lamps carries some third harmonic current even when the load is perfectly balanced.

The *American Electricians' Handbook* and other engineering data (See Ch 9, Table 8) lists the resistance of No. 6 Cu as .491 Ω per 1000′. When floor heights and distances on the architectural plans are measured, allowing a sufficient length of conductor for connections at either end of the run, the length of single conductor is found to be 104′ from Panel E to the switchboard in the basement.

The preferred method for calculating voltage drop (VD) for a 3ϕ circuit and % voltage drop (%VD) is by using the formulas:

$$VD = \frac{2 \times R \times L \times I \times .866}{1000}$$

where

VD = voltage drop (in V)
V_l = voltage loss (in V)
V_s = supply voltage (in V)
R = resistance of conductor (in Ω)
L = length of conductor (in feet)
I = current (in A)
$.866 = \frac{\sqrt{3}}{2}$ (used for 3ϕ only)
1000 = 1000′ or less of conductor

$$VD = \frac{2 \times R \times L \times I \times .866}{1000}$$

$$VD = \frac{2 \times .491 \times 104 \times 45 \times .866}{1000}$$

$$VD = \frac{3979.93}{1000}$$

$$VD = 3.9799 = \textbf{3.98 V}$$

$$\%VD = \frac{V_l}{V_s} \times 100$$

$$\%VD = \frac{3.98}{208} \times 100$$

$$\%VD = .01913 \times 100$$

$$\%VD = 1.913 = \textbf{1.91\%}$$

FPN 2 following 215-2 recommends that the voltage drop for feeders should be limited to 3% of the phase voltage at the farthest outlet. The recommendation for voltage drop for both feeders and branch circuits is 5%. In this calculation, the recommended voltage drop should not exceed 6.24 V (208 V × 3% = 6.24 V). The 3.98 V is well below the 3% recommendation. The feeder to Panel E consists of four No. 6 THW Cu conductors in 1″ RMC.

Calculations for Panel D

The loads on the circuits are:

Circuit	Load*	Circuit	Load*
1	1200	8	1152
2	1920	9	720
3	1200	10	1080
4	1920	11	1080
5	800	12	720
6	1152	13	720
7	1152		

* in VA

The circuits devoted to general lighting are 1, 2, 3, 4, 5, and 6. The general lighting circuits require a total of 8192 VA (1200 VA + 1920 VA + 1200 VA + 1920 VA + 800 VA + 1152 VA = 8192 VA).

Measurement shows the floor area supplied with general illumination from this panel to be 2496 sq ft . The minimum VA required is 9360 VA (2496 sq ft × 3.75 VA = 9360 VA). This value, being somewhat higher than the actual load on the circuits, is used in the feeder calculations.

The office has an area of 440 sq ft, requiring 1650 VA of illumination at the 3.75 VA rate (440 sq ft × 3.75 VA = 1650 VA). Table 220-3(a) lists office buildings at a minimum unit load of 3.5 VA per sq ft. For a continuous load, this minimum would have to be increased to 4.4 VA (1.25 × 3.5 VA = 4.375 = 4.4 VA).

Although this provision does not apply strictly to the one small office space here, it does indicate that office buildings and office areas need somewhat more lighting intensity than sales areas. Using 4.4 VA per sq ft, this area requires 1936 VA (440 sq ft × 4.4 VA = 1936 VA). The lighting load connected to Circuits 7 and 8 amounts to 2304 VA. Because the connected load of 2880 VA (2304 VA × 1.25 = 2880 VA) is larger than that required by Table 220-3(a) and 210-20(a), the higher value of 2880 VA is used.

Plug receptacle circuits 9, 10, 11, 12, and 13 take 4320 VA. The total load on the feeder is 16,560 VA (9360 VA + 2880 VA + 4320 VA = 16,560 VA). The current is 46 A (16,430 VA ÷ 360 = 46 A), for which the minimum size of THW conductors is No. 8.

The length of single conductor between Panel D and the switchboard is approximately 83′. The voltage drop for a No. 8 stranded conductor is too high. Use of No. 4 conductor results in an acceptable drop. Appd C, Table C8 shows that four No. 4 THW Cu conductors may be installed in 1¼″ RMC. Ch 9, Table 8 lists the resistance of No. 4 Cu as .308 Ω per 1000′.

$$VD = \frac{2 \times R \times L \times I \times .866}{1000}$$

$$VD = \frac{2 \times .308 \times 83 \times 46 \times .866}{1000}$$

$$VD = \frac{2036.73}{1000}$$

$$VD = 2.036 = \textbf{2.04 V}$$

$$\%VD = \frac{V_l}{V_s} \times 100$$

$$\%VD = \frac{2.04}{208} \times 100$$

$$\%VD = .0098 \times 100$$

$$\%VD = \textbf{.98\% VD}$$

FIRST FLOOR PLAN

Three lighting panels are shown on this floor, Panel B in the north wall, and Panels A and C in the south wall. Panel C is primarily confined to sup-

plying outlets in the entrance, canopy, and show window areas. Panels A and B are confined to the general store illumination. The wiring plan on this floor is similar to that on the second floor.

A single riser is indicated in the north wall. The feeder conduit is associated with Panel E on the second floor. The riser near the left end of the south wall, marked UP, is the same one shown on the second floor plan, marked DN.

The riser next to Panel A is a feeder conduit to Panel D, above. The next two risers, also marked UP, are explained on the Basement Floor Plan. The third riser, marked UP 10, is the other end of the conduit supplying the lighting unit at the head of the stairs on the second floor. The two risers in the east wall of the passenger elevator shaft are shown on this floor plan.

Calculations for Panel B

The loads on the circuits are:

Circuit	Load*	Circuit	Load*
1	1800	8	1920
2	1920	9	1556
3	1556	10	720
4	1800	11	1080
5	1920	12	1260
6	1556	13	1080
7	1800		

* in VA

The circuits devoted to general lighting are 1, 2, 3, 4, 5, 6, 7, 8 and 9. The general lighting circuits use 15,828 VA (1800 VA + 1920 VA + 1556 VA + 1800 VA + 1920 VA + 1556 VA + 1800 VA + 1920 VA + 1556 VA = 15,828 VA).

Measurement shows the floor area lighted by these circuits to be 3344 sq ft. The minimum acceptable VA for this area is 12,540 VA (3344 sq ft × 3.75 VA = 12,540 VA). Since the connected load is greater than this amount, the higher value is used in the feeder calculations.

Plug receptacle circuits 10, 11, 12, and 13 require 4140 VA. The load on the feeder is 19,968 VA (15,828 VA + 4140 VA = 19,968 VA). The current is approximately 56 A (19,968 VA ÷ 360 = 55.47 = 56 A). The minimum size of THW Cu conductor per Table 310-16 is No. 6.

The length of single conductor between Panel B and the switchboard is approximately 92′. The voltage drop in this length of No. 6 conductor equals 4.38 V. The resistance for a No. 6 Cu conductor is .491 Ω per Ch 9, Table 8.

$$VD = \frac{2 \times R \times L \times I \times .866}{1000}$$

$$VD = \frac{2 \times .491 \times 92 \times 56 \times .866}{1000}$$

$$VD = \frac{4381.32}{1000}$$

$$VD = \mathbf{4.38\ V}$$

$$\%VD = \frac{V_l}{V_s} \times 100$$

$$\%VD = \frac{4.38}{208} \times 100$$

$$\%VD = .021 \times 100$$

$$\%VD = \mathbf{2.1\%}$$

The feeder to Panel B consists of four No. 6 THW Cu conductors in 1″ RMC.

Calculations for Panel A

The loads on the circuits are:

Circuit	Load*	Circuit	Load*
1	300	7	1920
2	1000	8	1556
3	1800	9	540
4	1920	10	1260
5	1556	11	1080
6	1800		

* in VA

The circuits concerned with general lighting are 1, 2, 3, 4, 5, 6, 7, and 8. The general lighting circuits have a connected load of 11,852 VA (300 VA + 1000 VA + 1800 VA + 1920 VA + 1556 VA + 1800 VA + 1920 VA + 1556 VA = 11,852 VA).

Measurement shows the floor area lighted by these circuits as 3080 sq ft, for which the minimum VA must be not less than 11,550 VA (3080 sq ft × 3.75 VA = 11,550 VA). The connected load, being somewhat higher than this value, is used in feeder calculations.

Plug receptacle circuits 9, 10, and 11 take 2880 VA. The total load on the feeder is 14,732 VA (11,852 VA + 2880 VA = 14,732 VA). The current is approximately 41 A (14,732 VA ÷ 360 = 40.9 =

41 A) and calls for a minimum size of No. 8 Cu THW Cu conductor per Table 310-16.

The length of single conductor from Panel A to the switchboard is about 73′. Voltage drop in this No. 8 conductor is 4 V. The resistance for a No. 8 Cu conductor is .778 Ω per Ch 9, Table 8.

$$VD = \frac{2 \times R \times L \times I \times .866}{1000}$$

$$VD = \frac{2 \times .778 \times 73 \times 41 \times .866}{1000}$$

$$VD = \frac{4033.05}{1000}$$

$$VD = 4.03\ V = \mathbf{4\ V}$$

$$\%VD = \frac{V_l}{V_s} \times 100$$

$$\%VD = \frac{4}{208} \times 100$$

$$\%VD = .019 \times 100$$

$$\%VD = \mathbf{1.9\%}$$

The feeder to Panel A consists of four No. 8 THW Cu conductors in 1″ RMC.

Calculations for Panel C

The loads on the circuits are:

Circuit	Load*	Circuit	Load*
1	1200	8	1500
2	1000	9	1000
3	1500	10	300
4	1000	11	1800
5	1500	12	864
6	1200	13	900
7	1000		

* in VA

Except for front stair lighting, Circuit 10, and cove lighting, Circuit 12, this panelboard feeds only show window, entrance, and canopy areas.

The circuits for lighting the show windows are 1, 2, 3, 4, 5, 6, 7, 8, and 9. They require a total of 10,900 VA (1200 VA + 1000 VA + 1500 VA + 1000 VA + 1500 VA + 1200 VA + 1000 VA + 1500 VA + 1000 VA = 10,900 VA).

NEC 220-12(a) provides that a minimum of 200 VA per linear foot be assigned for show window lighting. The two show windows are each 22′ long, so that a minimum of 8800 VA (200 VA × 22′ × 2 = 8800 VA) shall be used in the feeder calculations. Here, the greater value of the 10,900 VA connected load is used.

The total load on Panel C is equal to 14,764 VA. The current is approximately 41 A (14,764 VA ÷ 360 = 41.01 = 41 A). No. 8 THW Cu is required per Table 310-16. The conductor is 73′ in length. Since Panel C is directly over the switchboard, voltage drop in the feeder conductors is negligible, and four No. 8 THW Cu conductors in 1″ RMC are satisfactory.

Since the Specifications state that the basement ceiling is unfinished, and since floors are of poured, reinforced concrete construction, it is necessary to set basement lighting outlet boxes and conduit on the first floor deck at the same time that first floor plug receptacle boxes are spotted there. Plans and Specifications should always be checked with a view to detect these situations.

BASEMENT FLOOR PLAN

The basement lighting panel is included in the main switchboard. A power panel, located in the boiler room, furnishes current to pump motors there and to a motor that operates the package elevator unit.

The basement contains a shipping and receiving space, an alteration department, a large stockroom area, employee toilets, public toilets, and a small office. Both employee toilets and public toilets have exhaust fans connected directly to respective lighting circuits so they are in operation at all times while light switches are turned ON. The Specifications furnish details regarding the fans, stating that they consume 130 VA each. Plug receptacle circuit 13, in the alteration department, has two special outlets described in the Specifications.

A pair of risers in the north wall are feeders to Panels B and E. In the south wall, risers at the left are associated with Panels D and A. Toward the other end, a conduit rises to the power center on the roof. A feeder conduit to Panel C is next in line, followed by one to an electric sign on the roof. The DN marking in the east wall of the freight elevator shaft refers to a conduit leading to the pit light, and the identical marking in the passenger elevator wall shows a run to its pit light.

Calculations for Basement Panel

The loads on the circuits are:

Circuit	Load*	Circuit	Load*
1	1200	8	1360
2	870	9	1100
3	1224	10	1464
4	1624	11	1620
5	1220	12	1260
6	1220	13	1500
7	1220	14	1260

* in VA

The total load is 18,142 VA (1200 VA + 870 VA + 1224 VA + 1624 VA + 1220 VA + 1220 VA + 1220 VA + 1360 VA + 1100 VA + 1464 VA + 1620 VA + 1260 VA + 1500 VA + 1260 VA = 18,142 VA). The current is approximately 50 A (18,142 VA ÷ 360 = 50.39 = 50 A) and a No. 6 conductor is large enough to supply the panel per Table 310-16.

Service Calculation

The service load may be obtained by adding all the feeder requirements, including boiler room power, sign, and roof power. A load of 4800 VA is to be allowed for boiler room power, 8640 VA for the sign, and 36,000 VA for roof power.

Starting with Panel E and continuing in the order of calculations, the total load is 16,200 VA + 16,560 VA + 19,968 VA + 14,732 VA + 14,764 VA + 18,142 VA + 8640 VA + 4800 VA + 36,000 VA, or 149,806 VA.

SERVICE CALCULATION	
Panel	Load*
E	16,200
D	16,560
B	19,968
A	14,732
C	14,764
Basement Lighting	18,142
Sign	8640
Boiler Room Power	4800
Roof Power	36,000
Total VA	149,806

* in VA

The total load for the service calculation is 149,806 VA. The current is 416 A (149,806 VA ÷ 360 = 416.13 = 416 A).

$$I = \frac{VA}{V \times \sqrt{3}}$$

$$I = \frac{149,806}{360}$$

$$I = \textbf{416 A}$$

Note: When making calculations for balanced 4-wire, 120/208 V, 3ϕ systems, divide the total VA by 360 (208 × 1.732 = 360) in order to obtain current per phase wire.

Table 310-16 shows that a 600 kcmil THW conductor will carry 420 A. Two smaller conductors are more easily handled than a single large one. NEC 310-4 permits conductors of No. 1/0 and larger to be run in parallel, provided they are of the same length, conductor material, circular-mil area, insulation type, and are terminated in the same manner. Where parallel conductors are run in separate raceways or cables, the raceways or cables shall have the same physical characteristics.

Two parallel conductors on each leg of the circuit would each carry one-half of the total current, or 208 A. With all conductors in the same conduit, a total of six current-carrying wires, the rating of each is reduced to 80% of the value indicated in Table 310-16. A No. 4/0 THW conductor has an ampacity of 230 A where there are not more than three current-carrying conductors in a raceway. For six conductors, each shall have a normal ampacity equal to 208 A divided by .8, or 260 A. A 300 kcmil THW conductor would be required in this case. But if two sets of No. 4/0 conductors are installed in separate conduits, the full ampacity of each conductor is restored, and there is considerable saving of copper.

NEC 220-22 states that the neutral feeder shall be large enough to carry the maximum unbalanced load. The *maximum unbalanced load* is the load that can be connected between the neutral and any ungrounded conductor, which means that loads that have no neutral component, such as power loads confined to the 3ϕ wires, are subtracted from the total load in order to determine this value of current. The total load in the present instance is 416 A and the 3ϕ power load 114 A, (boiler room power plus roof power). The maxi-

mum unbalanced load on the neutral is 302 A (416 A − 114 A = 302 A).

NEC 220-22 provides that a demand factor of 70% may be applied to that portion of the neutral load in excess of 200 A. It appears that the neutral load could be reduced to 200 A + 102 A (.7 × 102 A = 71.4 A = 72 A) or 272 A. NEC 220-22 states that no reduction of neutral capacity shall be allowed for that part of the load that consists of electric discharge lighting. Since electric discharge lighting in the present example accounts for at least 150 A, there is no excess over 200 A to which a 70% factor might be applied. The neutral load, then, is calculated on the basis of 302 A.

NEC 300-3(b) requires that a grounded conductor be run in the same raceway with each set of phase conductors. Either of the two neutral wires in this case carry one-half (151 A) of the unbalanced current (302 A ÷ 2 = 151 A). The next largest conductor size listed in Table 310-16 is No. 2/0 THW, which has an ampacity of 175 A.

NEC 230-30, Ex. permits an uninsulated copper grounded conductor to be installed underground in duct or conduit. It may appear that a larger ampacity might be assigned to an uninsulated conductor, so that a size smaller than No. 2/0 THW could be used. But 310-15(b)(3) states that where bare conductors are used with insulated conductors, their allowable ampacity shall be limited to that permitted for an insulated conductor of the same size. The present Specifications require that an insulated neutral conductor be used. The service consists of two sets of three No. 4/0 THW conductors and one set of No. 2/0 THW conductors in parallel.

Service Conduit Size

When all conductors in a raceway are of the same size, Appd C, Tables C1 – C12 determine the size of conduit required. Each table includes all sizes of conductors with all of the different types of conductor insulation listed. Each of the tables in Appd C is for a specific type of raceway. For different sizes of wire in a conduit, calculations are based upon Chapter 9, Table 1 and Tables 4 through 8.

Table 4 gives dimensions and percent of area of conduit and tubing. The other tables give dimensions for the various types of conductor coverings including rubber, thermoplastic, lead-covered, asbestosvarnished-cambric, cross-linked polyethylene, and bare.

In the store service, each conduit has three No. 4/0 THW conductors and one No. 2/0 THW conductor. Table 5 gives the cross-sectional area of a No. 4/0 THW conductor as .3718 sq in. and a No. 2/0 THW conductor as .2624 sq in. The area of all four conductors equals 1.3778 sq in. (3 × .3718 sq in. + .2624 sq in. = 1.3778 sq in.). Table 1 allows 40% of the conduit cross-sectional area to be used when there are three or more conductors of different sizes. Cross-sectional area of the conduit, therefore, is equal to 3.4445 sq in. (1.3778 sq in. ÷ .4 = 3.4445 sq in.). According to Table 4, the required size of conduit is 2½″.

480 V Lighting System

A method of lighting that has become increasingly popular in commercial and industrial establishments is the four-wire, 277/480 V, 3ϕ electric discharge system. Lighting units are connected between the neutral conductor and phase wires, imposing 277 V on ballast assemblies. Switching may be at the panelboard if CBs are listed and marked SWD per 240-83(d). Where wall switches are used and the voltage between adjacent switches exceeds 300 VA, a permanent barrier is required in outlet boxes between switches per 380-8(b).

Aluminum Wire and Conduit

In recent years, aluminum wire and conduit have competed with copper wire and steel conduit. Since aluminum conduit is made in the same standard trade sizes as steel, there are no NEC® problems. Aluminum conductors, however, vary with respect to copper in both allowable ampacity and ohms of resistance because ampacity for a given size is less than that of copper and resistance is higher. The sizes of wire and conduit required are calculated when THW Al feeder and service conductors are substituted for THW Cu conductors in the store building example.

The formula for calculating voltage drop for aluminum conductors is the same as the formula used for copper conductors. The resistivity (R) for copper and aluminum conductors is given in Chapter 9, Table 8.

The formulas for calculating voltage drop (VD) and %VD are:

$$VD = \frac{2 \times R \times L \times I \times .866}{1000}$$

where

VD = voltage drop (in V)

R = resistance of conductor (in Ω)

L = length of conductor (in ft)

I = current (in A)

$.866 = \dfrac{\sqrt{3}}{2}$ (used for 3ϕ only)

1000 = 1000′ or less of conductor

$$\%VD = \frac{V_l}{V_s} \times 100$$

where

$\%VD$ = percentage of voltage drop (in %)

V_l = voltage loss (in V)

V_s = voltage supply (in V)

100 = constant

The VD and %VD has been determined for each of the panels using Cu conductors. The formulas may be used to find VD and %VD for aluminum conductors. Ch 9, Table 8 gives the resistance values and circular mils used in the voltage drop formulas.

COPPER					
Panel	**R**	**L**	**I**	**Conductors**	**Conduit**
A	.778	73	41	4 No. 8 THW	1″
B	.491	92	56	4 No. 6 THW	1″
C	.778	73	41	4 No. 8 THW	1″
D	.308	83	46	4 No. 4 THW	1¼″
E	.491	104	45	4 No. 6 THW	1″

ALUMINUM					
Panel	**R**	**L**	**I**	**Conductors**	**Conduit**
A	.808	73	41	4 No. 6 THW	1″
B	.508	92	56	4 No. 4 THW	1¼″
C	.808	73	41	4 No. 6 THW	1″
D	.808	83	46	4 No. 2 THW	1¼″
E	.508	104	45	4 No. 4 THW	1″

FEEDER RISER DIAGRAM

SIGN PANEL

POWER CENTER ELEVATOR CIRCUITS AND VENTILATION EQUIPMENT

THREE NO. 1 THW, $1\frac{1}{4}$" C

FOUR NO. 8 THW, 1" C

D

E

FOUR NO. 4 THW, $1\frac{1}{4}$" C

FOUR NO. 6 THW, 1" C

C

FOUR NO. 8 THW, 1" C

FOUR NO. 8 THW, 1" C

A

B

FOUR NO. 6 THW, 1" C

BASEMENT LIGHTING PANEL

BOILER ROOM POWER

THREE NO. 6 THW, 1" C

120/208 V, 3φ, 4-WIRE SERVICE TWO SETS OF THREE NO. 4/0 THW AND ONE NO. 2/0 IN $2\frac{1}{2}$" GALVANIZED CONDUITS

THREE NO. 10 THW, $\frac{3}{4}$" C

ELECTRICAL DETAILS

FIXTURE ATTACHED TO OUTLET BOX

BOX TO BE ACCESSIBLE AFTER FIXTURE IS INSTALLED • 410-14(b)

CONDUIT TO BE SECURED WITHIN 3' OF BOX

FLEXIBLE CONDUIT TO BE SECURED WITHIN 12" OF BOX • 350

6' MAXIMUM FIXTURE WHIP • 350-18, Ex. 3

CONDUIT

JUNCTION BOX

INCANDESCENT UNIT

FIXTURES SECURED TO FRAMING MEMBER • 410-16(c)

THERMAL PROTECTION INTEGRAL WITHIN BALLAST REQUIRED • 410-73(e)

SWITCHBOARD ELEVATION

1" C

$1\frac{1}{4}$" C

$1\frac{1}{2}$" C

1" C

$1\frac{1}{4}$" C

METER

POWER 125

BASEMENT LIGHTING PANEL 60 A CB 14 CIRCUITS

MAIN CB 600 A FRAME

C 60

A 60

D 60

CT SECTION

B 60

E 60

30   30

BOILER ROOM

SIGN

FIXTURE SYMBOLS

Symbol	Description
	MORBRITE-4 LT.-8 FT. CAT. 1679 OR EQ.
	MORBRITE-4 LT.-4 FT. CAT. 1772 OR EQ.
m	MORBRITE-2 LT.-8 FT. CAT. 1677 OR EQ.
v	MORBRITE-2 LT.-4 FT. CAT. 1770 OR EQ.
	MORBRITE-1 LT.-6 FT. CAT. 1831 OR EQ.
	MORBRITE-1 LT.-4 FT. CAT. 1852 OR EQ.
l	MORBRITE-3 LT.-6 FT. CAT. 1915 OR EQ.
e	MORBRITE-1 LT.-8 FT. CAT. 1676 OR EQ.
	ALLRAY-200 W REC. CAT. 17234 OR EQ.
f	ALLRAY-300 W REC. CAT. 17329 OR EQ.
i	ALLRAY-300 W REC. CAT. 17240 OR EQ.
h	ALLRAY-150 W REC. CAT. 17236 OR EQ.
a	ALLRAY-300 W REC. CAT. 17265 OR EQ.
c	ALLRAY-100 W BRACKET CAT. 10090 OR EQ.
p	ALLRAY-100 W PORC. CAT. 11007 OR EQ.
n	ALLRAY-200 W SPOT CAT. 11103 OR EQ.
k	ALLRAY-100 W DRUM CAT. 12966 OR EQ.

STORE BUILDING
FIRST FLOOR PLAN

STORE BUILDING
BASEMENT FLOOR PLAN

SERVICES

Building – A structure which stands alone or is cut off from adjoining structures by fire walls or fire doors.

A. **225-32** Conductors supplying each building shall be provided with readily accessible means for disconnecting ungrounded conductors from the source of supply. The disconnecting means for garages, etc., for residential use may be 3- and 4-way snap switches. See 225-36, Ex.

B. **230-3** Service conductors shall not feed one building or structure through another building or structure.

C. **230-6** Conductors installed under at least 2″ of concrete beneath a building or enclosed by concrete or brick at least 2″ thick or in a transformer vault shall be considered outside a building.

D. **230-43** Wiring methods for service conductors are limited to: (1) open wiring on insulators; (2) IGS cable; (3) RMC; (4) IMC; (5) EMT; (6) ENT; (7)

SE cable; (8) wireways; (9) busways; (10) auxiliary gutters; (11) RNMC; (12) cablebus; (13) MC cable; (14) MI cable; (15) FMC or LTFMC, 6′ maximum length, with EBJ; (16) LTFNMC.

E. **230-46** Service-entrance conductors shall only be spliced by clamps or bolted connectors.

F. **230-54(f)** Drip loops are required to prevent moisture and are used to connect service-entrance conductors to service-drop conductors below level of service head.

G. **250-32(a)(b)** Where a grounded system serves two or more buildings from the same service, each building shall have a neutral to its own grounding electrode except if the second building has a single-branch circuit and no equipment that requires grounding or an EGC is run to the second building.

SERVICE EQUIPMENT

A. **230-72(a)(b)** If supplied by two or more services on the same side of the building, the service equipment shall be grouped and marked for identification. Service for fire pumps is permitted to be in another location.

B. **240-24(b)** In a multiple-occupancy building, each occupant shall have access to overcurrent devices except as permitted by 240-24(b)(1)(2).

C. **230-93** If necessary to prevent tampering, the service equipment for a special load may be locked.

D. **230-92** If service equipment is locked or inaccessible, overcurrent devices of lower rating must be installed in an accessible location.

SERVICE DISCONNECTING MEANS

A. **230-75** If the switch or CB does not interrupt the grounded conductor, other means of disconnection shall be provided in the cabinet or switchboard.

B. **230-76** The service disconnecting means shall be operable by hand and shall consist of either a manually-operable switch or CB or an electrically-operated switch or CB also operable by hand.

384-3(c) Switchboard or panelboards used as service equipment shall have means within the service disconnect section for connecting the supply neutral to the metal frame and sized per 250-28(d).

CABLE LIMITERS

240-8: Current-Limiting Overcurrent Protective Device – An overcurrent device that, when interrupting currents in current-limiting range, reduces current in faulted circuit to a value less than obtainable in same circuit if replaced with a solid conductor of comparable impedance.

240-11: Current Limiting Device – An overcurrent device that, when interrupting currents in a current-limiting range, will reduce current in the faulted circuit to a value less than obtainable in the same circuit if replaced with a solid conductor having comparable

CABLE PROTECTION AND FAULT ISOLATION OF MULTIPLE CABLES PER PHASE

EMERGENCY CIRCUIT TAPS AHEAD OF MAINS

230-82 In general, equipment shall not be installed on the line or supply side of service disconnecting means. See Special Applications (1 – 7).

230-91 The service overcurrent device shall be immediately adjacent to the service disconnecting means or an integral part of it.

240-40 In circuits over 150 V to ground, a disconnecting means shall be installed on the line side of all fuses, where accessible to non-qualified personnel. See Special Applications for (1) cable limiters and (2) group operations in 430-112, Ex. and 424-22(c).

240-60(b) Current-limiting fuseholders shall be designed so that they only accept current-limiting fuses.

GROUNDING SEPARATELY DERIVED SYSTEMS – 250-30

A. 240-3 Overcurrent protection of conductors shall be in accordance with their ampacities per Tables 310-16 – 310-19 and notes to these tables.

B. 240-21 Overcurrent protection shall be connected in each ungrounded circuit conductor at the point of supply.

C. 250-20(d) A premises wiring system with power derived from a generator and having no direct electrical connection to supply conductors originating in another system shall be grounded per 250-30.

D. 250-28 The bonding jumper is sized per 250-28(d) for derived phase conductors. The bonding jumper connects the circuit conductor of the system that is required to be grounded to the system noncurrent-carrying equipment enclosures except as permitted in 250-24(a)(4). This connection is to be made on the supply side of separately derived system and ahead of any system disconnect or overcurrent device.

E. 250-30(a)(2) The GEC is sized per 250-66 for derived phase conductors. The GEC connects the circuit conductor of the system that is required to be grounded to the grounding electrode. Except as permitted by 250-24(a)(3), this connection is to be made at the source of separately derived system and ahead of any disconnect means or overcurrent device.

F. 250-30(a)(3) The grounding electrode shall be the nearest of the following: (1) nearest effectively grounded structural metal member of structure; (2) nearest effectively grounded metal water pipe; or (3) other electrodes per 250-50 and 250-52 where electrodes specified by (1) or (2) are not available.

ABC FIRE EXTINGUISHERS

TRASH • WOOD • PAPER

A ORDINARY COMBUSTIBLES

LIQUIDS • GREASE

B FLAMMABLE LIQUIDS

MOTORS • TRANSFORMERS

C ELECTRICAL EQUIPMENT

An ABC fire extinguisher can be used for fires of ordinary combustibles, flammable liquids, or electrical equipment. Remain 8′ from the fire and direct the spray at the base of the fire.

GROUND-FAULT PROTECTION OF EQUIPMENT – 230-95

GROUND-FAULT
SENSOR TYPE GT

N
φ
φ
φ

WINDOW AREA
FOR CONDUCTORS

GROUND-FAULT
RELAY TYPE GP

B

D

GROUND TRIP OF
CIRCUIT INTERRUPTER

F

MONITOR
PANEL

GROUND FAULT DETECTION SYSTEM

Square D Company

APPLICATION – CB

The ground current powered ground-fault detection system is designed for use on power systems which include a grounded conductor (neutral or grounded phase). When circuit conditions are normal, currents from all the phase conductors and neutral (if used) total zero, and the sensor produces no signal.

If a ground fault occurs, the current total equals the ground-fault current, and the sensor provides a signal proportional to the ground-fault current. This signal powers the ground-fault relay, which activates the ground trip mechanism on the CB. See 215-10 and 240-13.

SERVICE DISCONNECT
RATED 1000 A OR MORE

E

BOLT-LOC SWITCH

A

F

SENSOR

N
A
B
C

LOAD

277/480 V
WYE SERVICE

C

120 V

LIMIT
SWITCH

STORED
ENERGY
MECHANISM

ELECTRIC
TRIP COIL

B

GFR

INTERPOSING
RELAY

APPLICATION – SWITCH AND FUSE

Under normal operation, there is no output from the ground-fault sensor and therefore the GFR contact is open. If a ground-fault current of sufficient magnitude occurs, the GFR contact will close to energize the interposing relay, which in turn will energize the electric trip coil of the BOLT-LOC switch to open the circuit.

A. **110-9** Service equipment shall be suitable for available fault current or its supply terminals.

B. **110-10** Overcurrent devices shall be selected and coordinated so as to clear a fault without extensive damage to electrical components of the circuit.

C. **230-95** Ground-fault protection of equipment is required for grounded wye services of more than 150 V to ground but not more than 600 V phase-to-phase for each service disconnect rated 1000 A or more.

D. **230-95(a)** Operation of ground-fault protection shall cause service disconnect to open all ungrounded wires of faulted circuit. Ground-fault protection shall have maximum setting of 1200 A and a maximum time delay of one second for ground-fault currents of 3000 A or more.

E. **230-95(b)** For switch and fuse combination, the fuses shall be capable of interrupting any current higher than the interrupting capacity of the switch during a time when the ground-fault protective system will not cause the switch to open.

F. **230-95(c)** Ground-fault protection system shall be performance tested on-site in accordance with manufacturer's instructions when first installed and a record of this test shall be made available to the AHJ.

GROUNDING ELECTRODE CONDUCTOR (GEC)

A. 250-28(a); 250-102(a) MBJs and EBJs shall be of copper or other corrosion-resistant material and shall be sized per Table 250-66 and 250-122.

B. 250-62; 250-64 The GEC shall be of copper, aluminum, or copper-clad aluminum, and selected to be resistant to any corrosive conditions or be suitably protected against corrosion. It shall be solid or stranded, insulated, covered, or bare and continuous without splice or joint except: (1) busbars; (2) where more than one service (per 230-40, Ex. 2), the GEC, sized per 250-66, serving the largest service entrance conductors, may be tapped for GECs for other enclosures. This tap is permitted to be sized per 250-66 for the largest phase conductor serving respective separate enclosures, and then only permitted where connection extends within the enclosure.

C. 250-64(a) GECs No. 4 and larger shall be attached to the surface, also a No. 5 if not exposed to physical injury. Aluminum grounding conductors shall not be run in direct contact with masonry or earth or where subject to corrosive conditions. If run outside, aluminum grounding conductors shall not be installed within 18″ of the earth.

CONDUCTOR PROTECTION – POLE

A. 230-49 Underground service conductors carried up a pole shall be protected by suitable means to a height of at least 8′ above ground. See 300-5.

B. 300-5(d) Conductors emerging from the ground shall be suitably protected to a height of 8′ and at least 18″ below ground. Where subject to physical damage, RMC, IMC, Schedule 80 PVC, or equivalent are acceptable.

HIGH-LEG – IDENTIFICATION

A. 215-8 The 4-wire delta phase conductor having the highest voltage to ground (high-leg) shall be orange in color or identified by tagging or other effective means where the neutral is present.

B. 384-3(e) In a 4-wire delta-connected system, with the midpoint of one phase winding grounded, the high-leg shall be orange or marked by other effective means.

C. 384-3(f) The high-leg shall be connected to B (center) phase in switchboard or panelboard.

SERVICE – UNDERGROUND

A. 230-8 Underground raceway shall be sealed. See 300-5.

B. 230-46 SE conductors are permitted to be spliced with clamps or bolted connectors.

C. 230-49 Underground service conductors shall be protected against physical injury as required by 300-5.

230-53 Where exposed to weather or embedded in masonry, service raceways shall be raintight and arranged to drain.

D. 250-92(a) 2. Bonding is required around the enclosure. Bonding type locknuts or bushings or other approved means shall be used.

E. 300-5(c) If raceway is under a building and contains underground cable, it shall extend beyond the outside walls of building.

FLAT CONDUCTOR CABLE (FCC) – 328

TRANSITION BOX

TRANSITION ASSEMBLY

CARPET SQUARES

TRANSITION BOX COVER

36″

36″

328-2: Flat Conductor Cable (FCC) – A cable with three or more flat copper conductors edge-to-edge and separated and enclosed by an insulating material.

328-2: FCC System – A complete wiring system (including FCC, connectors, terminators, adapters, boxes, and receptacles).

A. 328-1 Type FCC cable is a field-installed wiring system using associated accessories for branch circuits and installed under carpet squares.

B. 328-3 FCC system shall comply with 210, 220, 240, 250, and 300.

328-6(a) Maximum branch-circuit voltage is 300 V between ungrounded conductors or 150 V to ground.

328-6(b) General-purpose circuits shall not exceed 20 A. Individual circuits shall not exceed 30 A.

C. 328-10 Type FCC cable shall be covered with carpet squares that have release-type adhesive and not larger than 36″ × 36″.

D. 328-16 All system components shall be firmly anchored by an adhesive or mechanical anchoring system identified for this use.

E. 328-11 All connections are to be made with identified conductors. All bare cable ends shall use listed insulating ends. All cable connectors and ends shall be sealed against dampness and liquid spillage.

F. 328-15 Transition from one system to FCC system shall be made with transition assembly identified for this use.

USES PERMITTED – 328-4

(a) Branch circuits

(b) Floors (made of concrete, ceramic, wood, etc.)

(c) Walls (in surface metal raceways)

(d) Damp locations

(e) Heated floors shall be identified for such use when above 30°C (80°F)

USES NOT PERMITTED – 328-5

(1) Outdoors or in wet locations

(2) Where subject to corrosive vapors

(3) In hazardous locations

(4) In residences, schools, and hospitals

MULTIOUTLET ASSEMBLY

MULTIOUTLET ASSEMBLY

BUILDING FINISH OR BASEBOARD

DRY PARTITION

A. 220-3(b)(8) Each 5′ or fraction thereof of fixed multioutlet assembly shall be considered as one outlet of 180 VA capacity, except where a number of appliances may be connected simultaneously, then each 1′ or fraction will be considered one outlet. *Note:* This provision does not apply to dwelling units or guest rooms of motels or hotels.

B. 353-3 May be used through dry partition where no outlet is concealed and provisions are made for removing the cap or cover.

USES PERMITTED – 353-2(a)

(a) In dry locations

USES NOT PERMITTED – 353-2(b)

(1) Where concealed

(2) Where subject to physical damage

(3) Over 300 V or more between conductors unless assembly is metal at least .040″ thick

(4) Where subject to corrosive vapors

(5) In hoistways

(6) In hazardous locations

HEAVY-DUTY LAMPHOLDERS – RATING

30 A CIRCUIT

HEAVY-DUTY 660 W LAMPHOLDERS

A. 210-21(a) Lampholders connected to circuits having a rating greater than 20 A shall be of the heavy-duty type and shall be rated not less than 660 W, if of a medium type, and not less than 750 W if any other type.

720-5 For circuits of less than 50 V, standard lampholders rated not less than 660 W are required.

RECEPTACLE RATINGS

SINGLE RECEPTACLE

20 A PLUG RECEPTACLES

20 A CIRCUIT

A. 210-7(a) Receptacles installed on 15 A and 20 A circuits shall be of the grounding type.

B. 210-21(b)(1) A single receptacle installed on an individual branch circuit shall have a rating not less than the branch circuit rating.

C. 210-21(b)(3) Plug receptacles shall have a rating not less than the load served. When connected to circuits having two or more outlets, they shall conform to the following:

Circuit*	Receptacle Rating
15	15
20	15 or 20
30	30
40	40 or 50
50	50

* in A

OVERCURRENT DEVICES AND AVAILABLE FAULT CURRENT

Non-Selective System – A fault on an individual branch circuit not only opens the branch-circuit overcurrent protective device, but also opens the feeder overcurrent protective device.

Selective System – A fault on an individual branch circuit opens only the branch-circuit overcurrent protective device and does not affect the feeder overcurrent protective device.

NON-SELECTIVE SYSTEM

SELECTIVE SYSTEM

A. 110-9 Overcurrent devices shall have sufficient rating for the available voltage and the current that is to be interrupted.

B. 110-10 The overcurrent protective devices, total impedance, component short-circuit withstand ratings, and other characteristics of the circuit to be protected shall be selected and coordinated to permit protective devices to clear a fault without extensive damage to electrical components of the circuit.

C. 110-9 Service equipment and branch protection shall have an interrupting capacity suitable for the available fault current at the protective device.

D. 240-12, FPN Coordination is properly localizing a fault condition to restrict outages to affected equipment by using the proper selective fault-protection devices.

E. 517-17(a)(b) In health care facilities where ground-fault protection is provided for service and feeders, they shall be selective, coordinated, and performance-tested when first installed.

CURRENT-LIMITING FUSES IN MAIN SERVICE SWITCH

FUSED OVERCURRENT DEVICES TO APARTMENT PANELS (LP1 HERE) WITH PROPERLY-SIZED OVERCURRENT DEVICES

APPROXIMATELY 60,000 A MAXIMUM LINE-TO-NEUTRAL AVAILABLE SHORT-CIRCUIT CURRENT

PANEL IN INDIVIDUAL APARTMENT USING FUSES OR CBs HAVING 5000 TO 10,000 A.I.C.

METERS AND FUSED DISCONNECTS FOR FEEDERS TO INDIVIDUAL TENANTS

PAD-MOUNTED TRANSFORMER

APPROXIMATELY 41,000 A L-L AND 61,400 A L-N SHORT-CIRCUIT CURRENT AT TRANSFORMER SECONDARY

167 kVA, 120/240 V, 1 φ, 1.7% IMPEDANCE

SYSTEM A

What is the short-circuit current in System A?

Step 1. Calculate the transformer's full-load current (1φ):

$$I = \frac{kVA \times 1000}{E}$$

$$I = \frac{167 \times 1000}{240}$$

$$I = 696 \text{ A (FLA)}$$

Step 2. Find the multiplier:

$$\frac{100}{Trans \text{ % } Impedance}$$

$$\frac{100}{1.7} = 58.8$$

Step 3. Find the short-circuit current at transformer secondary:

Trans FLA × multiplier

696 × 58.8

40,925 A short-circuit L-L current

Note 1. The L-N fault current of transformer (1φ center-tapped) secondary is approximately 1.5 × L-L value, i.e., 40,925 A × 1.5 = 61,388 A L-N. (Multiplier 1.5 varies from 1.33 to 1.67.)

Note 2. For 3φ systems, insert 1.73 in the formula and proceed with Steps 2 and 3.

$$I = \frac{kVA \times 1000}{E \times 1.73}$$

| **L-L** – Line-to-Line |
| **L-N** – Line-to-Neutral |

CIRCUIT BREAKERS

NONADJUSTABLE CIRCUIT BREAKER

UP IS ON POSITION

6' - 7" MAXIMUM

FLOOR

A. 240-3(b) If the allowable ampacity of a conductor does not correspond with a standard rating of an overcurrent device, the next larger rating may be used, provided that the current is not greater than 800 A.

B. 240-81 CBs shall be clearly marked ON or OFF. Up is ON position.

C. 240-82 A CB shall be designed so that alteration of its operating characteristics is difficult.

D. 240-83(d) Where CBs are used to switch 120 V and 277 V fluorescent fixtures, they shall be marked SWD and listed for such switching duty.

E. 380-8(a) Center of operating handle shall not be over 6'-7" from floor.

F. 380-11 CBs used as a switch shall conform to 240-81 and 240-83.

CONDUCTORS – OVERCURRENT PROTECTION

NO. 1 RH Cu CONDUCTOR, 130 A

FUSED DISCONNECT

150 A FUSES

A. 240-3 Conductors other than flexible cords and fixture wires shall be protected at their ampacities per 310-15. See 240-3 (a – g).

B. 240-3(b) If the allowable ampacity of a conductor does not correspond with a standard rating of an overcurrent device, the next larger rating may be used, provided that the current is not greater than 800 A and the conductor is not part of a multioutlet branch circuit supplying receptacles for cord- and plug-connected loads.

C. 210-19(a) Branch-circuit conductors shall have an ampacity not less than the branch circuit and not less than the maximum load served.

D. 240-21 Overcurrent protection is required where conductors receive their supply.

CORDS AND FIXTURE WIRE – OVERCURRENT PROTECTION

APPLIANCE

FLEXIBLE CORD

240-4(a) Flexible cords (including tinsel) shall be protected at their ampacities per Table 400-5(a).

240-4(b)(1) Cord listed with specific appliance or portable lamps shall be connected to a standard branch circuit as follows:

Cord	Circuit*
Tinsel or No. 18 +	20
No. 16 +	30
20 A +	40
30 A +	50

* in A

240-4(b)(2) Fixture wire connected to 120 V or more standard branch circuit shall be protected as follows:

Cord	Circuit*
No. 18 (to 50')	20
No. 16 (to 100')	20
No. 14 +	20 or 30
No. 12 +	40 or 50

* in A

240-4(b)(3) Listed extension cordsets with No. 16 or larger conductors are permitted on a 20 A branch circuit.

SNAP SWITCHES

CONTROLS RESISTIVE, INDUCTIVE, TUNGSTEN-FILAMENT, AND MOTOR LOADS
• 380-14(a)

MARKING
• 380-15

CONTROLS RESISTIVE, INDUCTIVE, AND TUNGSTEN-FILAMENT LOADS
• 380-14(b)

AC GENERAL-USE

AC-DC GENERAL-USE

CONTROLS DIRECT CONNECTION TO 20 A OR LESS AL CONDUCTORS
• 380-14(c)

CONTROLS NONINDUCTIVE AND INDUCTIVE LOADS NOT EXCEEDING SWITCH'S AMPERE AND VOLTAGE RATING
• 380-14(d)

CO/ALR

AC SPECIFIC-USE FOR 347 V

380-14(a – d) AC general-use snap switches controlling inductive loads shall not exceed the rating of the load at the applied voltage. They shall have a rating not exceeding 50% of the ampere rating of the switch at the applied voltage. Where rated 20 A or less and connected to aluminum wire, they shall be marked CO/ALR and be listed for the purpose. Snap switches include: (a) AC general-use; (b) AC/DC general-use; (c) CO/ALR; and (d) AC specific-use for 347 V.

380-15 Switches shall be marked with their current and voltage. If the switch is HP-rated, it shall be so marked.

TEMPORARY WIRING

Ⓐ TEMPORARY CIRCUIT
GROUND WIRE
Ⓑ
Ⓒ
GROUNDING TYPE PLUG AND RECEPTACLE W/GFCI PROTECTION
MOTOR-OPERATED DEVICE
SWITCH
CORD
SUPPLY CONDUIT

A. 305-4(a – j) Services shall conform to 230. Feeders and branch circuits shall be protected at rated ampacity. Receptacles shall be the grounding type and connected to a grounding conductor. Receptacles are not permitted on lighting branch circuits. Suitable disconnecting switches or plug connectors required to disconnect all ungrounded conductors of each temporary circuit. Lamps shall be protected by a suitable fixture or lampholder with guard. A box is not required for splices or junction connections when using cable assemblies, cords, or cables. See 110-14(b) and 400-9.

B. Grounding shall be per 250.

C. 305-6 All temporary wiring installations shall be provided with either GFCI protection per 305-6(a) or by use of an assured equipment grounding program per 305-6(b).

THREADED BOX SUPPORTS

18" MAXIMUM WITH DEVICES
3' MAXIMUM— NO DEVICES
Ⓑ
Ⓐ
LESS THAN 100 CU IN. BOX

A. 370-23(e) Threaded boxes or fittings not over 100 cu in. in size which contain no devices are considered fastened if two or more conduits are threaded into the box and are firmly supported within 3' of the box.

B. 370-23(f) Threaded boxes or fittings not over 100 cu in. which contain devices are considered supported if two or more conduits are threaded into the box and are firmly secured within 18" of the box.

SWITCHBOARDS

Ⓐ
Ⓑ WEATHERPROOF ENCLOSURE
DRY LOCATION
EXPOSED LIVE PARTS
WET LOCATION
QUALIFIED PERSON

A. 384-5 Switchboards with exposed live parts shall be located in permanently dry locations accessible only to qualified persons.

B. 384-6 A switchboard in a wet location or outside a building shall have a weatherproof enclosure per 373-2(a).

PANELBOARDS – 384, PART C

CIRCUIT DIRECTORY

1	LGTS-105, 106	10	LGTS-139, 140
2	LGTS-107	11	LGTS-141
3	LGTS-109, 110	12	LGTS-142, 143
4	LGTS-120 N	13	EXIT SIGNS
5	LGTS-120 M	14	EXIT LGTS
6	LGTS-120 S	15	SPARE
7	WASHROOMS	16	SPARE
8	LGTS-124	17	SPARE
9	LGTS-137, 138	18	SPARE

Panelboard – A single panel or group of panels which may be assembled as a single panel, including buses, overcurrent devices for controlling light, heat, or power circuits. Panelboards shall be accessible only from the front after installation.

384-14: Lighting and Appliance Branch-Circuit Panelboard – A panelboard with 10% of its overcurrent devices rated at 30 A or less, for which neutral connections are provided.

I-T-E PANELBOARD

TYPE: CDP-7   SERIES: 8 A   VOLTAGE: 208Y/120

AMPS: 225   SYSTEM: 3 PH-4 W

DATE: 225   S.O.: 70690-2

USE I-T-E ELECTRICAL PRODUCTS EQ OR ET CIRCUIT BREAKERS

I-T-E Electrical Products
A division of Siemens-Allis, Inc.
Atlanta GA, USA

TO RESET, MOVE HANDLE TO EXTREME
OFF POSITION THEN THROW ON

MARKINGS

A. 384-13 The panelboard rating shall not be less than the minimum feeder capacity per 220. They shall be marked with the voltage, current rating, number of phases, and manufacturer's name or trademark. A circuit directory shall be provided on the face or inside the door.

B. 384-15 Not over 42 overcurrent devices for a lighting and appliance panelboard, other than the mains, are permitted in one cabinet. A two-pole CB counts as two overcurrent devices, a three-pole CB as three.

C. 384-16(a) Each lighting and appliance branch-circuit panelboard shall be protected on the supply side by no more than two main CBs or two sets of fuses with a combined rating that does not exceed the panelboard's rating.

D. 384-16(a), Ex. 1 Individual protection is not required for a lighting and appliance branch-circuit panelboard if the feeder's overcurrent protection does not exceed the panelboard's rating.

384-16(a), Ex. 2 For existing installations, individual protection is not required if the lighting and appliance branch-circuit panelboard is in an individual residential occupancy.

E. 384-16(c) Panelboards with snap switches rated 30 A or less shall have overcurrent protection not in excess of 200 A.

384-16(d) The load on an overcurrent device shall not exceed 80% of its rating where the load is continued for 3 hr or more. See 384-16(d), Ex.

F. 384-17 Panelboards in damp or wet locations shall comply with 373-2(a).

384-18 Panelboards shall be mounted in enclosures designed for the purpose.

384-19 Except as permitted for services, panelboard fuses shall be installed on the load side of switches.

384-20 Metal panelboard cabinets and frames shall be grounded per 250 and 384-3(c).

FLEXIBLE CORDS AND CABLES — 400

USES PERMITTED — 400-7(a)

(1) PENDANTS
(2) FIXTURES
(3) APPLIANCES OR PORTABLE LAMP CONNECTIONS
(4) ELEVATOR CABLES
(5) CRANE HOISTS
(6) STATIONARY EQUIPMENT (ALLOWING FREQUENT INTERCHANGE)
(8) APPLIANCE REMOVED FOR MAINTENANCE
(7) PREVENTION OF NOISE OR VIBRATION TRANSMISSION
(9) DATA PROCESSING EQUIPMENT
(10) MOVING PARTS CONNECTION
(11) TEMPORARY WIRING

USES NOT PERMITTED — 400-8

(1) SUBSTITUTE FOR BUILDING WIRING
(2) PASSED THROUGH WALLS, CEILINGS, OR FLOORS
(3) PASSED THROUGH DOORWAYS, WINDOWS, OR SIMILAR OPENINGS
(4) ATTACHED TO BUILDING SURFACES
(5) CONCEALED BEHIND BUILDING WALLS, CEILINGS, OR FLOORS
(6) INSTALLED IN RACEWAYS UNLESS PERMITTED ELSEWHERE IN NEC®

400-6(a) Types S, SC, SCE, SCT, SE, SEO, SEOO, SJ, SJE, SJEO, SJEOO, SJO, SJT, SJTO, SJTOO, SO, SOO, ST, and STOO flexible cords and G, PPE, and W flexible cables shall be marked on the surface every 24″ with the type, size, and number of conductors.

400-7(a) Flexible cords and cables shall be used for: (1) pendants; (2) fixtures; (3) appliances or portable lamp connections; (4) elevator cables; (5) cranes and hoists; (6) stationary equipment (allowing frequent interchange); (7) prevention of noise or vibration transmission; (8) appliance removed for maintenance; (9) data processing equipment; (10) moving parts connections; or (11) temporary wiring.

400-8 Flexible cords and cables shall not be used: (1) as a substitute for building wiring; (2) where passed through walls, ceilings, or floors; (3) where passed through doorways, windows, or similar openings; (4) where attached to building surfaces; (5) where concealed behind building walls, ceilings, or floors; (6) where installed in raceways unless permitted elsewhere in the NEC®.

400-9 Flexible cords shall be used only in continuous lengths, not spliced or tapped except junior hard-service and hard-service flexible cord No. 14 and larger if spliced per 110-14(b).

400-11 Flexible cords used in show windows and show cases shall, in general, be Types AFS, S, SE, SEO, SEOO, SJ, SJE, SJEO, SJEOO, SJO, SJOO, SJT, SJTO, SJTOO, SO, SOO, ST, STO, or STOO.

400-12 Flexible cords or cables shall not be smaller than the sizes in Table 400-4.

10′ TAP RULE

TAP TO 150 A CB = NO. 1/0 THWN Cu
TAP TO LUGS OF PANEL = NO. 4/0 THWN Cu

150 A CB
225 A PANELBOARD

10′ MAXIMUM

TO OTHER LOADS

800 A FEEDER CIRCUIT

A. 240-21(b)(1) The tap conductors shall not exceed 10′.

B. 240-21(b)(1)(a) The tap conductor's ampacity shall not be less than the combined load and not less than the rating of the device or OCPD supplied.

C. 240-21(b)(1)(b) The tap conductors shall not extend beyond the panelboard, etc. supplied.

D. 240-21(b)(1)(c) The tap conductors shall be protected in a raceway.

E. 240-21(b)(1)(d) The line side OCPD shall not exceed 10 times the ampacity of the tap conductor.

25′ TAP RULE

TAP AMPACITY
$800 \text{ A} \times \frac{1}{3} = 266.7 \text{ A}$
266.7 A = NO. 300 kcmil THWN Cu

25′ MAXIMUM

TO OTHER LOADS

800 A FEEDER CIRCUIT

A. 240-21(b)(2) The tap conductor shall not exceed 25′.

B. 240-21(b)(2)(a) The tap conductor's ampacity shall not be less than 1/3 of the rating of the feeder OCPD.

C. 240-21(b)(2)(b) The tap conductors shall terminate in a single OCPD which limits the load to the ampacity of the tap conductors.

D. 240-21(b)(2)(c) The tap conductors shall be protected in a raceway.

25′ TRANSFORMER FEEDER TAP RULE

SERVICE RACEWAY AND CONDUCTORS

SERVICE EQUIPMENT

600 A FEEDER CIRCUIT

OCPD

PRIMARY CONDUCTORS		
TAP AMPACITY		

$600 \text{ A} \times \frac{1}{3} = 200 \text{ A}$
200 A = NO. 3/0 THWN Cu

SECONDARY CONDUCTORS

TAP AMPACITY
$\frac{240 \text{ V}}{480 \text{ V}} = .5$
$.5 \times 600 \text{ A} \times \frac{1}{3} = 100 \text{ A}$
100 A = NO. 3 THWN Cu

TO OTHER LOADS

25′ MAXIMUM

TRANSFORMER
480 V, 3φ PRIMARY
240 V, 3φ SECONDARY

MAIN

PANEL

A. 240-21(b)(3)(a) The primary conductor's ampacity shall not be less than 1/3 of the rating of the feeder OCPD.

B. 240-21(b)(3)(b) The ampacity of conductors times the ratio of secondary-to-primary voltage is at least 1/3 of the rating of the feeder OCPD.

C. 240-21(b)(3)(c) The total length of the primary and secondary conductors shall not exceed 25′.

D. 240-21(b)(3)(d) The tap conductors shall be protected in a raceway.

E. 240-21(b)(3)(e) The secondary conductors shall terminate in a single OCPD which limits the load current to no more than the conductor's ampacity.

OVER 25′ FEEDER TAP RULE

SERVICE RACEWAY AND CONDUCTORS

SERVICE EQUIPMENT FOR HIGH-BAY MANUFACTURING BUILDING OVER 35′ IN HEIGHT

TO OTHER LOADS

25′ TOTAL HORIZONTAL LENGTH

100′ TOTAL LENGTH

OCPD

A. 240-21(b)(4)(c) The primary conductor's ampacity shall not be less than 1/3 of the rating of the feeder OCPD.

B. 240-21(b)(4)(d) The tap conductors terminate in a single CB or set of fuses.

C. 240-21(b)(4)(e) The tap conductors shall be protected in a raceway.

D. 240-21(b)(4)(f) The tap conductors shall be continuous without splices.

E. 240-21(b)(4)(g) The tap conductors shall be at least No. 6 Cu or No. 4 Al.

F. 240-21(b)(4)(h) The tap conductors shall not penetrate walls, floors, or ceilings.

G. 240-21(b)(4)(i) The tap location shall not be less than 30′ from the floor.

BRANCH-CIRCUIT TAP RULES

NO. 10 AWG

SURFACE-MOUNTED BOX

GROUNDED THROUGH CONDUIT AND SCREWS
• 250-146(a)

NO. 12 AWG NOT OVER 18″

Ⓐ

INTERNALLY GROUNDED TO STRAP

A. **210-19(d), Ex. 1** Branch-circuit tap conductors (for other than cooking loads as per 210-2) shall be at least No. 14 AWG and shall be rated not less than 15 A when tapped from less than 40 A circuit and not less than 20 A when tapped from 40 A or 50 A circuit when loads consist of: (a) lampholder or fixture where tap is not over 18″ long; (b) fixture with tap conductors per 410-67; (c) individual outlets, other than those for receptacles, less than 18″; (d) industrial, infrared lamp heating appliances; and (e) nonheating leads of de-icing and snow-melting cables and mats. See 240-4.

BONDING JUMPERS (EBJs)

EXPANSION JOINT

GROUND CLAMP

CONDUIT

GROUND CLAMP

CONDUIT

Ⓐ

Ⓑ

6′ MAXIMUM LENGTH EBJ

A. **250-98** Expansion fittings and telescoping conduits shall be made electrically continuous by bonding jumpers or other means.

B. **250-102(e)** An EBJ is permitted on the outside of conduit if not over 6′ long.

EQUIPMENT GROUNDING — FIXED EQUIPMENT

GROUNDED LIGHT FIXTURE

Ⓐ

Ⓐ

GFCI RECEPTACLE
• 210-8(a)(1)

BATHROOM SINK

A. **250-110(2)** (General Rule) Exposed noncurrent-carrying metal parts of fixed equipment likely to become energized and located in wet or damp locations shall be grounded.

GROUNDING — BOXES

BOX

CONDUIT

CONDUIT

THREADS TO BE CLEANED

GROUNDING SCREW

Ⓑ

GROUND WIRE

Ⓓ

Ⓒ

WIRE FOR CONNECTION TO GROUND TERMINAL ON RECEPTACLE

Ⓐ

GROUNDING CLIP

NM CABLE WITH GROUND WIRE

GROUND WIRES SPLICED TOGETHER CRIMP TYPE OR TWIST-ON DEVICE

250-97 Metal conduits or cable systems containing conductors of more than 250 V to ground shall use one or more of the methods specified in 250-94 to assure electrical continuity.

A. **250-8** EGCs shall be connected by listed pressure connectors, clamps, exothermic welding, or other listed means. Sheet metal screws shall not connect EGC to enclosure. Connections or fittings that depend solely on solder are not permitted.

B. **250-96(a)** To ensure electrical continuity, any nonconducting coating shall be removed from conduit or cable armor before fastening to metal enclosure.

C. **250-148** Where more than one grounding conductor enters a box, the electrical connection shall be such that the removal of a device shall not interrupt the grounding continuity.

D. **250-148(a)** The connection shall be made between a grounding conductor and a metallic box, by means of a grounding screw used for no other purpose, or by a listed grounding device.

PERSONAL PROTECTIVE EQUIPMENT

DANGER
HARD HAT REQUIRED IN THIS AREA

DANGER
EAR PROTECTION AREA

DANGER
EYE PROTECTION AREA

CAUTION
SAFETY SHOES REQUIRED IN THIS AREA

CAUTION
PROTECTIVE GLOVES REQUIRED AT THIS OPERATION

Always wear the appropriate personal protective equipment.

SUPPORTING CONDUCTORS IN VERTICAL RACEWAYS — 300-19

(1) WEDGES

(2) INSULATING SUPPORT

(3) DEFLECTION

(4) EQUALLY EFFECTIVE METHODS

NOT LESS THAN 90°

CABLE SUPPORT

VERTICAL RACEWAY

NO. 2/0 Cu CONDUCTORS

BOX

80' MAXIMUM

ADDITIONAL BOXES WITH SUPPORTS REQUIRED IF OVER TABLE 300-19(a) VALUES

CONDUCTOR SUPPORTS— TABLE 300-19(a)		
Wire Size	Al or Cu Clad Al*	Cu*
No. 18 — No. 8	100	100
No. 6 — No. 1/0	200	100
No. 2/0 — No. 4/0	180	80
250 kcmil — 350 kcmil	135	60
400 kcmil — 500 kcmil	120	50
600 kcmil — 750 kcmil	95	40
Over 750 kcmil	85	35

* in feet

SUPPORT METHODS

A. **300-19(a)** Conductors in vertical raceways shall be supported if the vertical rise exceeds the values in Table 300-19(a). One cable support shall be provided at the top of the vertical raceway. Intermediate supports shall be provided as needed to limit supported conductor length to values in Table 300-19(a).

300-19(b) Support methods for conductors shall be by one of the following:

(1) Wedges inserted in the top ends of the raceways

(2) Boxes at required intervals with insulating supports installed

(3) In junction boxes by deflecting cables not less than 90° and carrying them horizontally not less than twice the cable diameter. Support spacing is reduced 20% for this method.

(4) Equally effective methods

WIRING WITHIN FLOURESCENT FIXTURES

BALLAST

THW CONDUCTORS

SECONDARY WIRES FROM BALLAST TO ELECTRIC-DISCHARGE LAMP

1000 V OR LESS LAMP

A. **300-3(c) a.** Secondary wiring to electric-discharge lamps of 1000 V or less may occupy the same enclosure as branch-circuit conductors if insulated for the secondary voltage of the ballast.

B. **410-33** Branch-circuit conductors within 3″ of a ballast shall have a temperature rating of at least 90°C, such as RHH, THHN, THHW, FEP, FEPB, SA, XHHW, and THW.

RACEWAYS EXPOSED TO DIFFERENT TEMPERATURES

CONDUIT

CONDUIT SEALED AT THIS POINT

REFRIGERATED SPACE

A. **300-7(a)** To prevent circulation of air through interior conduit runs whose ends are exposed to widely different temperatures, conduit shall be sealed off.

300-7(b) Expansion joints shall be provided where necessary to compensate for thermal movement.

CONTINUITY — RACEWAYS AND BOXES

LOCKNUTS AND BUSHINGS OR EQUAL

250-96(a) Where a metal raceway serves as a grounding conductor, it shall be electrically continuous or bonded to assure ability to safely conduct possible fault currents.

250-102(e) The EBJ shall be inside the conduit with circuit wires except if it is not over 6′, then it may be outside the conduit if routed with the conduit.

300-10 Interior metal raceways and enclosures shall be effectively joined throughout the whole installation to provide electrical continuity.

AIR PLENUMS — WIRING METHODS

INTERMEDIATE RMC, EMT, MI, TYPE MC CABLE, IMC, OR ALS CABLE

DUCTS OR PLENUMS USED FOR ENVIRONMENTAL AIR

300-22(a) No wiring of any kind shall be installed in ducts used to transport dust, loose stock, flammable vapors, or commercial cooking.

300-22(b) Where necessary to run a wiring system through air handling ducts or plenums, types MI or MC cables, EMT, FMC, IMC, or RMC may be used. Physically adjustable equipment and devices in ducts may be wired with FMC or LTFMC in lengths not exceeding 4′.

300-22(c) Wiring methods in hollow spaces above lowered ceilings that are also used for supply or return plenums for environmental air are limited to types MI, MC, AC cables, and other factory-assembled multiconductor control or power cable specifically listed for the use. Other cables and conductors shall be in EMT, IMC, RMC, FMC, or where accessible, surface metal raceway, metal wireway, or solid-bottom metal cable tray with solid metal cover.

300-22(c), Ex. Stud or joist space in dwelling units where used as air handling permits normal wiring to pass through if perpendicular to space.

300-22(c)(1) Listed prefabricated cable assemblies of manufactured metallic wiring systems where listed for this use shall be permitted.

300-22(c)(1), Ex. LTFMC in single lengths not over 6′ shall be permitted.

424-58 Electric duct heaters shall be identified as suitable for the installation.

424-65 The electric dust heater controller shall be accessible and the disconnecting means shall be within sight from the controller.

UNDERFLOOR RACEWAYS — 354

USES PERMITTED — 354-2(a)

- Beneath concrete or other flooring material or in offices, where laid flush and covered with linoleum or equivalent floor covering.

USES NOT PERMITTED — 354-2(b)

(1) Where subject to corrosive vapors

(2) In hazardous locations except as permitted by 504-20 and in Class I, Division 2 locations as permitted by 501-4(b), Ex.

354-1 Underfloor raceways shall comply with the wiring methods of 300.

354-3 Raceway coverings shall comply with 354-3(a – d).

A. **354-3(a)** Half-round and flat-top raceways less than 4″ wide shall be covered with at least ¾″ of concrete or wood.

B. **354-3(b)** Flat-top raceways from 4″ to 8″ wide with a minimum 1″ spacing between raceways shall be covered with at least 1″ of concrete.

354-3(c) Trench-type raceways with removable covers are permitted to be laid flush with the floor.

C. **354-3(d)** In offices, approved metal flat-top raceways less than 4″ wide are permitted to be laid flush with the concrete floor when covered with linoleum at least 1/16″ thick.

D. **354-4** No conductor larger than that for which the raceway is designed shall be installed.

E. **354-5** The cross-sectional area of all conductors or cables shall not exceed 40% fill of the raceway.

F. **354-6** Splices and taps are permitted only in junction boxes.

354-6, Ex. Splices and taps are permitted in trench-type flush raceways with a removable cover. The conductors, splices, and taps shall not exceed 75% fill at that point.

354-7 Conductors shall be removed from the raceway when an outlet is abandoned.

G. **354-8** Raceways shall be centered on junction boxes and be held firmly in place.

H. **354-9** A marker shall be installed near each end of each straight run of raceway.

I. **354-10** All dead ends of raceways shall be closed.

J. **354-13** Junction boxes shall be level with the floor and sealed. Metal junction boxes shall be used with metal raceways.

K. **354-14** Inserts shall be leveled and sealed. Metal inserts shall be used with metal raceways.

L. **354-15** FMC is required to make connections between raceways and distribution centers and wall outlets when not in concrete, RMC, IMC, EMT, or approved fittings. PVC or ENMT is permitted if provision for termination of EBC is provided.

CELLULAR METAL FLOOR RACEWAYS – 356

356-1: Cellular Metal Floor Raceway – The hollow spaces of cellular metal floors, with fittings, approved as enclosures for electric conductors.

Cell – A single, enclosed tubular space in a cellular metal floor with the axis being parallel to the axis of the metal floor member.

Header – A transverse raceway for electric conductors that provides access to particular cells.

USES NOT PERMITTED – 356-2

(1) Where subject to corrosive vapor

(2) In hazardous locations except as permitted by 504-20 and in Class I, Division 2 locations as permitted by 501-4(b)

(3) In commercial garages with exceptions

356-3 Cellular metal floor raceway shall comply with the wiring methods of 300.

A. 356-4 Conductors larger than No. 1/0 require special permission before installation.

B. 356-5 The cross-sectional area of all conductors or cables shall not exceed 40% fill of the raceway.

C. 356-6 Splices and taps are permitted only in headers or junction boxes.

356-7 Conductors shall be removed from the raceway when an outlet is abandoned.

D. 356-8 Markers shall be installed for future locating of cells.

356-9 Junction boxes shall be level with the floor and sealed. Metal junction boxes shall be used with metal raceways.

E. 356-10 Inserts shall be leveled and sealed. Metal inserts shall be used with metal raceways.

356-11 FMC is required to make connections between raceways and distribution centers and wall outlets when not in concrete, RMC, IMC, EMT, or approved fittings. PVC or ENMT is permitted if provision for termination of EBC is provided.

356-12 Adequate mechanical and electrical continuity of the system shall be secured. Conductors shall be completely enclosed. All interior surfaces shall be free of burrs and sharp edges.

CELLULAR CONCRETE FLOOR RACEWAYS – 358

358-2: Cell – A single, enclosed, tubular space in a precast cellular concrete slab floor with the direction of the cell being parallel to the direction of the floor member.

Header – A transverse metal raceway for electric conductors with access to predetermined cells, permitting the installation of electric conductors from a distribution center to the floor cells.

USES NOT PERMITTED – 358-4

(1) Where subject to corrosive vapor

(2) In hazardous locations except as permitted by 504-20 and in Class I, Division 2 locations as permitted by 501-4(b)

(3) In commercial garages, other than to supply ceiling outlets or extensions to the area below the floor but not above.

358-3 Cellular concrete floor raceways shall comply with the applicable wiring methods of 300.

A. 358-5 The header shall be at installed in a straight line at a 90° angle to the cells. It shall be mechanically secured. The end joints shall be closed and sealed. It shall be electrically continuous and bonded to the enclosure of the distribution system.

B. 358-6 Listed metal raceways and fittings shall be used to make connections.

358-7 Junction boxes shall be level with the floor and sealed. They shall be metal and mechanically and electrically continuous with the header.

358-8 Markers shall be installed for future locating of cells.

C. 358-9 Inserts shall be leveled and sealed. They shall be metal and fitted with grounded receptacles.

D. 358-10 Conductors larger than No. 1/0 require special permission before installation.

E. 358-11 The cross-sectional area of all conductors or cables shall not exceed 40% fill of the cell or header.

F. 358-12 Splices and taps are permitted only in headers or junction boxes. Loop wiring is not considered to be a splice or a tap.

358-13 Conductors shall be removed from the raceway when an outlet is abandoned.

WIREWAYS

362-1: Metal Wireway – A sheet metal trough with a hinged or removable cover that houses and protects wires and cables laid in place after the wireway has been installed.

362-14: Nonmetallic Wireway – A flame-retardant nonmetallic trough with a removable cover that houses and protects wires and cables laid in place after the wireway has been installed.

METAL WIREWAYS

USES PERMITTED – 362-2(a)

1. For exposed work
2. In concealed spaces per 640-24
3. In hazardous locations. See 501-4(b), 502-4(b), and 504-20.

A. 362-3 Wireway installation shall comply with the wiring methods of 300.

B. 362-4 No conductor larger than for which the wireway is designed shall be installed.

C. 362-5 There shall be no more than 30 current-carrying conductors at any cross section. The sum of the cross-sectional areas of the conductors shall not exceed 20% of the cross-sectional area of the wireway. See Note 8(a) of Notes to Ampacity Tables of 0 to 2000 Volts. See 362-5, Ex. 1 – 3.

USES NOT PERMITTED – 362-2(b)

- Where subject to physical damage or corrosive vapor

362-7 Splices and taps shall not fill the wireway to more than 75% at any cross section.

D. 362-8(a) Horizontal support shall not exceed 5′.

E. 362-8(b) Vertical support shall not exceed 15′.

362-9 Unbroken lengths of wireway are permitted to pass through walls.

362-10 All dead ends of wireway shall be closed.

F. 362-12 Wireways shall be marked by the manufacturer.

NONMETALLIC WIREWAYS

USES PERMITTED – 362-15

(1) Only for exposed work. See 640-4, Ex. c

(2) Where subject to corrosive vapors

(3) In wet locations if listed for the purpose

G. 362-17 Wiring installation shall comply with the wiring methods of 300. Where grounding is required by 250, a separate EGC is used.

H. 362-18 No conductor larger than for which the wireway is designed shall be installed.

I. 362-19 The sum of the cross-sectional area of the conductors shall not exceed 20% of the cross-sectional area of the wireway. See 310-15(b)(2) a.

USES NOT PERMITTED – 362-16

(1) Where subject to physical damage

(2) In hazardous locations

(3) Where exposed to sunlight unless listed for the purpose

(4) Where subject to ambient temperatures other than those for which it is listed

(5) For conductors with insulation temperatures above those for which the wireway is listed

362-21 Splices and taps shall not fill the wireway to more than 75% at any cross section.

J. 362-22(a) Horizontal supports shall not exceed 3′.

K. 362-22(b) Vertical supports shall not exceed 4′.

362-24 Unbroken lengths of wireway are permitted to pass through walls.

L. 362-27 Wireways shall be marked by the manufacturer.

AUXILIARY GUTTERS

Auxiliary Gutter – A metal or nonmetallic enclosure with a hinged or removable cover designed to supplement wiring space for conductors or busbars.

A. 374-2 Auxiliary gutters shall not extend over 30′ beyond the equipment supplemented.

B. 374-3(a) Sheet metal auxiliary gutter support shall not exceed 5′.

C. 374-3(b) Nonmetallic auxiliary gutter support shall not exceed 3′ and at each end or joint.

D. 374-4 Covers shall be securely fastened.

E. 374-5(a) Sheet metal auxiliary gutters shall not contain over 30 current-carrying conductors at a cross section and the cross-sectional area of the conductors shall not exceed 20% of the cross-sectional gutter area. See 374-5(a)(1 – 4).

F. 374-5(b) The cross-sectional area of conductors in a nonmetallic auxiliary gutter shall not exceed 20% of the cross-sectional gutter area.

G. 374-6(a) Thirty or less current-carrying conductors in a sheet metal auxiliary gutter shall not be derated.

H. 374-6(b) All current-carrying conductors in a nonmetallic auxiliary gutter shall be derated per 310-15(b)(2) a.

374-8(a) Splices and taps shall not fill the gutter to more than 75% of its cross-sectional area.

374-9(a) Electrical and mechanical continuity shall be secured.

PULL AND JUNCTION BOXES

STRAIGHT PULLS

A. 370-28(a) The minimum dimensions of pull and junction boxes for No. 4 or larger conductors and cable containing No. 4 or larger conductors shall be per 370-28(a – d).

B. 370-28(a)(1) For straight pulls, the box length shall be not less than 8 times the diameter of the largest raceway.

C. 370-28(a)(2) For angle or U pulls, the distance between opposite walls shall be not less than 6 times the largest raceway diameter plus the diameters of other raceways on the same wall.

370-28(b) All conductors shall be cabled in pull and junction boxes over 6′ in any dimension.

D. 370-28(c) All pull and junction boxes shall be covered.

370-28(d) If permanent barriers are installed, each section shall be considered as a separate box.

ANGLE PULLS

CONDUCTORS — NUMBER IN CONDUIT

A. Ch 9, Table 1 permits conduit or tubing with more than two conductors to be filled to 40%. Appd C, Table C1 lists six No. 12 RHW or RHH conductors for ¾″ EMT.

B. Ch 9, Table 1 permits conduit or tubing with more than two conductors to be filled to 40%. Appd C, Table C1 lists 16 No. 12 THWN or THHN conductors for ¾″ EMT. Where there are three or more current-carrying conductors in a raceway, the ampacity shall be reduced per 310-15(b)(2) a.

C. 346-6 The number of conductors in a conduit shall be per the percentage of fill permitted in Ch 9, Table 1.

Chapter 9, Notes to Tables

Note 1 See Appd C for specific conduit or tubing for permitted conduit fill for same size conductors.

Note 2 Table 1 does not apply to short sections of conduit or tubing but to the complete raceway system.

Note 3 EGCs are included in the conduit fill calculations. Actual conductor dimensions shall be used to calculate fill.

Note 4 Conduits or tubing nipples, not over 24″, are permitted. Derating is required by Note 8(a) of Notes to Ampacity Tables of 0 to 2000 Volts does not apply. Fill is permitted to be at 60%.

Note 5 Actual dimensions of conductors, such as multiconductor cables, shall be used if not included in Ch 9.

Note 6 Use Tables 5 and 5A for dimensions of conductors of different size installed in the same raceway. Use Table 4 for specific conduit or tubing area.

Note 7 When calculating fill for conductors of the same size, the next higher number shall be used when the decimal is 0.8 and larger.

Note 8 Dimensions for bare conductors are per Ch 9, Table 8.

Note 9 Multiconductor cables shall be treated as a single conductor for conduit fill.

SURFACE RACEWAYS — 352

A. 352-1(a)(b) Surface metal raceway shall be used only in dry locations. They shall not be used:

(1) Where subject to severe physical damage unless approved

(2) Where over 300 V unless .040″ thick

(3) Where subject to corrosive vapors

(4) In hoistways

(5) Concealed, except under raised floors for data processing systems per 645-5(d) 2.

352-5 Surface raceway may extend through dry walls, partitions, or floors if the length is unbroken.

B. 352-6 Combination raceways are permitted for signal, power, and lighting circuits if in different compartments and identified throughout the system by sharply contrasting colors of the interior finish.

MANUFACTURED WIRING SYSTEMS — 604

604-1 Applies to field-installed wiring using factory manufactured subassemblies for branch circuits, remote control circuits, signaling and communication circuits in accessible areas. A drop ceiling with lift-out panels is considered accessible – See Definitions.

604-4 Manufactured wiring systems are permitted to be installed in accessible locations, dry locations, plenums, and spaces used for environmental air where listed and installed per 300-22.

604-5 Manufactured wiring systems shall not be used where the conductors or cables are limited by 333 and 334.

604-6(a)(1)(2) Wiring systems shall be AC or MC cable with No. 10 or No. 12 Cu insulated conductors with a bare or insulated Cu EGC sized equivalent to the ungrounded conductors. FMC or listed LTFMC is permitted with No. 10 or No. 12 Cu insulated conductors with a bare or insulated Cu EGC, sized equivalent to ungrounded conductors. See Exceptions.

604-6(b) Receptacles and connectors shall be locking type, uniquely polarized, and identified for the purpose and part of listed assembly for appropriate system.

604-6(c) Component parts shall be listed for appropriate system.

604-7 Unused outlets shall be capped to close connector openings.

COVE LIGHTING

A. 410-9 Coves shall have ample wiring space and be fully accessible for maintenance.

FIXTURES — SHOW WINDOWS

A. 210-62 At least one receptacle outlet shall be provided for each 12' of show window.

B. 410-7 No externally-wired fixtures other than the chain type shall be permitted in show windows. See 400-11.

FIXTURE MOUNTING

A. 410-76(b) Surface-mounted fixtures containing ballasts shall be listed for mounting on low-density cellulose fiberboard or spaced not less than 1½" therefrom. See FPN for definition of combustible low-density cellulose fiberboard sheets, panels, and tiles.

FIXTURE SUPPORTS

410-14(b) When an electric-discharge lighting fixture is mounted over a box, an opening is required in the back of the fixture so the box is accessible.

A. 410-16(a) Fixtures shall be attached to and supported by the outlet box. Fixtures weighing more than 50 lb shall be supported independently of the outlet box unless the box is listed for the weight to be supported.

410-16(b) Connections between fixture conductors and circuit conductors shall be accessible for inspection without requiring disconnection of any part of the wiring except when connected by attachment plugs and receptacles.

FIXTURES AS RACEWAYS

410-31 Fixtures shall not be used as raceways for circuit conductors.

A. 410-31 Listed fixtures identified for the use may be used as raceways for circuit conductors.

B. 410-32 End-to-end assemblies or fixtures connected by recognized wiring methods are permitted to carry conductors supplying the fixtures.

410-73(e) Fluorescent fixtures installed indoors shall have thermal protection integral within the ballast. Replacement ballasts shall also have integral protection. See Exceptions.

CORD-CONNECTED SHOWCASES

A. 410-29 A single showcase, not permanently installed, is permitted to be cord-and-plug connected to a permanent receptacle. Up to six such showcases are permitted to be coupled together by flexible cord and locking-type connections but one showcase of the group shall connect to a permanent receptacle.

B. 410-29(a) Flexible cord shall be hard-service type, with conductors not smaller than branch-circuit conductors and ampacity equal to the OCPD, and with an EGC.

410-29(b) Receptacles, connectors, and plugs shall be a listed grounding type rated at 15 A or 20 A.

C. 410-29(c) Flexible cord shall be secured to underside of the showcase so that: (1) the wiring is not exposed to mechanical damage; (2) the separation between showcases is not over 2" and the separation between the first case and the supply receptacle is not over 12"; and (3) the free lead at the end of a line of showcases has a female fitting not extending beyond the showcase.

D. 410-29(d) No other electrical equipment may be connected to the showcase.

410-29(e) For cord-connected showcases, the secondary circuits of electric-discharge lighting ballasts are limited to one showcase.

410-77(a) Auxiliary equipment which is not part of the light fixture shall be enclosed in accessible permanently-installed metal cabinets.

CONNECTION OF FLUORESCENT FIXTURES

410-14(a) Fluorescent fixtures supported independently of an outlet box shall be connected by metal raceway, nonmetallic raceways, AC cable, MI cable, metal-clad cable, or nonmetallic cable except cord connections permitted per 410-30(b)(c).

A. 410-14(a) Cord-connected fixtures are permitted per 410-30(b) and 410-30(c).

B. 410-30(c)(1) Listed fixtures are permitted to be cord-connected directly below the outlet box or busway if the cord is visible throughout its length outside the fixture and is not subject to strain or physical damage.

410-74 Fixtures designed for direct current shall be so marked.

410-78 Autotransformer ballasts which raise voltage to more than 300 V shall be supplied only by a grounded system.

SCREW-SHELL LAMPHOLDERS

A. 410-47 Screw-shell lampholders shall not be installed for use as plug receptacles. Where supplied by a circuit having a grounded conductor, the conductor shall connect to the screw-shell.

410-49 Lampholders shall be weatherproof for installation in wet or damp locations.

B. 410-50 The outer metal shell and cap shall be of insulating material. The lining shall not extend more than ⅛″ beyond the metal shell.

LAMPHOLDER
Ⓐ
FEMALE PLUG
$\frac{1''}{8}$
METAL SHELL
Ⓑ CAP

ELECTRIC DISCHARGE LIGHTING – OVER 1000 V

Ⓑ
TERMINALS
Ⓕ
SWITCH WITHIN SIGHT Ⓓ
TRANSFORMER OVER 1000 V Ⓐ
SWITCH OPENS ALL UNGROUNDED PRIMARY CONDUCTORS Ⓒ

PRECAUTIONARY LABEL
Ⓚ
CAUTION X000 VOLTS
SUPPORTS
CONDUIT OR EQUAL
LAMPS WITH NO EXPOSED LIVE PARTS Ⓔ

Ⓙ
Ⓖ Ⓗ Ⓘ
TRAVEL
ADJACENT COMBUSTIBLE MATERIAL
SHOW WINDOW

TRANSFORMER
LOW-VOLTAGE WINDING
HIGH-VOLTAGE WINDING
NO. 14 INSULATED CONDUCTORS
TRANSFORMER
GROUND
NEON TUBING
LOW-VOLTAGE WINDING
HIGH-VOLTAGE WINDING

410-18(a) Where circuit operates at over 150 V to ground, all noncurrent-carrying metal parts shall be grounded except tie wires, screws, clips, etc. at least 1½″ from lamp terminals.

A. 410-80(b) Equipment with open-circuit voltage exceeding 1000 V shall not be installed in or on dwellings.

B. 410-80(c) The terminal of an electric-discharge lamp connected to a circuit of over 300 V is considered a live part.

C. 410-81(a) Fixtures or lamp installations shall be connected singly or in groups by an externally-operable switch or CB that opens all ungrounded primary conductors.

D. 410-81(b) Switch or CB shall be within sight of fixture or lamps or else capable of being locked in open position.

E. 410-82 Lamps and lampholders are to be designed so that there will be no exposed live parts when lamps are inserted or removed.

F. 410-83 The secondary open-circuit voltage shall not be over 1500 V with a test allowance of an additional 1000 V. The secondary current rating shall not be over 120 mA if the open-circuit voltage is over 7500 V. It shall not be over 240 mA if the open-circuit voltage is 7500 V or less.

G. 410-84 Transformers shall be enclosed and listed.

410-85 High-voltage transformer windings shall not be connected in series or in parallel.

H. 410-86(a) Transformers shall be accessible after installation.

I. 410-86(b) Transformers shall be installed as near to the lamps as practicable.

J. 410-86(c) Transformers shall be located so that adjacent combustible materials are not subjected to temperature in excess of 90°C (194°F).

410-88 Conductors shall be installed per 600-32.

K. 410-91 Each fixture or circuit of tubing over 1000 V shall have a precautionary label. The voltage indicated is open-circuit voltage.

TAPS FROM BRANCH CIRCUITS

Branch-Circuit Conductor – The circuit conductor(s) between the final overcurrent device protecting the circuit and the outlet(s).

FEEDER

208 V, 3φ, 3-WIRE CIRCUIT (NO NEUTRAL)

MULTIWIRE BRANCH CIRCUIT

120/208 V, 3φ, 4-WIRE, CIRCUIT

OUTLETS

(A) PANELBOARD

UNGROUNDED 1φ, 208 V CIRCUITS

GROUNDED NEUTRAL CONDUCTOR

A. **210-4(a)** All multiwire branch-circuit conductors shall originate in same panelboard.

210-4(b) In dwelling units where more than one receptacle on the same yoke (split receptacle) is fed by a multiwire branch circuit, means shall be provided at the panelboard to simultaneously disconnect all ungrounded conductors.

210-4(c) Multiwire branch circuits are permitted by 210 and shall supply only line-to-neutral load, except: (1) where they supply only one piece of utilization equipment; (2) where branch-circuit overcurrent protective device simultaneously opens all ungrounded conductors of the multiwire circuit.

210-4(d) If more than one nominal voltage system exists in a building, each ungrounded system conductor shall be identified by phase and system.

210-10 Ungrounded conductors for feeders and branch circuits may be tapped from ungrounded wires of circuits which have identified neutrals. Switching devices in these circuits shall have a pole in each conductor. For branch circuits, all poles shall open simultaneously where they serve as the disconnecting means for appliances. See 422-31, 410-48, 410-54(b), 424-20, 426-51, 430-85, and 430-103.

215-7 Two-wire DC circuits and AC circuits with two or more ungrounded conductors are permitted to be tapped from the ungrounded conductors of circuits with a grounded neutral. Switching devices in each tapped circuit shall have a pole in each ungrounded conductor.

FEEDER AND BRANCH-CIRCUIT LOAD CALCULATIONS

215-2, FPN 2 The voltage drop for power, heating, or lighting feeders is sized to not exceed 3%. The voltage drop for the feeder plus the branch circuit is sized to not exceed 5%. See 210-19(a), FPN 4.

210-19(a) The branch-circuit rating shall be not less than the noncontinuous load plus 125% of the continuous load. The minimum branch-circuit conductor size, without correction factors, shall be equal to or greater than the noncontinuous load plus 125% of the continuous load.

Table 220-11, Note 1 Demand factors in Table 220-11 do not apply to computed loads on feeders to areas in hospitals, hotels, and motels where the entire lighting load is used generally at one time.

L₁

NEUTRAL CONDUCTOR

N

3-WIRE DC OR 1φ AC CIRCUIT WITH A NEUTRAL CONDUCTOR

L₂

220-22 The feeder neutral is sized for maximum unbalance between neutral and phase conductors, except the neutral for household ranges, ovens, and dryers is permitted to be reduced to 70% of the load as calculated per Table 220-18 and Table 220-19. A further demand of 70% is permitted for the portion of the maximum unbalanced load in excess of 200 A. This also applies to 4-wire, 3φ and 5-wire, 2φ systems. This 70% demand factor does not apply for that portion of the feeder load which consists of nonlinear loads.

COMMON NEUTRAL FOR FEEDERS

A

B

N

2 SETS OF 3-WIRE FEEDERS WITH COMMON NEUTRAL

A

B

N

3 SETS OF 3-WIRE FEEDERS WITH COMMON NEUTRAL

A

B

C

N

2 SETS OF 4-WIRE FEEDERS WITH COMMON NEUTRAL

A

B

C

D

N

2 SETS OF 5-WIRE FEEDERS WITH COMMON NEUTRAL

215-4(a) A common neutral may be used with two or three sets of 3-wire feeders or two sets of 4-wire or 5-wire feeders.

220-22 The neutral should be sized to carry the maximum unbalance of the load as determined by 220.

225-7(b) The neutral conductor shall have an ampacity not less than the maximum net calculated load current between the neutral and all ungrounded conductors connected to any one phase of the circuit.

MANUFACTURED BUILDINGS

545-3: Manufactured Building – A building of closed construction, normally factory-made or assembled for installation or assembly and installation on building site. This does not include mobile homes or recreational vehicles.

A. 545-4(a) All wiring methods in the NEC® are permitted for use, and other wiring systems and fittings specifically approved for use with manufactured buildings are also permitted. Wiring devices with integral enclosures are permitted if a sufficient length of conductor is provided for replacement of devices.

B. 545-4(b) No. 10 and smaller concealed cables are permitted to be installed without supports, such as staples or straps, as required in normal construction, if cables are secured at cabinets, boxes, and fittings and protected from physical damage.

C. 545-5 Service-entrance conductors shall be installed per 230.

D. 545-6 Service-entrance conductors are to be installed at the building site, except where the point of attachment is known before manufacture.

E. 545-7 Service equipment shall be located at nearest, readily-accessible point, either inside or outside of the building per 230-70.

545-8 Protection of equipment and conductors that are exposed during manufacturing, transit, and erection at building site is required.

F. 545-9(a) Boxes other than those in Table 370-6(a) are permitted if listed, identified, and tested.

545-9(b) Boxes under 100 cu in. for mounting in closed construction shall have anchors or clamps for rigid and secure installation.

G. 545-10 A switch or receptacle with an integral enclosure is permitted when tested, identified, and listed.

545-11 Bonding and grounding of prewired panels and building components shall conform to 250, Parts E, F, and G.

H. 545-12 Means shall be provided to route the GEC from the service equipment to the point of attachment to the grounding electrode.

545-13 Where on-site interconnection of modules or other building components is necessary, fittings and connectors when tested, identified, and listed are permitted to be used. These fittings are to be equal to the wiring method and may be concealed at the time of on-site assembly.

AGRICULTURAL BUILDINGS

A. 547-3 Electrical equipment and devices shall be installed to function at their full rating without overheating.

B. 547-4 Wiring methods to be UF, NMC, Cu SE, RNMC, LTFNMC, or other cables or raceways suitable for the location with approved fittings.
547-4(b) All cables shall be secured within 8″ of each cabinet, box, or fitting.

C. 547-4(d) Flexible connections shall be made with dust-tight flexible connectors, LTFMC, or hard usage flexible cord, with listed and identified fittings.

D. 547-5 Devices, such as switches, fuses, CBs, controllers, pushbuttons, relays, etc., shall be installed in dustproof, waterproof, watertight, or corrosion-resistant enclosures designed to keep out dust, water, and corrosive elements, and shall have telescoping or close-fitting covers.

E. 547-5(a)(b) All boxes containing devices, taps, joints, or terminal connections shall be dust- and watertight and made of corrosion-resistant material.

F. 547-6 Rotating electrical machinery and motors shall be designed to limit the entrance of dust, moisture, or corrosive particles, or to be totally enclosed.

G. 547-7(a) Lighting fixtures shall be installed to minimize the entrance of dust, foreign matter, moisture, and corrosive material.

H. 547-7(b) Lighting fixtures exposed to physical damage shall have suitable guards.

I. 547-7(c) Lighting fixtures shall be watertight if exposed to water from condensation and/or cleansing water or solution.

SOLAR PHOTOVOLTAIC SYSTEMS – 690

A. **690-2: Array** – An assembly of modules or panels mounted on a support structure to form a power-producing unit.

690-2: Interactive System – A system that operates in parallel with and is permitted to deliver power to a normal utility service source connected to the same load.

690-2: Inverter – Equipment used to change voltage level or waveform, or both, of electrical energy.

B. **690-2: Module** – The smallest complete, environmentally-protected assembly of solar cells, exclusive of tracking, designed to generate DC power under sunlight.

C. **690-2: Panel** – A collection of modules, secured together, wired, and designed to provide a field-installable unit.

690-2: Photovoltaic Output Circuit – Conductors between photovoltaic source circuits and the inverter or DC utilization equipment.

690-2: Photovoltaic Power Source – One or more arrays which generate DC power at system voltage and current.

690-2: Photovoltaic Source Circuit – Conductors between modules and from modules to common junction point(s) of the DC system.

D. **690-2: Solar Cell** – Basic photovoltaic device that generates electricity when exposed to light.

690-2: Solar Photovoltaic System – The system used to convert solar energy into electrical energy suitable for connection to a utilization load.

690, Part H covers storage batteries. 690-71 deals with installation, 690-72 deals with the state of charge.

690-1 Solar photovoltaic systems may interact with other electric power production sources or may stand alone. Either AC or DC output for utilization is permitted.

E. **690-5** Dwelling units with roof-mounted photovoltaic arrays shall be provided with ground-fault protection to reduce fire hazards.

F. **690-7(d)** In one-family and two-family dwellings, live parts in source and output circuits over 150 V to ground, while energized, shall not be accessible to other than qualified persons.

690-7(d), FPN See 110-27 for guarding of live parts and 210-6 for voltage to ground and between conductors.

690-15 Means shall be provided to disconnect solar photovoltaic equipment from all ungrounded conductors of all sources.

G. **690-18** A means shall be provided to disable an array or portions of an array to protect persons against shock while installing, servicing, or replacing an array.

Cells form modules.
Modules form arrays.
Arrays form panels.

Name _____

Date _____

Trade Competency Test 1

NEC®	Answer	

_____ T F **1.** In a fuseholder, the screw shell shall be connected to the line side of the circuit.

_____ _____ **2.** Header ducts used with cellular concrete raceways shall be constructed of _____.

_____ T F **3.** Taps not over 18″ long for individual lampholders may be smaller than the branch circuit conductors.

_____ _____ **4.** Equipment intended to break current at fault levels shall have a(n) _____ rating sufficient for the nominal circuit voltage and current that is available at the line terminals of the equipment.

_____ _____ **5.** Dead ends of underfloor raceways shall be _____.

_____ _____ **6.** One of the approved methods of identifying the high-leg of a 4-wire, delta-connected secondary system is by using a _____ wire.

    A. white   C. red
    B. green   D. neither A, B, nor C

_____ T F **7.** Openings around electrical penetrations through fire-resistant-rated walls shall be firestopped using approved methods.

_____ _____ **8.** The minimum feeder allowance for show-window lighting shall be _____ VA per linear foot.

    A. 100   C. 300
    B. 200   D. 500

_____ _____ **9.** Service conductors of 600 V or less buried under _____″ or more of concrete are considered outside the building.

_____ _____ **10.** Conductors from discontinued outlets in an underfloor raceway installation shall be _____ from the raceway.

_____ _____ **11.** Interior raceways installed between areas differing widely in temperature shall be _____.

_____ _____ **12.** _____ are required at or near the end of each run of underfloor raceway to locate the last insert.

_____ _____ **13.** The total load on a branch circuit shall not exceed its _____.

_____ _____ **14.** The unit load for a store is listed as _____ VA per sq ft.

    A. 2   C. 3
    B. 2.5   D. 3.75

_____ _____ **15.** If the allowable current-carrying capacity of a conductor does not correspond to the rating of a standard size overcurrent protection device, the next larger size may be used provided the current does not exceed _____ A.

_____ _____ **16.** Temporary wiring methods shall be acceptable only if _____ based on the conditions of use and any special requirements of the temporary installation.

17. Threaded boxes not over 100 cu in. in size that contain devices need no additional support if two or more conduits, which are properly supported within _____″ of the box, are threaded into the box from the same side.

18. Where permissible, the demand factor applied to that portion of an unbalanced neutral feeder load in excess of 200 A is _____%.
    A. 40   C. 70
    B. 60   D. 80

19. In other than dwellings, each 5′ or fraction of multioutlet assembly shall normally be considered as a load of _____ VA.

T   F   20. The neutral feeder conductor shall be the same size as the ungrounded conductors for a resistance load.

21. Neon installed on dwelling units shall not exceed _____ V.

22. Metal auxiliary gutters shall be supported throughout their entire length at intervals not exceeding _____′.

23. The unit load for a hall in other than a one-family dwelling or an apartment is listed as _____ VA per square foot.

T   F   24. Lighting outlets installed in other than dwelling units shall not be connected to 50 A branch circuits.

25. A CB having an interrupting rating other than _____ A shall be marked with the rating of a standard size overcurrent device.

26. A(n) _____ is required where underground cable leaves conduit providing physical protection.

27. Bonding shall be provided where necessary to ensure electrical _____ and the capacity to safely conduct any fault current which is likely to be imposed.

28. A minimum size of No. _____ Cu EGC is required for a 30 A automatic overcurrent device ahead of the equipment.

29. CBs used to switch 120/277 V fluorescent fixtures shall be approved for the purpose and marked _____.

T   F   30. In interior wiring, a common neutral feeder wire shall not be run for two or three sets of 3-wire feeder circuits.

31. CBs shall clearly _____ whether in open or closed position.

32. A 15 A receptacle on a standard 15 A or 20 A branch circuit shall not supply a portable appliance load in excess of _____ A.
    A. 9   C. 12
    B. 10   D. 15

T   F   33. Flexible cords and cables are permitted to be attached to building surfaces over 6′-7″ from the floor or a platform.

34. The _____ of an underfloor raceway shall determine the largest conductor that can be installed.

35. A unit load for each outlet supplying heavy-duty lampholders that are not used for general illumination in industrial occupancies shall be _____.
    A. 1½ A   C. 180 VA
    B. 5 A   D. 600 VA

Trade Competency Test 2

NEC®		Answer	

_____   T   F   **1.** Splices are sometimes permitted in service conductors.

_____   _____   **2.** Auxiliary gutters shall not extend more than _____ ′ beyond the equipment they supplement.

_____   _____   **3.** EBJs are permitted to be routed outside a conduit system if not over _____ ′ in length.

_____   _____   **4.** The cross-sectional fill of a wireway shall not exceed _____ %.

_____   _____   **5.** Every commercial building must have a(n) _____ A circuit for a sign outlet.

_____   T   F   **6.** The maximum rating of a 250 V cartridge fuse is 600 A.

_____   _____   **7.** Underground service conductors carried up a pole shall be protected from mechanical injury to a height of at least _____ ′.

    A. 8   C. 12

    B. 9   D. 15

_____   _____   **8.** Metal enclosures for a GEC using a water pipe electrode for a 120/208 V service with three 250 kcmil Cu conductors and a No. 3/0 neutral shall be _____ .

    A. rigid conduit only   C. bonded

    B. not less than ¾″ in dia.   D. electrically continuous

_____   T   F   **9.** Bonding jumpers required for grounding interior equipment on the load side of the service shall be selected from Table 250-94.

_____   _____   **10.** The area of conductors and splices in a gutter shall not exceed _____ % of the cross-sectional area.

_____   T   F   **11.** Fuses shall never be field connected in parallel.

_____   _____   **12.** The size of a copper GEC using a water pipe electrode for a 120/208 V service with three 250 kcmil Cu conductors and a No. 3/0 neutral shall be No. _____ .

    A. 2   C. 6

    B. 4   D. 8

_____   _____   **13.** Panelboards equipped with snap switches rated at 30 A or less shall have overcurrent protection not in excess of _____ A.

    A. 100   C. 200

    B. 150   D. 300

_____   T   F   **14.** The service OCPD shall be an integral part of, or shall be located adjacent to, the service disconnecting means.

_____   _____   **15.** The current carried continuously in bare copper bars in metal auxiliary gutters shall not exceed _____ A per sq in.

_____ _____ **16.** No. 1/0 Cu conductors in vertical raceways shall be supported at intervals not exceeding _____'.
  A. 50   C. 100
  B. 75   D. 125

_____ _____ **17.** Three-way and four-way switches shall be wired so that all switching is done in the _____ conductors.

_____ T   F **18.** In general, the disconnecting means shall be within sight of the controller.

_____ _____ **19.** The wiring between 3-way switches and outlets in metal enclosures shall be run with all _____ in the same enclosure.

_____ _____ **20.** The largest standard cartridge fuse rating is _____ A.
  A. 100   C. 1200
  B. 600   D. 6000

_____ _____ **21.** Screw-shell lampholders shall be installed for use as _____ only.

_____ T   F **22.** Switchboards with exposed live parts shall be located in permanently dry locations accessible only to qualified persons.

_____ _____ **23.** A nipple having a maximum length of 24" can be filled to _____% of its internal cross-sectional area.

_____ _____ **24.** Transformers for electric-discharge lighting systems of more than 1000 V shall be _____ after installation.

_____ _____ **25.** Sheet metal auxiliary gutters shall be supported throughout their entire length at intervals not exceeding _____'.

_____ _____ **26.** A 25' tap taken from a 100 A feeder shall have an ampacity of not less than _____ A.
  A. 15   C. 25
  B. 20   D. 33⅓

_____ _____ **27.** When applying the 100' tap rule, the horizontal run is limited to _____' or less.

_____ _____ **28.** Tap conductors shall have at least one-third ampacity of supply conductors when applying the _____' tap rule.

_____ _____ **29.** In straight pulls, the length of a box shall be not less than _____ times the diameter of the largest conduit.

_____ _____ **30.** When applying the 25' tap rule from the secondary side of the transformer for industrial installations, OCPDs shall not _____ output.

_____ _____ **31.** Branch-circuit conductors routed within 3" of a ballast in a fixture shall be rated at least _____°C.

_____ _____ **32.** In U-pulls, the size of the box shall be _____ times the diameter of the largest conduit plus all remaining conduits on the same wall.

_____ _____ **33.** No. 12 conductors enclosed in vertical raceways shall be supported every _____'.

_____ _____ **34.** Not over _____ overcurrent devices for a lighting and appliance panelboard, other than the mains, are permitted in one cabinet.

_____ _____ **35.** In a building where water pipe is available, the minimum size of the GEC for an ungrounded system with No. 3/0 Cu service conductors shall be No. _____.
  A. 2 Cu   C. 6 Cu
  B. 4 Cu   D. 4 Al

Trade Competency Test 3

Name _____

Date _____

NEC®	Answer	
_____ T F		**1.** Conduit nipples not over 24″ long may be filled to 60% of their cross-sectional area.
_____	_____	**2.** Nonmetallic wireways shall not be used where exposed to _____ unless listed and marked as suitable for the purpose.
_____	_____	**3.** Coves that enclose lighting units shall have space fully accessible for _____.
_____		**4.** The secondary OCPD of a transformer may consist of _____ or less CBs or fuses.

    A. two   C. four
    B. three   D. six

_____ _____   **5.** In cellular concrete floor raceways, conductors larger than No. _____ shall not be installed unless by special permission.

_____ _____   **6.** An autotransformer ballast that raises the voltage above _____ V shall be supplied by a grounded system.

_____ _____   **7.** The largest conductor permitted in ⅜″ FMC is No. _____.

    A. 10   C. 14
    B. 12   D. 16

_____ _____   **8.** No wiring method of any kind shall be installed in _____ used with commercial cooking.

_____ T   F   **9.** Conductors in underfloor raceway systems may be spliced or tapped at individual outlets.

_____ _____   **10.** When applying the 100′ tap rule, the junction box containing the tap shall be at least _____′ from the floor.

_____ _____   **11.** Underfloor raceways over 4″ but less than 8″ wide, and spaced less than 1″ apart, shall be covered with concrete to a depth of not less than _____″.

    A. ¾   C. 1½
    B. 1   D. 2

_____ T   F   **12.** Unbroken lengths of surface metal raceway may be extended through dry floors.

_____ _____   **13.** A lighting and appliance panelboard shall not have more than _____ OCPDs.

    A. 42   C. 50
    B. 46   D. neither A, B, nor C

_____ _____   **14.** If a change occurs in the size of the ungrounded conductor, a similar change is permitted in the _____ conductor.

_____ _____ **15.** The combined cross-sectional area of all conductors or cables in an underfloor raceway shall not exceed _____%.

A. 20   C. 40
B. 30   D. 50

_____ _____ **16.** Where more than one voltage system exists in a building, each ungrounded conductor of a multiwire branch circuit shall be identified by _____ and system.

_____ T   F **17.** A switchboard in a damp or wet location shall have a waterproof enclosure.

_____ T   F **18.** Header ducts used with cellular concrete floor raceways may be run diagonally across the cells, provided they run in straight lines.

_____ _____ **19.** Cord-equipped fixtures can be suspended directly below an outlet box if the cord _____ in a grounding-type attachment plug.

_____ _____ **20.** Underfloor raceways not over 4″ wide shall be covered by concrete or wood to a thickness of not less than _____″.

A. ¾   C. 1½
B. 1   D. 2

_____ _____ **21.** Transformers shall have a secondary circuit current rating of not more than _____ mA.

_____ _____ **22.** A feeder neutral conductor on a pure resistive load shall have a demand factor of _____% applied for all loads above 200 A.

_____ T   F **23.** The full cross-sectional area of cellular metal floor raceway may be used if AC cable is used.

_____ _____ **24.** Fixtures that are designed to be supplied by direct current shall be _____.

_____ _____ **25.** Expansion joints in raceway systems shall be installed where necessary to compensate for _____ movement.

_____ _____ **26.** The largest size conductor allowed in a cellular metal floor raceway without special permission is _____.

A. No. 1/0   C. 250 kcmil
B. No. 3/0   D. 500 kcmil

_____ _____ **27.** Raceway systems that are exposed to widely different temperatures shall be _____.

_____ T   F **28.** Surface metal raceways are permitted for use in dry or wet locations.

_____ _____ **29.** In cellular metal floor raceways, splices and taps shall be made only in _____ access units or junction boxes.

_____ _____ **30.** A fuse or OCPD shall be connected in _____ with each ungrounded conductor.

_____ _____ **31.** The center of the grip of an operating handle in its highest position shall not exceed _____ from the floor.

_____ _____ **32.** Each fixture of each secondary circuit of tubing rated over _____ V shall have a warning label.

_____ _____ **33.** An outlet tap for a fixture shall not extend over _____″.

_____ T   F **34.** No conduit shall be run through a duct used for dust removal.

Commerical Locations

Trade Competency Test 4

NEC®   **Answer**

_____   _____   **1.** Panelboards with overcurrent protection not above 200 A may contain snap switches of _____ A or less.

_____   T   F   **2.** Plastic inserts may be used with metal underfloor raceways.

_____   _____   **3.** Any 15 A or 20 A receptacle shall not supply a cord-and-plug connected load which exceeds _____% of the rating of the receptacle.

_____   _____   **4.** If the ampacity of a conductor does not correspond with a standard overcurrent protection device, the next larger size up to _____ A can be used.

      A. 800   C. 1200
      B. 1000   D. 1500

_____   T   F   **5.** No wiring system of any type shall be installed in a duct used for ventilation of commercial cooking equipment.

_____   _____   **6.** The OCPD of a phase converter is computed at _____% of the 1ϕ input FLA.

_____   T   F   **7.** SE conductors can be spliced where the wiring method is changed.

_____   _____   **8.** Only _____ mains are permitted in a lighting and appliance panelboard.

_____   T   F   **9.** Conductors are considered outside the building when encased in at least 2″ of concrete.

_____   _____   **10.** Splices and wires within any portion of an auxiliary gutter shall not fill the cross-sectional area of the gutter by more than _____%.

      A. 20   C. 60
      B. 50   D. 75

_____   T   F   **11.** The cross-sectional area of conductors in any section of sheet metal auxiliary gutter shall not exceed 20% of its area.

_____   _____   **12.** The use of a multioutlet assembly shall be permitted in _____ locations.

_____   T   F   **13.** AC-DC general-use snap switches controlling ordinary inductive loads shall have an ampere rating twice that of the load.

_____   _____   **14.** The maximum number of OCPDs in a lighting and appliance branch-circuit panelboard shall not exceed _____.

      A. 30   C. 42
      B. 36   D. 48

_____   T   F   **15.** SE conductors shall not feed one building or structure through another building or structure.

16. The general rule requires grounding receptacles to be installed on _____ A and _____ A circuits.

   A. 15; 20   C. 25; 30

   B. 20; 25   D. neither A, B, nor C

17. Listed extension cord sets with No. _____ or larger conductors can be connected to a 30 A branch circuit.

T   F   **18.** Handles of CBs shall be clearly marked ON or OFF.

19. Fixture wire used on a parking lot lighting pole can extend up to _____′ for No. 18 conductors.

20. The load for show windows is computed at a unit load of _____ VA per linear foot.

21. A fixture whip can be used if its length does not exceed _____′.

22. All 15 A and 20 A, 120 V receptacles used on construction sites shall be _____-protected.

   A. GFCI   C. GP

   B. GPF   D. neither A, B, nor C

23. FMC does not require support for lengths of _____′ or less at terminals where flexibility is required.

24. Receptacles located in floors shall be installed in floor boxes _____ for such applications.

25. Angle-pull dimensional requirements apply to junction boxes only when the size of conductor is equal to or larger than No. _____.

   A. 1/0   C. 4

   B. 3/0   D. 6

26. Electric mixers traveling in and out of open-type mixing tanks are considered _____ utilization equipment.

27. Flexible cords enclosing No. _____ and larger conductors are permitted to be spliced.

28. The total load on any OCPD in a panelboard shall not exceed _____% of its rating if in normal use for over 3 hours.

29. For existing installations in residential occupancies, _____ protection is not required in a lighting and appliance panelboard used as service equipment.

30. A 120 V CB used to switch fluorescent fixtures shall be marked _____.

   A. SWD   C. SWF

   B. SWR   D. FLS

31. The maximum setting of ground-fault protection of equipment shall be _____ A.

32. The general rule requires conductors used in flexible cords to be No. _____ or larger.

33. The maximum number of quarter bends permitted in a run of FMC is _____.

T   F   **34.** The general rule allows overcurrent protection devices to be inaccessible to occupants of a building.

T   F   **35.** Where GFCI protection of personnel for temporary electrical power is required, listed cord sets incorporating this protection shall be permitted.

An *industrial location* is a building where goods are manufactured. Common service sizes are 120/240 V, 1φ; 120/208 V, 3φ; 277/480 V, 3φ; 4160 V, 3φ; and 12,470 V, 3φ.

POWER INSTALLATION FOR INDUSTRIAL PLANT

The print shows only power outlets that are supplied by a 265/460 V, 3φ, 4-wire wye distribution system. Lighting and receptacle outlets are supplied by a separate 120/208 V, 3φ, 4-wire system and appear on a separate print. Specifications call for THW Cu conductors throughout the project.

MOTOR CIRCUITS

Article 430 explains when to use the tables at the end of 430 and when to use the motor nameplate rating to determine the ampacities and ratings of the various components of motor circuits. NEC 430-6(a)(1) states that where the current rating of a motor is used to determine the ampacity of conductors or the ampere ratings of switches, controllers, branch-circuit overcurrent protection devices,

etc., the values given in Tables 430-147, 430-148, 430-149, and 430-150, including notes, shall be used instead of the motor nameplate rating.

If only the full-load current is marked on the nameplate, the horsepower rating shall be determined by the corresponding value given in Tables 430-147 through 430-150. In the selection of separate motor-running overcurrent protection which will protect motors, motor-control apparatus, and motor branch-circuit conductors against excessive heating due to overloads and failure to start, the full-load current rating that appears on the motor nameplate shall be used.

NEC 430-22(a) states that branch-circuit conductors supplying a single motor shall have an ampacity not less than 125% of the full-load current of the motor. NEC 430-32(a)(1) specifies that continuous-duty motors rated more than 1 HP shall be protected against overload, in general, by a device that trips at not more than 125% of the full-load current of the motor.

INDUSTRIAL PLANT

CIRCUITS
THREE NO. 6 CONDUCTORS IN 1″ CONDUIT
3 #6-1″
TO PANEL
THREE CONDUCTORS
M — MOTOR

SYMBOLS	
PANEL	▨
SWITCHBOARD	▭
MOTOR	Ⓜ
JUNCTION BOX	Ⓙ
SPECIAL LOAD	▲
SECONDARY RESISTOR	⊠

NEC 430-52(b) requires that overcurrent protection devices protecting branch-circuit conductors shall be capable of carrying motor-starting current. Motor-starting current usually varies between 150% and 300% of the full-load current, depending on the type of motor. Table 430-152 lists information necessary to calculate the maximum rating or setting of motor branch-circuit, short-circuit, and ground-fault protective devices. In rare cases, starting current runs somewhat higher than values given there. NEC 430-52 provides for this by allowing larger overcurrent protection devices, which permit the motor to start.

Application

A1 on the print shows a 40 HP, 460 V, 3φ squirrel-cage motor which has no Code letter marked on the nameplate, a marked service factor of 1.15, a full-load current of 49 A, and which is started by means of an autotransformer compensator. Table 430-150 gives the full-load current as 52 A. Under 430-22(a), the branch-circuit conductors shall be able to carry 65 A (52 A × 1.25 = 65 A). Table 310-16 lists the nearest size of THW Cu wire as No. 6. Appd C, Table C8 permits up to five No. 6 THW conductors in 1″ RMC.

Table 430-152 lists all types of motors including squirrel-cage and synchronous motors. Row 2 of Table 430-152, "squirrel-cage, other than Design E" applies here. Column 2 shows the branch-circuit nontime-delay fuse rating as 300% of the full-load current. For a motor current of 52 A, circuit protection should be 156 A (52 A × 3 = 156 A). The nearest standard fuse or non-adjustable CB rating that does not exceed the 300% is 150 A per 240-6(a).

Motor running overcurrent protection, per 430-32(a)(1), should be not greater than 61.25 A (49 A × 1.25 = 61.25 A). This protection may be obtained with an adjustable device set at 61.25 A or by the nearest size fuse or non-adjustable CB, which in this case is 60 A.

Vertical Milling Machine A5 is driven by a 20 HP motor, Code letter C, a marked service factor of 1.15, a full-load current of 25 A, and is protected with an ITCB. Table 430-150 gives the full load current as 27 A. NEC 430-22(a) calls for a No. 10 conductor (27 A × 1.25 = 33.75 A). Appd

C, Table C8 specifies a ½″ RMC. Motor-running overcurrent protection per 430-32(a)(1) should be not greater than 31.25 A (25 A × 1.25 = 31.25 A). The nearest standard fuse or non-adjustable CB is rated at 30 A per 240-6(a). Table 430-152 gives 250% of the full-load current for branch-circuit protection, or 67.5 A (27 × 2.5 = 67.5 A). This allows a 70 A ITCB for branch-circuit protection.

The 7½ HP, Code letter F, full-voltage starting motor on Radial Drill A4 has a nameplate current of 10 A and a marked service factor of 1.15. Motor-running protection is 12.5 A (10 A × 1.25 = 12.5 A), for an adjustable protector per 430-32(a)(1). Branch-circuit protection is 300% of 11 A, or 33 A (11 A × 3 = 33 A), which permits a 35 A NTDF per Table 430-152.

Pipe-Threading Machine A21 uses a 20 HP wound-rotor motor. The size of circuit conductors, conduit, and motor-running overcurrent protection are the same as for the 20 HP squirrel-cage motor of Vertical Milling Machine A5, but the branch-circuit protection is different. Referring to Table 430-152, select the wound-rotor motor which allows 150% of the full-load current, or 40.5 A (27 A × 1.5 = 40.5 A). A 45 A NTDF or a 45 A time-limit CB may be used.

The motor nameplate lists a secondary current of 50 A. NEC 430-23(a) states that for continuous-duty motors, the current-carrying capacity of wires between the motor secondary (rotor) and the controller shall not be less than 125% of the full-load secondary current. This value is 62.5 A (50 A × 1.25 = 62.5 A), the nearest conductor being No. 6. Table 430-23(c) specifies ampacities of conductors between the controller and resistors. In most cases, high-temperature wire is necessary because of the heat that is generated in resistor grids.

Lathe D3 is driven by a 10 HP, 230 V, DC motor with a full-load current of 37 A and a marked service factor of 1.15. Table 430-147 gives the full-load current as 38 A, requiring two No. 8 conductors in ¾″ conduit. The motor-running overcurrent protection is 46.25 A (37 A × 1.25 = 46.25 A) and the branch-circuit protection is 57 A (38 A × 1.5 = 57 A) per Table 430-152, which permits a 60 A NTDF.

Elevator Motor

A23 is a 15 HP, high-reactance, 30 and 60 minute rated motor with full-voltage starting. Table 430-22(b), Note states that the duty cycle on elevator motors shall be classed as intermittent duty. An *intermittent duty motor* is a motor that operates for alternate intervals of load and no load; load and rest; or load, no load, and rest. NEC 430-33 provides that the ampacity of the branch-circuit conductors for intermittent duty motors shall be considered as protected against running overcurrent by a branch-circuit device which does not exceed the values specified in Table 430-152.

NEC 430-22(b) states that the ampacity of the branch-circuit conductors for intermittent duty motors shall be not less than the percentage of the nameplate current rating of the motor as shown in Table 430-22(b) unless the AHJ grants special permission for conductors of a smaller size. In this case, the ampacity of the branch-circuit conductors for elevator motor A23 need be only 90% of the nameplate full-load current of the motor.

The full-load current rating of the elevator motor on A23 is listed as 21 A. The ampacity of the conductors is rated on the basis of 18.9 A (21 A × .9 = 18.9 A). A No. 12 conductor is needed. Table 430-152 shows that the branch-circuit protection may be 150% of the full-load current, or 31.5 A (21 A × 1.5 = 31.5 A) which permits a 35 A non-time-delay fuse or ITCB.

Feeder Supplying Two or More Motors

NEC 430-24 states that conductors supplying two or more motors shall have an ampacity of not less than 125% of the full-load current rating of the highest rated motor in the group, plus the sum of the full-load current of the remaining motors. Thus, for two motors, one having a full-load current rating of 20 A and the other 14 A, the feeder ampacity shall be not less than 39 A (20 A × 1.25 = 25 A + 14 A = 39 A). A No. 8 THW Cu wire is required.

NEC 430-62(a) provides that overcurrent protection for conductors supplying two or more motors shall be not greater than the branch-circuit overcurrent for the largest rating or setting plus the full-load currents of the other motors. With a 250% protective requirement for the larger motor in the preceding example, the feeder protective device should be set at not more than 64 A (20 A × 2.5 = 50 A + 14 A = 64 A). The nearest standard fuse or CB that does not exceed this value is 60 A per 240-6(a).

Crane Motors

Bridge Crane A22 has three motors, a 20 HP and two 7½ HP units. The nameplates specify 15-minute ratings. Noncontinuous-duty motors are commonly used on this type of service, the ratings being stated according to the nature of the application as 15 minutes, 30 minutes, or 60 minutes.

Table 610-14(a) has ampacities for conductors supplying this type of motor, the permissible currents being larger than those allowed by Table 310-16. Columns of Table 610-14(a) list values according to types of insulation and ambient operating temperatures, the first of each set of two columns dealing with 60-minute ratings, the second with 30-minute ratings. A footnote states that allowable ampacities of conductors for 15-minute motors shall be equal to the 30-minute ratings increased by 12%.

In the present instance, the 75°C columns are applicable. Since the motors are 15-minute units, the sizes of the conductors are determined on the basis of the 30-minute rating increased by 12%. The full-load current of the largest motor is 27 A, and the others are 11 A each. According to the table, a No. 14 THW Cu conductor has a 30-minute current-carrying capacity of 26 A. A 12% increase gives this conductor a capacity of 29 A (26 A × 1.12 = 29.12 = 29 A), making it suitable for use with the larger motor.

NEC 610-14(c) states that a No. 16 conductor may be used with crane motors, if otherwise suitable. The 30-minute rating of No. 16 is given as 12 A, and the allowable 12% increase raises the 15-minute rating to 13.4 A (12 A × 1.12 = 13.44 = 13.4 A), a value satisfactory for the two small motors.

NEC 610-14(e)(2) states that the ampacity of the power supply conductors on the crane shall be not less than the combined short-time, full-load ampere rating of the largest motor or group for any single crane motor, plus 50% of the short-time full-load ampere rating of the next largest motor or group. Adding the 27 A of the largest motor to 50% of 22 A (full-load current of the other motors) equals 38 A (27 A + 11 A = 38 A). A No. 8 THW conductor may be used per Table 310-16. However, the specifications call for a No. 6 THW conductor here. NEC 610-14(d) lists the sizes of bare contact conductors which, in the present case where the runway is longer than 60′, is No. 2.

Metalworking Machine Tools

NEC 670-3(a) states that the nameplate shall give necessary data, including the full-load current and the ampere rating of the largest motor. Per 670-4(a), the ampacity of supply conductors shall be not less than the full-load current marked on the nameplate, plus 25% of the full-load current rating of the highest rated motor as indicated on the nameplate. This requirement is equivalent to the general rule stated in 430-24.

Automatic Gear Machines C1 and C2 fall within the metalworking/machine tool classification. The total nameplate rating is 10.2 A and the highest rated motor is 4.8 A. Per 670-4(a), the supply conductors shall have a current-carrying capacity of not less than 11.4 A (4.8 A × .25 = 1.2 A + 10.2 A = 11.4 A). The feeder to Panel C consists of three No. 14 conductors in ½″ conduit. Short-circuit protection for each feeder per 430-62(a) is 19.8 A (4.8 A × 3 = 14.4 A; 2.6 A + 1.8 A + 1 A = 5.4 A; 14.4 A + 5.4 A = 19.8 A). A 15 A NTDF is acceptable.

Synchronous Motor

An air compressor in the switchboard room is driven by a 100 HP synchronous motor operating at a leading power factor of 80%. Table 430-150 gives the full-load current at unity power factor as 101 A. A footnote states that this figure shall be multiplied by 1.25 when the motor operates at 80% power factor, the current here becoming 126 A (101 A × 1.25 = 126.25 = 126 A). Circuit wires shall have an ampacity of at least 158 A (126 A × 1.25 = 157.5 = 158 A) per 430-22(a). Table 310-16 lists No. 2/0 THW Cu wire at 175 A. Appd C, Table C8 shows that 1½″ RMC is required.

A footnote to Table 430-152 states that for synchronous motors used in applications of this nature, branch-circuit protection need not be greater than 200% of the full-load current. For a current of 126 A, the overcurrent protection device could be set as low as 252 A (126 A × 2 = 252 A). By using an ITCB, a 300 A CB is needed. Two No.

10 conductors in ½″ conduit, originating at Panel D, supply direct current excitation for the revolving field of the synchronous motor.

X-Ray Unit

The No. 10 conductors specified for this device are larger than might seem to be required. Their size is dictated by a desire to hold the voltage drop to a minimum.

Resistance Welders

NEC 630-31(a)(1) states that the ampacity of conductors supplying a resistance welder shall be not less than 70% of the rated primary current for automatically fed welders and 50% for manually operated ones. NEC 630-31(b) requires that rated ampacity of conductors which supply two or more such welders shall be not less than the sum of the above value for the largest welder plus 60% of such values for the smaller welders.

NEC 630-32(a)(b) demands that each welder be protected by an overcurrent protection device set at not more than 300% of the rated primary current, and that a feeder supplying one or more units shall be protected by a device set at not more than 300% of the conductor rating.

Unit B1 is a 5 kVA spot welder rated at 11 A. Unit B2 is a 10 kVA spot welder rated at 22 A. Both are manually operated. Conductors supplying B1 may be No. 14. Those for B2 may also be No. 14. B1 may be protected by a 30 A fuse or a non-adjustable CB and B2 by a 60 A fuse or non-adjustable CB. The feeder should not be smaller than 14.3 A (22 A × 5 = 11 A; 11 A × .5 × .6 = 3.3 A; 11 A + 3.3 A = 14.3 A). The question of voltage drop must be considered if the units are to operate efficiently. Two No. 10 conductors are run in ½″ conduit from the junction box to Panel B. Two No. 12 conductors extend from the small welder to the junction box.

Transformer Arc Welders

NEC 630-11(b) provides that the conductor rating for a group of welders shall be based on the nature of the operation. The maximum demand for a similar group of welders used on an identical application is 50% of the total capacity of the units. The total capacity is 45 kVA (5 kVA + 10 kVA + 15 kVA + 15 kVA = 45 kVA). One half of this amount is 22.5 kVA (45 kVA ÷ 2 = 22.5 kVA), and the current taken when they are balanced across a 460 V, 3φ supply line is equal to about 30 A per leg. By applying the formula $I = \dfrac{VA}{V \times \sqrt{3}}$, the current is found to be 28 A (I = 22,500 VA ÷ 797 = 28.23 = 28 A).

NEC 630-12(a) requires that each welder have overcurrent protection not greater than 200% of the rated primary current. NEC 630-12(b) requires that feeder protection be not greater than 200% of the conductor rating.

Motor-Generator Arc Welder

B7, a single-operator, motor-generator arc welder, has a duty cycle of 80%. Per 630-11(a), conductors supplying such a unit shall have an ampacity of 10 A, which is equal to 91% of the nameplate rating (11 A × .91 = 10.01 = 10 A). The smallest permissible conductor per Table 310-16 is No. 14 Cu.

Feeder Calculations

The load on the sub-feeder from Panel C to Panel A totals about 53 A. Three No. 6 THW Cu conductors are used in 1″ RMC. The load on the sub-feeder from Panel E to Panel A is approximately 21.6 A. Specifications call for a No. 10 Cu feeder. Panel A by its own units, including 25% of the largest motor-running current, equals approximately 544 A. The total load on Panel A equals approximately 544 A plus the full-load current of the motors connected to Panels C and E, for a total feeder load to Panel A of 618 A (544 A + 52 A + 22 A = 618 A). The print shows an 800 A busway from the main switchboard to Panel A.

Feeder conductors for Panel B supply the following loads: 14.3 A, 1φ to resistance welders B1 and B2; 28 A, 3φ to transformer arc welders B3, B4, B5, and B6; 10 A, 3φ to motor-generator arc welder B7; 11 A, 1φ to the X-ray unit B8; and 7.6 A, 3φ to the grinder B9.

When this load is balanced across the three phases as equally as possible, the most heavily loaded conductor carries about 70 A. No. 4 THW Cu conductors would normally be used, but because of the nature of the equipment, especially the X-ray unit, three No. 1 conductors are

specified. Chapter 9, Table C8 shows that 1¼″ RMC is needed for these conductors.

Service Calculations

The service load is equal to the full-load motor currents of all of the motors plus all other equipment on Panels A and B plus the full-load current of the 25 HP motor-generator, plus that of the 100 HP synchronous motor, plus 25% of the full-load current of the synchronous motor. The total is:

Panel A =	525 A
Panel B =	70 A
25 HP MG Set =	34 A
100 HP Synchronous Motor =	126 A
+ 25% of Largest Motor =	32 A
	787 A

This value is larger than that for the conductors listed in Table 310-16, so two or more conductors in parallel are required to be used.

One half of 787 A equals 393.5 A (787 A ÷ 2 = 393.5 A). Per Table 310-16, the nearest listed conductor is 600 kcmil THW Cu. The service may consist of two sets of three 600 kcmil THW Cu conductors in two runs of 3″ RMC per Appd C, Table C8. For ease of handling, smaller conductors can be used. Table 310-16 gives the ampacity of 250 kcmil THW Cu as 255 A. With six conductors in one conduit, the ampacity of each is reduced to 80% of 255 A, or 204 A (255 A × .80 = 204 A). Two such conductors on each leg furnish a capacity equal to 408 A (204 A + 204 A = 408 A). Appd C, Table C8 permits six 250 kcmil THW Cu conductors in 3″ RMC. The service then consists of two sets of six 250 kcmil THW Cu conductors in 3″ conduits.

Note: This service configuration is one of many possible alternatives. Supply conductors could be run in three or more sets of conduit, to avoid penalty with respect to ampacity, or even in the form of busduct if conditions warranted its use.

POWER INSTALLATION FOR RESTAURANT

The power installation is to be confined to the kitchen area, with the exception of a 5 HP exhaust fan on the roof. The service is 4-wire delta with 3φ power at 240 V between phase wires and 1φ power at 120/240 V between two phase wires and a neutral conductor, as indicated on the small diagram to the right of the print.

The print covers only the power installation. Lighting outlets appear on a separate drawing. This procedure is often used where either the power or the lighting is involved to avoid any confusion which may result from crowding both power and lighting onto the same sheet.

The panelboard is divided into two sections, A and B. Section A supplies 3ϕ loads and large 1ϕ loads which do not require a neutral conductor. Section B supplies any 1ϕ loads which require a neutral conductor. THW Cu conductors are used.

Circuit Analysis

Circuits A1 and A2 are connected to 21 kVA, 3ϕ ranges, with the nameplate current rating being 53 A. These loads, in normal operation, are likely to continue for long periods of time. NEC 210-20(a) requires that conductors serving such loads shall carry not more than 80% of the listed current. The conductors supplying each range are rated at a minimum ampacity of 67 A (53 A ÷ .8 = 66.25 = 67 A). Table 310-16 rates No. 4 THW Cu conductors at 85 A. Appd C, Table C8 indicates that 1″ RMC is needed for three No. 4 Cu conductors.

Deep fat fryers A3 and A4 are rated at 10 kVA, 1ϕ, with a current rating of 44 A. These loads are similar to those of the ranges in that they continue for long periods of time so that the 80% factor is used. The conductors have a normal ampacity of 55 A (44 A ÷ .8 = 55 A). Two No. 6 Cu conductors in ¾″ conduit will meet the requirements.

Circuit A5 supplies a 5 HP, 3ϕ blower motor on the roof with a nameplate current of 14.8 A and a service factor of 1.15. A riser leading to this unit may be seen at the left side of the panelboard. NEC 430-6(a) states that, except for hermetic motors, the ampacity of conductors, branch-circuit overcurrent devices, etc., shall be determined from the motor current tables at the end of 430 instead of the current ratings on motor nameplates. Table 430-150 lists the current of the 5 HP, 240 V, 3ϕ motor as 15.2 A.

NEC 430-22(a) requires that branch-circuit conductors supplying a single motor shall have an ampacity of 125% of 15.2 A, or 19 A (15.2 A × 1.25 = 19 A). No. 12 THW Cu conductors are rated at 20 A per Table 310-16. Appd C, Table C8 shows that three No. 12 Cu conductors may be run in ½″ RMC.

NEC 430-52 and Table 430-152 require that the initial rating of a dual-element fuse, used for branch-circuit protection, shall not exceed 175% of the full-load current of the motor listed in Table 430-150. The rating of the dual-element fuse for the 5 HP motor should not exceed 26.6 A (15.2 A × 1.75 = 26.6 A).

NEC 430-52(c)(1) requires the use of a 25 A fuse. NEC 430-83 states that a controller shall have a horsepower rating not less than that of the motor. NEC 430-109 states a similar requirement for the disconnecting means. A 25 A CB satisfies the needs of a controller per 430-83(c) and of a disconnect per 430-110(a). The rating of a manually-operable switch on the roof of the building is 5 HP.

Per 430-32(a)(1), each continuous-duty motor rated at more than 1 HP shall be protected against running overcurrent by a separate overcurrent protection device rated at not more than 125% of the full-load current of the motor, or else contain an integral thermal protector. If the blower motor does not have built-in thermal protection, an overcurrent protection device shall be supplied (perhaps in a magnetic switch used to start and stop the motor). The rating of the device should be 18.5 A (14.8 A × 1.25 = 18.5 A).

NEC 430-52 states that the maximum value of branch-circuit protection shall be limited to the value given in Table 430-152 (300% for NTDFs). NEC 430-58 states that a circuit breaker is permitted to be used for motor branch-circuit protection if it conforms to 430-110(a), which in turn requires it to have an ampacity of 115% of the full-load current of the motor. The 5 HP motor has a full-load current of 17.5 A (15.2 A × 1.15 = 17.48 = 17.5 A). If a standard 20 A CB at the panel carries starting current of this motor without tripping, it satisfies the need for motor overcurrent protection per 430-52 and 430-58.

Circuit A6 supplies a 1½ HP, 3ϕ dishwasher motor with a built-in thermal device. Table 430-150 gives the current as 5.2 A, so that No. 14 Cu conductors are large enough. Per 430-83(c) 1., a controller located at the operating location may be a general-use switch, which has an ampere rating at least twice that of the full-load current of the motor.

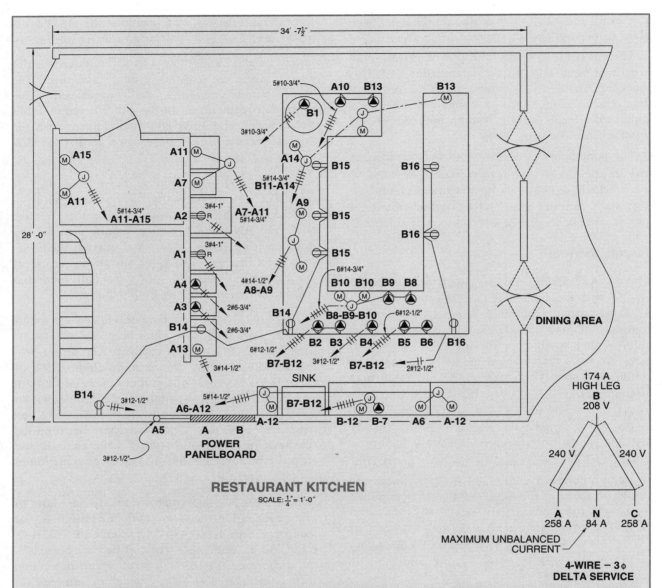

RESTAURANT KITCHEN

SCALE: $\frac{1}{4}$" = 1'-0"

MOTOR LOAD ON PANEL A	
A5	Roof Fan – 5 HP, 240 V, 3φ, 15.2 A
A6	Dishwasher – $1\frac{1}{2}$ HP, 240 V, 3φ, 5.2 A
A7	Food Mixer – 1 HP, 240 V, 3φ, 3.6 A
A8	Compressor – $\frac{3}{4}$ HP, 240 V, 3φ, 2.8 A
A9	Compressor – $\frac{3}{4}$ HP, 240 V, 3φ, 2.8 A
A11	Compressor – $\frac{1}{2}$ HP, 240 V, 1φ, 4.9 A
A11	Chef's Ref – $\frac{1}{2}$ HP, 240 V, 1φ, 4.9 A
A12	Disposer – $\frac{1}{2}$ HP, 240 V, 1φ, 4.9 A
A12	Potato Peeler – $\frac{1}{2}$ HP, 240 V, 1φ, 4.9 A
A13	Food Chopper – $\frac{3}{4}$ HP, 240 V, 3φ, 2.8 A
A14	Ice Machine – $1\frac{1}{2}$ HP, 240 V, 3φ, 5.2 A
A15	Hermetic Comp – 240 V, 3φ, 11 A

COMMERCIAL ELECTRIC COOKING AND KITCHEN EQUIPMENT ON PANEL A	
A1	Range – 21.00 kVA, 240 V, 3φ
A2	Range – 21.00 kVA, 240 V, 3φ
A3	Deep Fat Fryer – 10.00 kVA, 240 V, 3φ
A4	Deep Fat Fryer – 10.00 kVA, 240 V, 3φ
A10	Hot Table – 3.25 kVA, 240 V, 3φ

MOTOR LOAD ON PANEL B	
B10	Compressor – $\frac{1}{4}$ HP, 120 V, 1φ, 5.8A
B10	Compressor – $\frac{1}{4}$ HP, 120 V, 1φ, 5.8 A
B11	Carbonator – $\frac{1}{4}$ HP, 120 V, 1φ, 5.8 A
B12	Silver Burnisher – $\frac{1}{3}$ HP, 120 V, 1φ, 7.2 A

COMMERCIAL ELECTRIC COOKING AND KITCHEN EQUIPMENT ON PANEL B	
B1	Coffee Maker – 5 kVA, 120/240 V, 1φ
B2	Toaster – 2.5 kVA, 120/240 V, 1φ
B3	Toaster – 2.5 kVA, 120/240 V, 1φ
B4	Toaster – 2.5 kVA, 120/240 V, 1φ
B5	Waffle Iron – 2.5 kVA, 120/240 V, 1φ
B6	Waffle Iron – 2.5 kVA, 120/240 V, 1φ
B7	Sterilizer – 1 kVA, 240 V, 1φ
B8	Plate Warmer – 1 kVA, 240 V, 1φ
B9	Bun Warmer – .75 kVA, 120 V, 1φ
B13	Infrared Cooker – 3.25 kVA, 120/240 V, 1φ

Three No. 14 Cu conductors in ½″ conduit extend from the motor location to a junction box where they are joined by two No. 14 Cu conductors from a garbage disposer connected to A12. A ½″ conduit carries the five wires to a second junction box where two conductors from a potato peeler motor are tapped to Circuit A12 before the five wires continue to the panelboard.

Two motors are connected to Circuit A12 per 430-53(a), which states that motors not exceeding 1 HP each and having a full-load current not in excess of 6 A are permitted to be connected to a 20 A, 120 V, or less, circuit. Individual running overcurrent protection is required as these motors are permanently installed. See 430-32(b)(2). Usually, motors of this type are provided with a thermal protector integral with the motor which gives required protection.

Other motors are combined in this manner on Circuits B10, B11, and A11. Those on A11 are connected at the panelboard because they are run in separate conduits. The ¾ HP compressor motors on Circuits A8 and A9 may be combined on a single circuit but would require running protection, as they are automatically started per 430-32(c).

Circuit A10, for the hot table, and B13, for the infrared cooker, are taken to the panelboard as five No. 10 Cu conductors in ¾″ conduit. Circuit A10 is rated at 3250 VA or 13.5 A (3250 VA ÷ 240 V = 13.54 = 13.5 A) and Circuit B13 at 3500 VA or 14.6 A (3500 VA ÷ 240 V = 14.58 = 14.6 A). Since these are continuous loads, 210-20(a) requires that the total load on any overcurrent protection device not exceed 80% of its rating.

When there are from four to six current-carrying conductors in the same conduit supplying a continuous load, they are derated once as per 310-15(b)(2). No. 12 conductors are permissible for A10 and B13, but in this case a No. 10 THW Cu conductor was selected. Appd C, Table C8 permits six No. 10 THW Cu conductors in ¾″ RMC. Derating considerations also enter into calculations for toasters and waffle irons on Circuits B2, B3, B5, and B6.

Circuit A15, whose conduit is run exposed, supplies an 11 A sealed hermetic compressor motor. NEC 440-6(a) states that the full-load current marked on the compressor nameplate shall be used to determine the ampacity of the branch-circuit conductors. NEC 440-4(a) requires that the locked rotor current for such motors shall be marked on the nameplate. Locked rotor current is marked as 53 A. NEC 440-41(a) provides that controllers for sealed hermetic motors shall be selected on the basis of full-load current and locked rotor current.

Per Table 430-150, Column 3, the nearest current greater than 11 A, (the full-load current on the motor nameplate) is 15.2 A, which applies to a 5 HP unit. In Table 430-151, the nearest 3ϕ locked rotor current to the 53 A value on the nameplate of the motor is 58 A, applying to a 3 HP motor. NEC 440-41(a) states that where two different horsepower ratings are obtained from different tables, the higher of the the values shall be used. Here, the larger of them is used, necessitating a 5 HP controller. NEC 440-12(a) imposes a similar demand with respect to the disconnecting means.

Service Calculation

The load on Panel A is obtained by converting motor loads to VA and adding them to the VA of the cooking and heating appliances. A value equal to 25% of the largest motor rating is included per 430-24 as well as a demand factor permitted by Table 220-20 for commercial electric cooking and other kitchen equipment. The total load on Panel A is 72,178 VA, or 174 A.

Note: When total wattage or VA is to be divided by phase-to-phase (3ϕ) voltage times 1.732, the following values may be substituted:

- for 208 V × 1.732, use 360
- for 230 V × 1.732, use 398
- for 240 V × 1.732, use 416
- for 440 V × 1.732, use 762
- for 460 V × 1.732, use 797
- for 480 V × 1.732, use 831

MOTOR LOAD ON PANEL A

A5 Roof Fan — 5 HP, 240 V, 3ϕ, 15.2 A

VA = 15.2 × 416 =   6323 VA

A6 Dishwasher — 1½ HP, 240 V, 3ϕ, 5.2 A

VA = 5.2 × 416 =   2163 VA

A7 Food Mixer — 1 HP, 240 V, 3ϕ, 3.6 A

VA = 3.6 × 416 =   1498 VA

A8 Compressor — ¾ HP, 240 V, 3ϕ, 2.8 A

VA = 2.8 × 416 =   1165 VA

A9 Compressor — ¾ HP, 240 V, 3φ, 2.8 A

VA = 2.8 × 416 =	1165 VA

A11 Compressor — ½ HP, 240 V, 1φ, 4.9 A

VA = 4.9 × 240 =	1176 VA

A11 Chef's Ref — ½ HP, 240 V, 1φ, 4.9 A

VA = 4.9 × 240 =	1176 VA

A12 Disposer — ½ HP, 240 V, 1φ, 4.9 A

VA = 4.9 × 240 =	1176 VA

A12 Potato Peeler — ½ HP, 240 V, 1φ, 4.9 A

VA = 4.9 × 240 =	1176 VA

A13 Food Chopper — ¾ HP, 240 V, 3φ, 2.8 A

VA = 2.8 × 416 =	1165 VA

A14 Ice Machine — 1½ HP, 240 V, 3φ, 5.2 A

VA = 5.2 × 416 =	2163 VA

A15 Hermetic Comp — 240 V, 3φ, 11 A

VA = 11 × 416 =	4576 VA
Total Motor VA =	24,922 VA

220-14: + 25% of largest motor

6323 VA × .25 =	1581 VA
Total Motor Load on Panel A =	**26,503 VA**

COMMERCIAL ELECTRIC COOKING AND KITCHEN EQUIPMENT LOAD ON PANEL A

A1	Range — 21 kVA, 240 V, 3φ	21,000 VA
A2	Range — 21 kVA, 240 V, 3φ	21,000 VA
A3	Deep Fat Fryer — 10 kVA, 240 V, 3φ	10,000 VA
A4	Deep Fat Fryer — 10 kVA, 240 V, 3φ	10,000 VA
A10	Hot Table — 3.25 kVA, 240 V, 3φ	3250 VA
	Total =	65,250 VA

Table 220-20: 65,250 VA × .70 = 45,675 VA

Total Commercial Electric Cooking and Kitchen Equipment Load on Panel A =	**45,675 VA**
Total VA on Panel A =	**72,178 VA**

$$I = \frac{VA}{V \times \sqrt{3}}$$

$$I = \frac{72,178}{240 \times 1.73}$$

$$I = \frac{72,178}{416} = 173.5 = 174 \text{ A}$$

Panel A = 174 A

The load on Panel B, found in the same manner, is 20,226 VA or 84 A.

MOTOR LOAD ON PANEL B

B10 Compressor — ¼ HP, 120 V, 1φ, 5.8 A

VA = 5.8 × 120 =	696 VA

B10 Compressor — ¼ HP, 120 V, 1φ, 5.8 A

VA = 5.8 × 120 =	696 VA

B11 Carbonator — ¼ HP, 120 V, 1φ, 5.8 A

VA = 5.8 × 120 =	696 VA

B12 Silver Burn. — ⅓ HP, 120 V, 1φ, 7.2 A

VA = 7.2 × 120 =	864 VA
Total Motor VA =	2952 VA

220-14: + 25% of largest motor

864 VA × .25 =	216 VA
Total Motor Load on Panel B =	**3168 VA**

COMMERCIAL ELECTRIC COOKING AND KITCHEN EQUIPMENT LOAD ON PANEL B

B1	Coffee Maker — 5 kVA, 120/240 V, 1φ	5000 VA
B2	Toaster — 2.5 kVA, 120/240 V, 1φ	2500 VA
B3	Toaster — 2.5 kVA, 120/240 V, 1φ	2500 VA
B4	Toaster — 2.5 kVA, 120/240 V, 1φ,	2500 VA
B5	Waffle Iron — 2.5 kVA, 120/240 V, 1φ	2500 VA
B6	Waffle Iron — 2.5 kVA, 120/240 V, 1φ	2500 VA
B7	Sterilizer — 1 kVA, 240 V, 1φ	1000 VA
B8	Plate Warmer — 1 kVA, 240 V, 1φ	1000 VA
B9	Bun Warmer — .75 kVA, 120 V, 1φ	750 VA
B13	Infrared Cooker — 3.5 kVA, 120/240 V, 1φ	3500 VA
	Total =	23,750 VA

Table 220-20 : 23,750 VA × .65 = 15,438 VA

Total Commercial Electric Cooking and Kitchen Equipment on Panel B =	**15,438 VA**
Receptacles, 120 V: 180 VA × 9 =	**1620 VA**
Total VA on Panel B =	**20,226 VA**

$$I = \frac{VA}{V}$$

$$I = \frac{20,226}{240} = 84.275 = 84 \text{ A}$$

Panel B = 84 A

These values have been marked on the diagram at the right of the drawing, the load on phase B being 174 A, that on A or C is 258 A, and the current in N (neutral) is 84 A.

This method of adding 1φ current directly to 3φ current in order to determine the total current on a particular supply wire is the one customarily used and gives results sufficiently accurate for practical needs.

Table 310-16 gives the nearest ampacities for THW Cu conductors as No. 2/0 for Phase *B*, 300 kcmil for Phase *A* and Phase *C*, and No. 4 for *N*. Chapter 9, Table 5 lists cross-sectional areas as .2624 sq in. for No. 2/0, .5281 sq in. for 300 kcmil, and .0973 sq in. for No. 4.

The cross-sectional areas of conductors are given in Chapter 9, Table 5. These areas are multiplied by the number of conductors to determine the total cross-sectional conductor area.

The total cross-sectional conductor area equals:

No. 2/0 THW =	$0.2624 \times 1 = .2624$
300 kcmil THW =	$0.5281 \times 2 = 1.0562$
No. 4 THW =	$0.0973 \times 1 = .0973$
	Total 1.4159 sq in.

For 40% fill, the cross-sectional area of conduit must be at least 3.5398 sq in. (1.4159 sq in. ÷ .4 = 3.5398 sq in.). Chapter 9, Table 4 lists the nearest size of conduit as 2½″.

Per 250-24(b)(1) the size of a copper neutral conductor is No. 2. The increased size provides an adequate return path for fault current. The conduit size remains the same in this case.

As in the case of the store building, parallel service conduits could be used in order to reduce the size of the conductors. The CB may be a 400 A frame, with current settings suited to protect each of the 3ϕ conductors. See 240-6(c) for the conditions that shall be adhered to with adjustable trip CBs.

Table 250-66 states that a No. 2 Cu GEC is required for this application.

MOTOR CONTROLLERS – 430, PART G

430-81(a): Motor Controller – A switch or device normally used to start or stop a motor by making or breaking the motor circuit current.

TO POWER SOURCE
FUSIBLE DISCONNECT
CONTROLLER RATED FOR DESIGN E MOTOR
• 430-83(a)(1)
25 HP, 230 V, 3φ DESIGN E MOTOR

What is the minimum size controller required for the 25 HP motor?
430-83(a)(1): 25 HP x 1.4 = 35 HP
Controller = **35 HP**

TO POWER SOURCE
FUSIBLE DISCONNECT
GENERAL-USE SWITCH MAY BE USED AS CONTROLLER IF IT IS 2X RATING OF MOTOR FLC
• 430-83(c)(1)
2 HP, 230 V, 3φ MOTOR

TO POWER SOURCE
GENERAL-USE SNAP SWITCH SUITABLE FOR AC ONLY
• 430-83(c)(2)
9 A, 1 HP, 230 V, 1φ MOTOR

TO POWER SOURCE
ITCB RATED IN AMPS USED AS CONTROLLER
• 430-83(a)(2)

What is the minimum size general-use snap switch permitted?
Table 430-148: Motor rating = 10 A
430-83(c) 2.: $\frac{1}{.80}$ = 1.25
    10 A x 1.25 = 12.5 A
240-6(a): Next higher standard size = 15 A
Snap switch = **15 A**

TO POWER SOURCE
DISCONNECTING MEANS
CONTROLLER SHALL HAVE CONTINUOUS-DUTY FLC RATING NOT LESS THAN MOTOR NAMEPLATE CURRENT RATING
• 430-83(d)
TORQUE MOTOR

TO POWER SOURCE
DISCONNECTING MEANS
CONTROLLER NOT REQUIRED TO BE HP RATED
• 430-83(a)(b)
• 430-81(b)
$\frac{1}{8}$ HP, OR LESS, STATIONARY MOTOR

TO POWER SOURCE
DISCONNECTING MEANS
CONTROLLER NOT REQUIRED TO BE HP RATED
• 430-83(b)
• 430-81(c)
$\frac{1}{3}$ HP, OR LESS, PORTABLE MOTOR

SINGLE CONTROLLER FOR TWO MOTORS ON ONE MACHINE
• 430-87, Ex.
MOTOR TURNS SPINDLE
SPINDLE
TABLE
MOTOR OPERATES TABLE
METALWORKING MACHINE

$\frac{1}{2}$ HP, 120 V, 6 A FLC MOTOR
SINGLE OCPD
SINGLE CONTROLLER FOR MOTORS PROTECTED BY 1 OCPD
• 430-87, Ex. (b)
$\frac{1}{2}$ HP, 120 V, 5 A FLC MOTOR

ROOM   CONTROLLER
MOTORS WITHIN SIGHT FROM SINGLE CONTROLLER
• 430-87, Ex. (c)

430-81(b) For stationary motors of $\frac{1}{8}$ HP or less that are normally left running, the branch-circuit protective device is permitted to serve as the controller.

430-81(c) For portable motors of $\frac{1}{3}$ HP or less, the controller is permitted to be a plug and receptacle.

430-83(a) The horsepower rating of the controller shall not be less than the horsepower rating of the motor.

430-83(a)(1) Controllers for Design E motors rated over 2 HP shall be marked as rated for such use or shall have a horsepower rating not less than 1.4 times the motor rating for 3 HP through 100 HP motors, or not less than 1.3 times the motor rating of motors over 100 HP.

430-83(a)(2) A branch-circuit ITCB rated in amperes only is permitted to be used as a controller.

430-83(b) Devices permitted by 430-81(b)(c) are not required to be horsepower-rated.

430-83(c) For stationary motors of 2 HP or less and 300 V or less, the controller is permitted to be a general-use switch having an ampere rating at least twice the FLC rating of the motor. For AC circuits, a general-use snap switch suitable for use on an AC circuit shall be used to control a 2 HP or less and 300 V or less motor with a current rating not more than 80% of the ampere rating of the switch.

430-83(d) Motor controllers for torque motors shall have a continuous-duty, full-load current rating not less than the motor's nameplate current rating.

430-84 Controllers are not required to open all conductors unless they also serve as the disconnecting means per 430-111.

430-87 In general, each motor of 600 V or less shall have its own controller.

430-87, Ex. A single controller rated at not less than the total horsepower ratings of all the motors in a group is permitted providing a. the motors drive parts of the same machine; b. the motors have one overcurrent device per 430-53(a); or c. the group of motors is in a single room within sight from the controller.

SWITCHBOARDS – 384, PART B

110-26(a) The working space dimensions from live parts of electrical equipment operating at 600 V or less shall be 30″ wide from the floor or platform to the height required. Equipment of equal depth is permitted in the working space.

110-26(b) The working space shall not be used for storage.

110-26(c) At least one entrance shall provide access to the working space. For equipment rated 1200 A or more and over 6′ wide, a 24″ wide by 6½′ high door is required at each end. See Exceptions.

110-26(d) Illumination is required for all working spaces.

110-26(e) The minimum headroom of working spaces around switchboards, etc. is 6½′ or not less than the height of any equipment over 6½′ high.

384-5 Switchboards with any exposed live parts shall be located in permanently dry locations, under competent supervision, and accessible only to qualified persons.

384-6 Switchboards in damp or wet locations shall be installed per 373-2(a).

384-8(a) For switchboards which are not totally enclosed, the minimum clearance between the top and a combustible ceiling is 3′. The clearance can be reduced when a noncombustible shield is provided between the switchboard and the ceiling.

384-8(b) Clearances around switchboards shall be per 110-26.

384-9 Insulated conductors in switchboards shall be listed, flame retardant, and rated not less than the applied voltage.

250-112(a) Switchboard frames shall be grounded. Frames of 2-wire, DC switchboards which are effectively insulated do not require grounding.

PHASE CONVERTERS – 455

455-2: Phase Converter – An electrical device that converts 1ϕ power to 3ϕ power.

455-4 The nameplate shall indicate:

(1) Manufacturer's name
(2) Rated input and output voltage
(3) Frequency
(4) Rated 1ϕ input FLA
(5) Rated minimum and maximum single load in kVA or HP
(6) Maximum total load in kVA or HP
(7) For rotary units, 3ϕ amps at full load

455-5 An EGC shall be provided per 250-8.

455-6(a)(1)(2) Conductors on the 1ϕ side shall be sized at 125% of the phase converter's input FLA.

455-7(a)(b) The OCPD shall be sized at 125% of the phase converter's nameplate 1ϕ input FLA.

455-8(a) The disconnecting means shall be readily accessible and located in sight from the phase converter.

455-8(b) The disconnecting means shall be a HP-rated switch, CB, molded-case switch, and in some cases, an ampere-rated switch.

455-8(c) The disconnecting means shall be sized at 115% of the maximum rated 1ϕ input FLA. See Applications (1) and (2).

455-10 A terminal housing shall be provided on a phase converter per 430-12.

455-23 Capacitors installed for a motor load shall be a part of the rotary phase conversion system and shall be installed on the line side of the motor overload device.

BUSWAYS – 364

Busway – A grounded metal enclosure with factory-mounted, bare or insulated conductors, which are usually copper or aluminum bars, rods, or tubes.

USES PERMITTED – 364-4(a)

A. (1) In open areas and visible

(2) Totally enclosed, nonventilating busways with accessible joints are permitted behind panels when means of access are provided. The space behind access panels shall not be used for air handling unless it is for environmental air and conductors are insulated and have no plug-in connections.

USES NOT PERMITTED – 364-4(b)

(1) Where subject to physical damage or corrosive vapors

(2) In hoistways

(3) In hazardous locations, unless approved per 501-4(b)

(4) Outdoors or in wet or damp locations unless identified for the use

B. (5) Lighting and trolley busway shall not be installed less than 8′ above floor or working platform.

C. 364-5 Busways shall be supported at least every 5′.

D. 364-6 Unbroken lengths of busway are permitted to extend through dry walls. Busways are permitted to extend vertically through dry floors if unventilated and protected to at least 6′ above the floor.

364-7 All dead ends of busways shall be closed.

E. 364-8 Branches from busways shall be per 334, 344, 352, and 364. Cord and cable assemblies approved for hard usage and listed bus drop cable are also permitted. Where a nonmetallic raceway is installed, an EGC per 250-8 and 250-12 shall be utilized.

364-9 Overcurrent protection for busways shall be per 364-10 – 364-13.

364-10 Busways shall be protected against overcurrent per their allowable current ratings. If the rating does not correspond to a standard ampere rating, the next higher rated standard overcurrent device not exceeding 800 A is permitted.

F. 364-11 Overcurrent protection is required where busways are reduced in ampacity.

G. 364-11, Ex. For industrial establishments only, overcurrent devices may be omitted where the size is reduced if a smaller busway is not over 50′ long and at least ⅓ the rating of the next overcurrent device back on the line, provided it is not in contact with combustible material.

364-12, Ex. 2 When used to supply lighting units, the branch-circuit overcurrent device may be part of the fixture cord cap or may be made part of the fixture.

CABLEBUS – 365

365-1: Cablebus – An assembly of insulated conductors and terminations in an enclosed, ventilated protective metal housing.

A. 365-2(a) For 600 V or less, cablebus is permitted at any voltage or current for which the conductors are rated. It shall be exposed. Cablebus shall be identified for the use when installed outdoors or in corrosive, wet, or damp locations. It shall not be installed in hoistways or hazardous locations unless approved for the use. It may be used for branch circuits, feeders, and services.

365-2(b) Approved cablebus may be installed at over 600 V.

B. 365-3(a) Current-carrying conductors in cablebus shall have an insulation rating of 75°C (167°F) or higher.

C. 365-3(b) The ampacity of conductors shall be per Tables 310-17 and 310-19 or Tables 310-69 and 310-70 for over 600 V.

D. 365-3(c) Conductor size shall be per the cablebus design and not smaller than No. 1/0.

E. 365-3(d) Insulated conductors shall be supported at least every 3′ horizontally and 1½′ vertically.

F. 365-6(a) Cablebus shall be supported at least every 12′ unless the structure is designed for greater lengths.

G. 365-6(b) Cablebus in continuous sections is permitted to pass through partitions or walls when protected against physical damage and unventilated.

365-6(c) Cablebus is permitted to pass through dry floors and platforms except where firestops are required. It shall be totally enclosed and protected for up to 6′.

365-6(d) Cablebus is permitted to pass through wet floors and platforms except where firestops are required if curbs are present to prevent water flow and it is totally enclosed and protected for up to 6′.

365-7 Cablebus shall have approved fittings for: (1) changes in directions; (2) dead ends; (3) terminations for connected equipment; and (4) additional physical protection where required.

365-9 Cablebus shall be grounded per 250, excluding 250-86, Ex. 2.

H. 365-10 Each section of cablebus shall be marked to include the manufacturer's name and the maximum diameter, number, voltage rating, and ampacity of conductors. Markings shall be visible after installation.

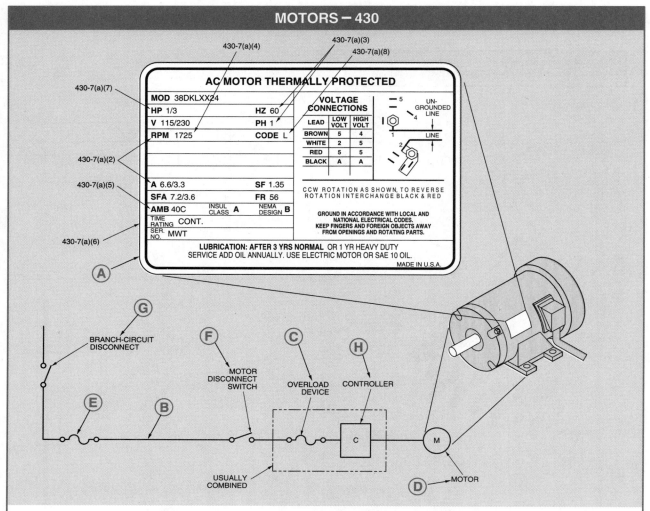

MOTORS — 430

A. **430-7** Motors shall be marked with specific information.

B. **430-22** The ampacity of branch-circuit conductors shall be not less than 125% of the motor current rating.

C. **430-32(a)(1)** Continuous-duty motors over 1 HP shall be protected against overload by a separate overload device rated at not more than 125% of the full-load current rating for motors with a service factor not less than 1.15 or a temperature rise not over 40°C, and at not over 115% for all other motors.

430-32(b)(1) Continuous-duty motors of 1 HP or less which are not permanently installed and are manually started and within sight from the controller are permitted to be protected by the branch-circuit, short-circuit, and ground-fault protective device.

430-32(b)(1), Ex. Any motor in 430-30(b)(1) is permitted on a 20 A, 120 V branch circuit.

430-32(c)(1) Continuous-duty motors of 1 HP or less which are automatically started shall be protected against overload by a separate overload device rated at not more than 125% of the full-load current rating for motors with a service factor not less than 1.15 or a temperature rise not over 40°C, and at not over 115% for all other motors.

D. **430-33** Any motor applications shall be considered as continuous duty unless the driven apparatus is such that the motor cannot operate continuously.

430-34 Where the overcurrent relay selected does not allow the motor to start, the next higher size overload relay is selected provided the trip current does not exceed 140% of the motor's full-load current rating for motors with a service factor of not less than 1.15 or a temperature rise not over 40°C, and 130% for all other motors.

430-35(a) A running overcurrent device may be shunted at starting of a manual-start motor if no hazard is introduced and the branch-circuit device of not over 400% is operative in the circuit during the starting period.

430-35(b) Shunting is not permitted if the motor is automatically started. See Exception.

430-40 Thermal cutouts and overload relays for motor-running protection not capable of opening short circuits shall be protected per 430-52 unless approved for group installation and marked with the maximum size of required protection.

E. **430-52** The motor branch-circuit overcurrent device shall be able to carry the starting current (150% – 300% per Table 430-152; absolute maximum 400% with NTDFs and Class CC fuses; and 225% with TDFs).

430-102(a) A disconnecting means shall be in sight from the controller location and shall disconnect the controller. See Ex. 1 and 2.

F. **430-102(b)** A disconnecting means shall be in sight from the motor location and the driven machinery location. See Exception.

430-107 One of the disconnecting means shall be readily accessible.

430-108 All disconnecting means shall comply with 430-109 and 430-110.

G. **430-109** The disconnecting means shall be a HP-rated switch, CB, molded-case switch, ITCB, combination controller, or manual motor controller shall be a listed device.

430-110(a) All disconnecting means shall have an ampere rating of at least 115% of the motor's FLC, taken from FLC tables per 430-6(a).

H. **430-111** A suitable switch or CB may serve as both the disconnecting means and the controller.

MOTOR DISCONNECTING MEANS – 430-109

100: Disconnecting Means – A device, or group of devices, by which the circuit conductors are disconnected from their source of supply.

100: General-Use Switch – A switch for use in general distribution and branch circuits. The ampere-rated switch is capable of interrupting its rated current at its rated voltage.

100: General-Use Snap Switch – A form of general-use switch constructed so that it can be installed in device boxes, etc.

100: Isolating Switch – A switch designed to isolate an electric circuit from the power source. It has no interrupting rating and is operated only after the circuit has been opened by other means.

FUSES OR CBs MAY SERVE
AS DISCONNECTING MEANS
• 430-109(b)

BRANCH-CIRCUIT
PANELBOARD

$\frac{1}{8}$HP, OR LESS,
STATIONARY MOTOR

TO POWER
SOURCE

CONTROLLER

20 HP, 230 V, 3 φ
DESIGN E MOTOR

DISCONNECTING MEANS SHALL BE
RATED FOR DESIGN E MOTOR
• 430-109(a)(1)

What is the minimum size disconnecting means required for the 20 HP, 230 V, 3φ Design E motor?

430-109(a)(1): 20 HP x 1.4 = 28 HP

Disconnecting means = **28 HP**

TO POWER
SOURCE

DISCONNECTING
MEANS

MOTOR
STARTER

**DO NOT
OPERATE
UNDER LOAD**

GENERAL-USE
OR ISOLATING
SWITCH PERMITTED
• 430-109(e)

SWITCH PLAINLY
MARKED

40 HP DC OR 100 HP
AC MOTOR

TO POWER
SOURCE

DISCONNECTING
MEANS

LIMITED MANUAL
MOTOR CONTROLLER
• 430-109(c)(3)

GENERAL USE SNAP SWITCH
RATED FOR AC USE ONLY
• 430-109(c)(2)

GENERAL-USE SWITCH
WITH AMPERE RATING AT
LEAST 2X MOTOR FLC
RATING MAY BE USED
• 430-109(c)(1)

2 HP, OR LESS, 300 V
STATIONARY MOTOR

430-109 The disconnecting means shall be a HP-rated switch, CB, or molded-case switch, and shall be a listed device.

430-109(a)(1) A motor-circuit switch for a Design E motor over 2 HP shall be marked for such use and shall have a horsepower rating not less than 1.4 times the motor rating for 3 HP – 100 HP motors, or not less than 1.3 for motors over 100 HP.

430-109(b) The branch-circuit overcurrent device (fuses or CBs) is permitted to serve as the switch for stationary motors of ⅛ HP or less.

430-109(c)(1) The disconnecting means is permitted to be a general-use snap switch with an ampere rating not less than twice the FLC rating of the motor for stationary motors of 2 HP and 300 V or less.

430-109(c)(2) On AC circuits, a general-use snap switch rated for AC use only may be used if the motor FLC rating is not more than 80% of ampere rating of switch.

430-109(c)(3) A listed manual motor controller with an HP rating not less than the motor rating and marked Suitable as a Disconnect may be used.

430-109(d) The disconnecting means is permitted to be a general-use switch for motors from 2 HP through 100 HP provided: (a) the motor drives a generator provided with overload protection; (b) the controller is provided with overload protection not exceeding 125% of the motor's FLC rating; and (c) separate fuses or an ITCB is set at not more than 150% of the motor's FLC rating.

430-109(e) The disconnecting means is permitted to be a general-use or isolating switch for stationary 40 HP DC or 100 HP AC motors where the switch is marked "Do not operate under load".

430-109(f) The disconnecting means is permitted to be a HP-rated plug and receptacle for cord-and-plug connected motors and other than Design E motors, and for a 2 HP or less Design E motor. See 422-22 and 440-63.

MOTOR CONTROL CIRCUITS – 430, PART F

430-71: Motor Control Circuit – The circuit of a control apparatus that carries electrical signals directing the controller's performance, but does not carry the main power current.

A. 300-3(c)(2) c. Excitation, control, relay, and ammeter conductors used in connection with the motor or starter may occupy the same enclosures as the motor-circuit conductors.

B. 430-72(a) A motor control circuit tapped from the load side of a branch-circuit, short-circuit, and ground-fault protective device shall be protected against overcurrent.

430-72(b)(1) (General Rule) Motor control conductor overcurrent protection shall not exceed the values listed in Table 430-72(b), Col A.

C. 430-72(b)(2) Control conductors that do not leave the controller are permitted to be protected by the motor branch-circuit protective device not exceeding the values listed in Table 430-72(b), Col B. Control conductors that extend beyond the controller are permitted to be protected by the motor branch-circuit protective device not exceeding the values listed in Table 430-72(b), Col C.

D. 430-72(b), Ex. 1 Motor branch-circuit overcurrent protective devices are permitted where the opening of a control circuit creates a hazard. (Example: The control circuit of a fire pump motor.)

430-72(b), Ex. 2 Transformer-fed, 2-wire primary, 2-wire secondary control conductors are considered protected if the motor branch-circuit overcurrent protective device protects secondary wires per Table 430-72(b). See 240-3(i).

430-72(c) Control circuit transformers are required to be protected per 450. See Ex. 1 – 5.

430-73 Control circuits shall have mechanical protection and be so arranged that an accidental ground will not start the motor or bypass the shutdown devices.

430-74(a) Control circuit shall be disconnected from the supply when the motor circuit is opened. If two separate devices are used, they shall be immediately adjacent. See Ex. 1 and 2.

430-74(b) Control transformers located inside the controller shall be on the load side of the disconnect for the control circuit.

WOUND-ROTOR MOTORS

Wound-Rotor Motor – A 3ϕ motor in which the rotor windings are connected to slip rings.

A. 430-23(a) For continuous duty, the conductors for the secondary of a wound-rotor motor to its controller shall be not less than 125% of the motor's FLC rating.

430-23(b) For other than continuous duty, the conductors for the secondary of a wound-rotor motor to its controller shall be not less than the values in Table 430-22(b).

B. 430-23(c) If the secondary resistor is separate from the controller, the ampacity of the conductors between the controller and resistor shall be not less than the values in Table 430-23(c).

C. 430-32(a)(1) The running overcurrent protection shall be not more than 125% for motors with service factors not less than 1.15 and motors with a temperature rise not over 40°C. For all other motors, the running overcurrent protection shall be not more than 115%.

D. 430-152 The motor branch-circuit, short-circuit, and ground-fault protective device shall be selected from Table 430-152. For NTDFs and dual-element fuses, the rating shall not exceed 150% of the motor's FLC rating.

PART – WINDING MOTORS

Part-Winding Motor – A motor arranged for starting by first energizing part of its primary (armature) winding and then energizing the rest of the winding in one or more steps.

A. 430-3 Where separate overcurrent devices are used with a part-winding motor, each half of the motor winding shall be individually protected per 430-32 and 430-37 with one half of the trip current specified.

B. 430-22(a) The size of the branch-circuit conductors for multispeed motors is based on the highest of the FLC ratings on the motor nameplate.

DIRECT CURRENT MOTORS

A. 430-22(a) Branch-circuit conductors shall be no less than 125% of the motor's FLC rating. See 430-22(a), Ex. 1 and 2.

430-22(a), Ex. 1 DC motors supplied from a rectified 1φ supply shall have conductor ampacity of: (a) 190% for 1φ half-wave rectifier; or (b) 150% for 1φ full-wave rectifier.

B. 430-32(a)(1) The running overcurrent protection shall be not more than 125% for motors with service factors not less than 1.15 and motors with a temperature rise not over 40°C. For all other motors, the protection shall be not more than 115%.

C. 430-82(c)(1) The rheostat shall be designed so the contact arm cannot be left on intermediate segments.

D. 430-82(c)(2) An under-voltage release is required on rheostats to interrupt the power supply where the motor's speed falls to ⅓ its normal rate.

430-88 The adjustable-speed control shall be arranged so the motor cannot start on a weakened field, except where so designed.

430-89 Separately excited DC motors, series motors, and motor generators and converters shall be provided with speed-limiting devices.

E. 430-152 The motor branch-circuit, short-circuit, and ground-fault protective device shall be selected from Table 430-152. For NTDFs and dual-element fuses, the rating shall not exceed 150% of the motor's FLC rating.

MOTOR FEEDER CALCULATIONS

A. 430-6(a)(1) For general motor applications, except for torque motors, Tables 430-147, 148, 149, and 150, including notes, shall be used to determine the conductor ampacity, ampere rating of switches, short-circuit and ground-fault protection, etc. The motor nameplate FLA is used to determine the overload protection. Where the motor is marked in amps but not HP, the HP rating is assumed to be the corresponding value given in Tables 430-147, 148, 149, and 150, interpolated if necessary.

B. 430-24 The minimum feeder ampacity is 125% of the FLC rating of the largest motor plus the full-load currents of all other motors. Where the motors are for types of service listed in 430-22(a), Ex. 1, similar procedures are followed, taking into consideration values obtained from Table 430-22(b). Where circuitry prevents all motors from operating at the same time, the conductor size is determined on the basis of the maximum load at any given time.

C. 430-62(a) The maximum feeder protection is determined by the largest branch-circuit protection plus the sum of the full-load currents of all other motors.

HERMETIC MOTORS

440-2: Hermetic Refrigerant Motor-Compressor – A combination compressor and motor enclosed in the same housing, without an external shaft or shaft seals, and the motor operating in the refrigerant.

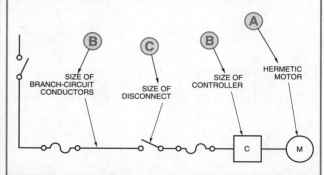

A. 440-4(a) The LRC shall be marked on the nameplate of all polyphase motors and 1φ motors over 9 A and 115 V or 4.5 A and 230 V.

B. 440-6(a) The data marked on the nameplate of the equipment shall be used to determine the rating of the controller. If no rated-load current is shown on the equipment nameplate, the rated-load current of the compressor shall be used. The full-load current marked on the equipment nameplate is used to determine the size of the branch-circuit conductors.

C. 440-12(a) The disconnecting means shall be sized based on the nameplate rated-load current or branch-circuit selection current, whichever is greater, and the LRC of the motor compressor per 440-12(a)(1)(2).

440-22(a) The motor is considered protected by the branch-circuit short circuit device rated for 175% of the rated-load current or the branch-circuit selection current, whichever is greater.

440-22(c) Where ratings shown on the manufacturer's heater table are less than those permitted by 440-22(a)(b), then the lower values listed on the manufacturer's heater table shall be used.

440-52(a)(1)(3) The overload setting shall be not over 140% of the motor's FLC for overload relays and 125% for other devices.

MOTOR CONTROLLERS

A. 430-87 (General Rule) Each motor of 600 V or less shall have its own controller.

B. 430-87, Ex. (a)(b)(c) A single controller is permissible if the motors are part of the same machine, or if in a group per 430-53(a), or if in a group in a single room in sight of the controller.

TWO OR MORE MOTORS ON ONE BRANCH CIRCUIT

A. 430-24 The minimum feeder current rating is 125% of the largest motor current plus the sum of the other motors.

B. 430-53(a) The maximum setting of the overcurrent device shall be 20 A, 120 V or less; 15 A, 600 V or less. Two or more motors not over 1 HP are permitted on the same 120 V, 20 A branch circuit or 600 V, 15 A branch circuit, if: 1. not over 6 A; 2. controller maximum branch circuit overcurrent device is not exceeded; 3. each overload is per 430-32.

C. 430-53(b) All motors have overload protection and the smallest motor is protected per 430-52, and where the branch-circuit device will not open under the most severe normal service conditions that might be encountered.

D. 430-53(c) Two or more motors of any size may be connected to one branch circuit if each has running overcurrent protection and all of the five requirements of 430-53(c) are met.

E. 430-53(d) Ampacity of taps shall not be less than 1/3 that of the branch circuit conductors, not over 25′ long, and protected from physical damage.

430-62(a) The maximum feeder protection is determined by the largest branch-circuit protection plus the sum of the full-load currents of all other motors. Use Table 430-152 or 440-22(a) for hermetic motors.

430-112 Individual disconnects are not required where provisions of 430-53(a) are met.

430-112, Ex. (a)(c) Individual disconnects are required unless the group is part of a single machine or if the group is in a single room in sight of the disconnect.

METALWORKING MACHINE TOOLS

A. 670-3(a) The nameplate of a metalworking machine tool shall show the full-load current ampere rating of the largest motor and the short-circuit interrupting capacity of the machine's overcurrent device if one is furnished.

B. 670-4(b) A machine tool is considered a single unit and shall have a disconnect which may be fed by branch circuits protected by CBs or fuses.

MISCELLANEOUS MOTOR RULES

Fluke Corporation

Motors shall be located where maintenance can be accomplished.

A. 430-12(a) Terminal housings shall be metal. Nonmetallic housings are permitted in other-than-hazardous areas.

B. 430-12(b) Table 430-12(b) gives sizes for wire-to-wire connections.

430-14(a) The motor shall be located where adequate ventilation is provided and where maintenance can be accomplished.

430-16 The motor shall be protected from the accumulation of dust.

430-43 An overload which can automatically restart a motor shall be approved for the motor and if motor restarts could not result in injury.

C. 430-132 Exposed live parts shall be guarded if over 50 V.

D. 430-142 Frames of stationary motors shall be grounded if: (1) metal enclosed wiring; (2) in wet location; (3) in hazardous location; or (4) over 150 V to ground. If the motor frame is not grounded, it shall be insulated from ground.

E. 430-145(a) Where the motor terminals of fixed motors are in metal raceways or metal enclosed cable, junction boxes for motor terminals are required. Metal raceways and Type AC cable connected shall be per 250.

F. 430-145(b) The junction box shall not be more than 6′ from the motor under conditions specified.

430-145(c) Instrument transformer secondaries and exposed metal parts of instruments or relays used in connection with motors shall be grounded per 250-170 – 250-178.

GENERATORS – 445

Generator – A device that is used to convert mechanical power to electrical power.

A. **445-1** Generators and associated wiring and equipment shall comply with 695, 700, 701, 702, and 705.

445-2 Generators shall be suitable for the locations where they are installed.

B. **445-3** The generator's nameplate shall show kVA or kW, amps, volts, frequency, power factor, maker's name, rpm, ambient temperature, etc.

C. **445-4(a)** Constant voltage DC generators (other than exciters) shall have overcurrent protection.

445-4(b) Two-wire, DC generators are permitted to have overcurrent protection in one conductor only if the overcurrent device is actuated by the entire current.

445-4(c) Generators of 65 V or less and driven by individual motors are considered to be protected by the overcurrent device protecting the motor if these devices operate when the generators are delivering not more than 150% of the FLC.

445-4(d) Control devices shall be arranged to disconnect balancer sets if they are part of the system.

445-4(e) Three-wire generators shall have overcurrent devices in each armature lead so arranged as to disconnect the generators and equalizers in the event of trouble.

445-5 The ampacity of conductors from the generator to the first overcurrent device shall not be less than 115% of the nameplate rating, except not less than 100% where design and operation prevent overloading.

445-6 Live parts of generators operating at over 50 V to ground shall not be exposed to accidental contact where accessible to unqualified persons.

445-7 Where necessary for the safety of attendants, insulating mats or platforms shall be provided per 430-133.

445-8 Bushings shall be used to protect wires passing through openings from sharp edges in enclosures, boxes, etc.

MOTOR OVERLOAD PROTECTION

430-34 Where the overload relay will not start the motor or carry the load, the next higher size overload relay is permitted if the trip current does not exceed 140% of the motor's FLC rating for motors with a service factor not less than 1.15 or motors with a temperature rise not over 40°C. For all other motors, the trip current shall not exceed 130%.

430-35(a) The overload protection may be shunted at the starting of a manual-start motor if no hazard is introduced and the branch-circuit device of not over 400% is operative in the circuit during starting.

430-35(b) The overload protection shall not be shunted if the motor is automatically started. See Exceptions.

430-36 Where fuses are used for motor overload protection in a 3-wire, 3φ AC system with one wire grounded, a fuse shall be used in the grounded conductor.

430-37 Three motor overload protective devices shall be used to protect 3φ motors unless protected by other approved means. The number of overcurrent devices other than fuses is governed by Table 430-37.

MOTORS OVER 600 V

A. **430-122** Controllers shall be marked with the maker's name, voltage, current, HP rating, and other necessary data and control voltage. See 430-8.

430-124 The ampacity of conductors shall not be less than the current at which the motor overload protective device is selected to trip.

B. **430-125(a)** Motors shall have overload protection and the motor, motor circuit conductors, and control apparatus shall have fault current protection. This protection shall be coordinated and automatically operated except a vital motor may use a supervised annunciator or alarm.

430-125(b)(1) Each motor shall be protected against overloads or failure to start by a thermal protector integral with the motor or external sensing devices or both.

430-125(b)(2) Secondary circuits of wound-rotor motors are considered to be protected by motor overcurrent protection.

430-125(b)(3) Operation of the overload device shall open all ungrounded conductors at the same time.

430-125(b)(4) Overload sensing devices shall not automatically restart the motor unless restarting is no hazard.

430-125(c)(1)(b) Fault current protection is required. It may be provided by suitable type and rated fuses that cannot be serviced while energized.

C. **430-125(c)(1)(c)** A CB is permitted for fault current protection. It may sense fault current by integral or external sensing elements.

D. **430-127** The disconnect for the controller shall be lockable in the open position.

CRANES AND HOISTS

A. **430-112, Ex. a.** Individual disconnects are not required where a number of motors drive a single machine.

B. **610-11** Supply conductors shall be in raceways, AC cable with insulated EGC, MI, or MC cable, except: (a) bar contact conductors; (b) short lengths of open conductors at resistors, collectors, etc.; (c) for flexibility, flexible stranded wire installed in FMC, LTFMC, LTFNMC, multiconductor cable, or approved PVC enclosure; (d) multiconductor cable to pendant pushbutton station; or (e) flexibility for moving parts.

610-14(a) Table 610-14(a) provides allowable ampacity for conductors.

610-14(d) Bare conductors No. 6 span 0' – 30'; No. 4 span 30' – 60'; No. 2 over 60'.

610-14(e)(2) The ampacity of supply conductors shall be not less than the sum of the short-time full-load current rating for the largest motor, or group, for a single crane motion plus 50% of that for the next larger motor or group.

C. **610-15** Where there is more than one motor, a common return conductor is permitted.

610-21(f) The grounded track may serve as a conductor if the supply is by way of an insulating transformer not over 300 V and other conductors are insulated. See Conditions (1 – 4).

D. **610-31** The disconnecting means shall be operable from the ground and in sight of the crane and runway contact conductors.

610-32 A lockable motor disconnect is required in leads from runway contact conductors or other power supply on all cranes and monorail hoists. See Exception. Where the disconnect is not readily accessible from the operating station, means are required at the operating station to open the power circuit of the crane or monorail hoist.

610-33 The minimum rating of the disconnecting means shall be 50% of the combined short-time ampere ratings of all motors and 75% of the sum of the short-time ampere ratings of motors for single crane operation.

610-41 An overcurrent device is required for the main contact conductors.

E. **610-42(a)** Each motor shall have branch-circuit protection.

610-42(b) Two motors operating as a unit may have a single overload protector for taps to brake coils.

610-55 A limit switch or other device is required on a hoist unit.

610-61 Grounding of metal parts is required per 250.

ELEVATORS, DUMBWAITERS, ESCALATORS, AND MOVING WALKS

A. **620-5** There shall be sufficient, clear working space around control panels to permit safe and convenient access to all live parts for maintenance and adjustment. The minimum space shall be per 110-26. Where the minimum working space cannot be provided for escalator and moving-walk panels and the panel is in the same space as the drive machine, the clearances may be waived where the entire panel and disconnecting means are arranged to be readily removed from the machine space and has flexible leads to all external connections.

B. **620-21(a)(3)(d)** Conductors may be grouped and bound together without being installed in a raceway if supported every 3' and free from physical damage.

C. **620-32** 50% fill is permitted in a wireway. Vertical runs require support at least every 15'. 362-5 shall not apply.

D. **620-33** 40% fill is permitted in raceways.

620-35 The 30' length restriction of 374-2 and number of wires restriction of 374-5 do not apply to auxiliary gutters.

620-36 Conductor of power, control, lighting, and signal systems not over 600 V may be run in the same raceway or cable if insulated for the maximum voltage found in the cable or raceway system.

OPERATING REQUIREMENTS

A. **620-13(a)** The ampacity of conductors for a single motor shall conform to 430-22(a)(b).

620-13(d) Conductors supplying several motors shall comply with the standard rule governing motor feeders unless a demand factor is allowed per 620-14.

620-21 In general, all elevator wiring shall be in RMC, IMC, EMT, RNMC, wireways, or in Types MC, MI, or AC cable. See Applications (a), (b), and (c).

B. **620-51** An identified disconnect is required for the power supply to each unit.

C. **620-51(a)** The disconnect shall be an enclosed, externally operable motor switch or a CB which can be locked in the open position.

620-51(b) The disconnect shall be readily accessible to qualified persons.

D. **620-52(b)** For power from more than one source and multiple disconnects, a clearly legible warning sign shall read: "WARNING: PARTS OF THE CONTROLLER ARE NOT DE-ENERGIZED BY THIS SWITCH." The sign shall be mounted on or adjacent to the disconnecting means.

620-61(b)(1) Elevators and dumbwaiters are protected on an intermittent-duty basis. Elevator motors shall be classified as intermittent-duty motors and protected against overload per 430-33.

620-61(b)(2) Escalator and moving-walk driving motors shall be classified as continuous-duty motors and shall be protected against overload per 430-32.

620-82 In general, operating equipment shall be grounded per 250.

620-91(c) Where emergency power is provided, a disconnect is required by 620-51. It shall disconnect both emergency and normal power.

VOLTAGE TRANSFORMERS

A. **450-3(c)** Voltage transformers shall be fused on the primary side when installed indoors. See 384-32.

TRAVELING CABLES AND CONTROL HOISTWAY

620-2: Control System – The overall system governing the starting, stopping, motion direction, acceleration, speed, and retardation of the moving member.

620-2: Motor Controller – The operative units of the control system consisting of the starter device(s) and power conversion equipment used to drive an electric motor or the pump for hydraulic control equipment.

620-2: Operation Controller – The electric device(s) for the part of the control system that initiates starting, stopping, and direction of motion in response to a signal from an operating device.

620-2: Operating Device – The car switch, pushbuttons, etc. used to activate the operation controller.

A. **620-3** Control and signal voltage limited to 300 V except for special limited-energy circuits. See Applications (a), (b), and (c).

B. **620-11(b)** Traveling cables used as flexible connections shall be of the types listed in Table 400-4 as elevator cable or be of other approved types.

C. **620-12(a)** Traveling cables: (1) for lighting circuits, minimum No. 14 or equivalent in parallel conductors No. 20 or larger; (2) for other circuits, minimum size shall be No. 20.

620-21(a)(1)(a) FMC, LTFMC, and LTFNMC are permitted in the hoistway or in the car for limit switches, interlocks, doors, etc.

D. **620-37(a)(c)** Only wiring directly related to elevator or related systems shall be permitted inside the hoistway and the machine room. Main power feeders supplying elevator or dumbwaiters shall be installed outside the hoistway unless necessary for elevators with driving machine motors located in the hoistway, on car, or on counterweight, or by special permission.

620-41 Traveling cables shall be suspended at the car and hoistway ends to reduce strain.

620-41(1)(2)(3) Traveling cables supported by: (1) steel supporting fillers; (2) looping cables around supports for unsupported lengths less than 100′; (3) means that automatically tighten around cable (for cables up to 200′) when tension is increased.

620-53 A single disconnecting means shall be provided for disconnecting all ungrounded car light and accessory power supply conductors for each unit.

620-81 Metal raceways and cable assemblies attached to elevators shall be bonded to the grounded metal frame of car they contact.

SECONDARY TIES

450-3(a), FPN 1 The overcurrent protection of conductors is per 240-3, 240-21, and 240-100.

A. 450-6 A secondary tie is a 600 V or less circuit between phases that connects two power sources or power supply points. As used in this section, the word "transformer" means one transformer or a bank of transformers operating as a unit.

450-6(a) The circuits shall be provided at each end with overcurrent protection per 240. See Applications (1), (2), (3), (4), and (5).

B. 450-6(a)(1) The rated ampacity of the tie shall not be less than 67% of the rated secondary current of the largest transformer connected to the secondary tie system if overcurrent protection is not provided per 240.

450-6(a)(2) Where the load is connected to the tie between transformer supply points and overcurrent protection is not provided per 240, the rated ampacity of the tie shall not be less than 100% of the rated secondary current of the largest transformer connected to the secondary tie system.

C. 450-6(a)(3) Limiters or CBs having comparable current characteristics may be used under conditions described in (a)(1) and (a)(2).

450-6(a)(4) Ordinarily, corresponding phase wires shall be joined together at tap points.

450-6(a)(4)(b) Loads shall be connected to individual conductors and without limiters where the total ampacity of the ties is not less than 133% of the rated secondary current of the largest transformer, and the total load of the taps does not exceed this value.

450-6(a)(5) If the operating voltage is over 150 V to ground, a disconnect switch is required at each end.

450-6(b) For secondary ties, a CB set at not more than 250% of the rated secondary current of the transformers shall be provided. Additionally, an automatic CB shall be provided.

IMPEDANCE OF TRANSFORMERS

$$Z = \frac{input\ voltage}{output\ rating}$$

$$Z = \frac{120}{2400}$$

$$Z = .05\ or\ 5\%$$

IMPEDANCE – The impedance that appears on the nameplate of a transformer is determined by applying the following steps:

1. Connect an ammeter across (in series) the secondary winding.
2. Connect a variable rheostat in the primary winding circuit.
3. Connect a voltmeter across the primary winding.
4. Apply a voltage to the primary circuit. Adjust the rheostat to increase the voltage across the primary winding until the secondary winding current reaches its rated current.

5. Read the voltage across the primary winding.
6. Divide the applied primary voltage by the rated primary winding voltage.

 For example, a transformer has a 2400 V, 10 A primary rating and a 240 V, 100 A secondary rating. A voltage of 120 V is needed across the primary winding for the ammeter in the secondary winding circuit to read 100 A. Dividing 120 by 2400 equals .05 or 5% impedance.

OVERCURRENT PROTECTION OF TRANSFORMERS (OVER 600 V) – 450-3(a)

450-3(a) Transformers over 600 V shall have primary and secondary protection per Table 450-3(a). Where the fuse or CB rating exceeds a standard rating, the next higher rating is permitted.

Table 450-3(a) Primary protection of 600% for CBs and 300% for fuses is permitted if not over 6% Z and 400% for CBs and 300% for fuses if over 6% but not over 10% Z. See Table 450-3(a)(1) for secondary protection. See Table 450-3(a), Notes 1 – 5.

Supervised installations require individual protection on the primary side. The continuous rating of the primary fuse shall not be over 250% of the transformer and for CBs, the rating shall not be over 300%.

Table 450-3(a), Note 1 The next higher fuse or CB rating is permitted if the fuse or CB does not correspond to the next higher standard size.

OVERCURRENT PROTECTION OF TRANSFORMERS (600 V OR LESS) – 450-3(b)

240-3 All conductors feeding in the primary or out of the secondary shall be protected at their rated ampacities. This protection shall be provided where the conductors receive their supply per 240-21.

Table 450-3(b) Each transformer of 600 V or less is required to have individual primary overcurrent protection. This overcurrent device is to be rated or set at not over 125% of the rated primary current except: (1) if the primary current is 9 A or more and 125% is not a standard size fuse or CB, then the next larger size device may be used. Where the primary is less than 2 A, the overcurrent device may be rated or set at not more than 300% of the rated primary amps; (2) the primary circuit protection may also protect transformer if of proper size; (3) if secondary is protected per Table 450-3(b).

Table 450-3(b), Notes 1 – 3 Individual primary overcurrent protection is not required if the primary feeder is protected at 250% of the primary rated current and the secondary is protected at not over 125% of the rated secondary current. Individual transformer primary protection is not required if the transformer has coordinated thermal overload protection by the manufacturer, is arranged to open the primary circuit, and if the primary overcurrent device is not over 6 times the rated primary current for transformers with not more than 6% Z and not more than 4 times the rated primary current for transformers having more than 6% but not more than 10% Z except: where the secondary current is 9 A or more and 125% is not a standard size fuse or CB, then the next larger device may be used. Where secondary is less than 9 A, the overcurrent device may be rated or set at not more than 167% of the rated primary current.

DRY-TYPE TRANSFORMERS – INSTALLED INDOORS

CASE

CORE

COILS

TERMINAL
ARRANGEMENT

MOUNTING
FRAME

ABB Power T&D Company Inc.

Dry-type transformers may be installed indoors when they are rated at not more than 112½ kVA and not more than 600 V, and the transformer is completely enclosed except for ventilating openings.

2″ x 12″
ROOF JOISTS

COMBUSTIBLE
MATERIAL

$\frac{5''}{8}$ PLYWOOD

12″ MINIMUM
CLEARANCE

Ⓐ

VENTILATION
OPENING

50 kVA, 480 V
DRY-TYPE
TRANSFORMER

FIRE-RESISTANT,
HEAT-INSULATING
MATERIAL

WALLBOARD

Ⓑ

NO MINIMUM
CLEARANCE

Ⓒ

NO MINIMUM
CLEARANCE

VENTILATION
OPENING

COMPLETELY
ENCLOSED

50 kVA, 480 V
DRY-TYPE
TRANSFORMER

FIRE-RESISTANT
ROOM

Ⓓ

TRANSFORMER OVER 112½
kVA WITH CLASS 155 OR
HIGHER INSULATION

WOOD
PANELING

NON FIRE-
RESISTANT
ROOM

12′
MINIMUM

Ⓔ

6′
MINIMUM

6′
MINIMUM

TRANSFORMER OVER 112½
kVA WITH CLASS 155 OR
HIGHER INSULATION

WOOD
PANELING

NON FIRE-
RESISTANT
ROOM

Ⓕ

COMPLETELY ENCLOSED
TRANSFORMER OVER 112½
kVA WITH CLASS 155 OR
HIGHER INSULATION

A. 450-21(a) Dry-type transformers installed indoors and rated at 112½ kVA or less shall be a minimum of 12″ from combustible material.

B. 450-21(a) The 12″ minimum is not required when transformers are separated from combustible material by fire-resistant, heat-insulating material.

C. 450-21(a) The 12″ minimum is not required when the transformer is 600 V or less and completely enclosed.

D. 450-21(b) Dry-type transformers over 112½ kVA rating shall be installed in fire-resistant rooms.

E. 450-21(b), Ex. 1 Transformers with Class 155 or higher insulation are permitted to be installed in a non fire-resistant room provided they are separated from combustible material by a minimum of 6′ horizontally and 12′ vertically.

F. 450-21(b), Ex. 2 Transformers with Class 155 or higher insulation and completely enclosed except for ventilation openings are permitted to be enclosed in a non fire-resistant room.

TRANSFORMERS IN PARALLEL

OVERCURRENT PROTECTION

Ⓐ

TRANSFORMERS CONNECTED IN PARALLEL

T1   T2

A. 450-7 Transformers may be connected in parallel and protected as a unit when their electrical characteristics are suitable, provided overcurrent protection for each transformer meets requirements of 450-3(a) or 450-3(b).

OIL-INSULATED TRANSFORMERS – INSULATED INDOORS AND OUTDOORS

OIL-INSULATED TRANSFORMER

450-26 (General Rule) A transformer vault is required for an oil-insulated transformer indoors.

450-26, Ex. 1 If the transformer is not over $112\frac{1}{2}$ kVA, the thickness of the vault wall need be only 4".

450-26, Ex. 2 A vault is not required if not over 600 V, not over 10 kVA in combustible area, or 75 kVA in fire-resistant area.

450-26, Ex. 3 Electric furnace transformers totaling not more than 75 kVA are permitted to be installed in a fire-resistive location if provisions are made to prevent the spread of an oil fire.

450-26, Ex. 4 A vault is not required if the transformer is in a suitable detached building accessible only to qualified persons.

450-27 Oil-insulated transformers installed outdoors shall be installed so combustible material, combustible buildings, parts of buildings, door and window openings, fire escapes, etc. are safeguarded from transformer oil fires when the transformer is installed, attached to, or adjacent to a building or combustible material, or on roofs.

AUTOTRANSFORMERS

208 V SUPPLY 7.5 kVA   Ⓐ

36.06 A   36.06 A

$I = \frac{VA}{E}$

$I = \frac{1000}{208}$

$I = \textbf{4.81 A}$

$kVA = \frac{E \times I}{1000}$

$kVA = \frac{208 \times 36.06}{1000}$

$kVA = \frac{7500.48}{1000}$

$kVA = \textbf{7.5 kVA}$

4.81 A

32 V
31.25 A

H1   H2 H3   H4   X1   X2 X3   X4

208 V

$kVA = \frac{E \times I}{1000}$

$kVA = \frac{240 \times 31.25}{1000}$

$kVA = \frac{7500}{1000}$

$kVA = \textbf{7.5 kVA}$

$I = \frac{VA}{E}$

$I = \frac{1000}{32}$

$I = \textbf{31.25 A}$

Ⓑ

31.25 A   240 V LOAD 7.5 kVA

A. 210-9 Autotransformers are not permitted to supply branch circuits unless the grounded conductor is electrically connected to both supply and load.

B. 210-9, Ex. 1 Autotransformer is permitted to extend or add individual branch circuit to the existing installation for equipment load without an electrically continuous grounded conductor when transforming nominal 208 V to 240 V or 240 V to 208 V.

210-9, Ex. 2 Autotransformers in industrial occupancies are permitted to supply nominal 600 V loads from nominal 480 V systems and nominal 480 V loads from nominal 600 V systems, without connection to a similar grounded conductor, if installed by a qualified person.

ASKAREL OR NONFLAMMABLE TRANSFORMERS – INSTALLED INDOORS

ASKAREL-INSULATED TRANSFORMER INSTALLED INDOORS

PRESSURE-RELIEF VENT

Ⓐ

450-24 Fluid-insulated transformers insulated with a dielectric fluid identified as nonflammable are permitted indoors and outdoors. When over 35,000 V and indoors, a vault is required. In this section a nonflammable dielectric fluid does not have a flash or fire point and is not flammable in air.

A. 450-25 Askarel-insulated transformers 25 kVA or over shall have pressure-relief vent. If more than 35,000V, they shall be installed in an approved vault. Where askarel-insulated transformers are installed in a poorly-ventilated area, means shall be provided to absorb any gas, or a chimney or flue shall be connected to carry such gases outside buildings.

HIGH-VOLTAGE INSTALLATION IN APPROVED TRANSFORMER VAULT

A. **110-34(e)** Bare conductors are insulated by elevation; 8'-6" up to 7500 V. Minimum depth of working space to 9000 V is 4'. See Table 110-34(a).

230-202(b) For services exceeding 600 V, wiring methods shall be per 710-4.

B. **230-203** Warning signs that read "DANGER HIGH VOLTAGE KEEP OUT" shall be provided wherever unauthorized persons may come in contact with energized parts.

C. **300-37** Conductors above ground shall be in RMC, IMC, PVC, cable trays, cablebus, MI cable, or other identified raceways. Bare conductors and busbars permitted where accessible only to qualified persons.

300-50 Undergound conductors to be suitable for voltage and prevailing conditions and, if buried, per 310-7. Over 600 to 22 kV conductors run underground in approved direct burial cable meeting requirements of 250-2(d) shall be at least 30" deep. See Table 300-50 and exceptions for other wiring methods and unusual conditions.

450-8(c) Energized parts shall be guarded per 110-17 and 110-34.

450-10 Exposed noncurrent-carrying metal parts of transformer installations shall be grounded.

D. **450-42** Walls and roof of vaults shall be of fire-resistive construction with a minimum fire resistance of 3 hours. See Ex. for 1-hour rating. Concrete floors shall be at least 4" thick.

E. **450-43(a)** Vault doors shall have a minimum fire rating of 3 hr. The door sill shall be at least 4" high. Vault doors shall be locked, accessible only to qualified persons, and out-swinging.

F. **450-45(c)** Natural ventilation shall be no less than 3 sq in. per kVA. For under 50 kVA, the minimum shall be 1 sq ft.

450-47 Foreign pipes or ducts shall not pass through a transformer vault.

G. **450-48** Material shall not be stored in transformer vaults.

490-24 For minimum clearance of live parts, see Table 490-24. Listed equipment is exempt.

SURGE ARRESTERS

A. **280-3** Where used at a point on a circuit, a surge arrester shall be connected to each ungrounded conductor.

280-11 Surge arresters are permitted indoors or outdoors and accessible only to qualified persons unless listed for accessible locations.

B. **280-12** The conductor from the surge arrester to the line and ground shall be as short as possible and contain no sharp bends.

C. **280-21** A minimum of No. 14 Cu or No. 12 Al is required for line and ground connecting conductors. The grounding conductor shall be one of the following: 1. grounded service conductor; 2. grounding electrode conductor; 3. service grounding electrode; or 4. equipment grounding bus in service equipment.

D. **280-25** Grounding connections shall be per 250 except as given in 280.

PROTECTION REQUIREMENTS

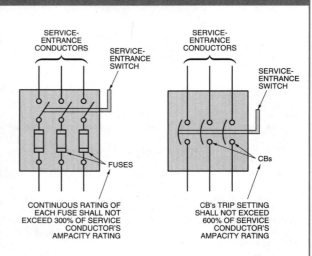

230-208 Fuses used for short-circuit protection shall not exceed three times the ampacity of the conductor. For CBs, the trip setting shall not exceed six times the ampacity of the conductors.

TRANSFORMER FEEDER TAP — 25' RULE

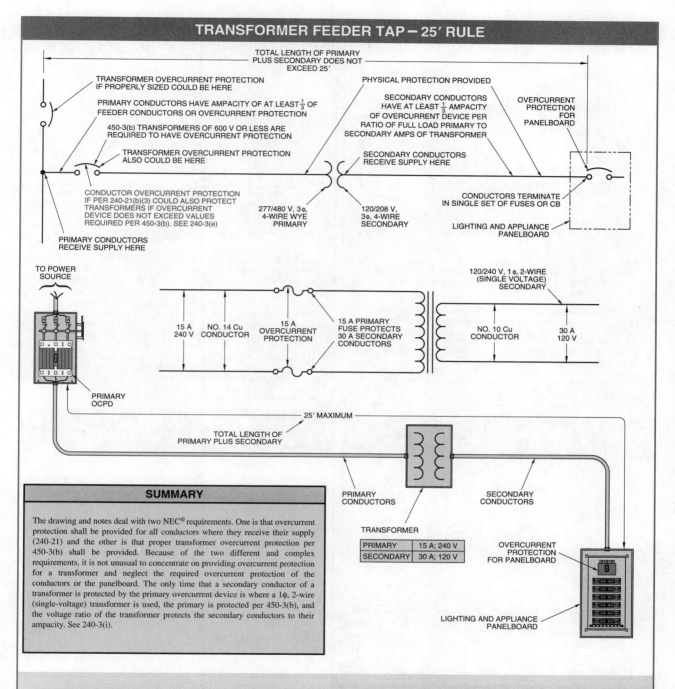

TOTAL LENGTH OF PRIMARY PLUS SECONDARY DOES NOT EXCEED 25'

TRANSFORMER OVERCURRENT PROTECTION IF PROPERLY SIZED COULD BE HERE

PHYSICAL PROTECTION PROVIDED

PRIMARY CONDUCTORS HAVE AMPACITY OF AT LEAST $\frac{1}{3}$ OF FEEDER CONDUCTORS OR OVERCURRENT PROTECTION

SECONDARY CONDUCTORS HAVE AT LEAST $\frac{1}{3}$ AMPACITY OF OVERCURRENT DEVICE PER RATIO OF FULL LOAD PRIMARY TO SECONDARY AMPS OF TRANSFORMER

OVERCURRENT PROTECTION FOR PANELBOARD

450-3(b) TRANSFORMERS OF 600 V OR LESS ARE REQUIRED TO HAVE OVERCURRENT PROTECTION

TRANSFORMER OVERCURRENT PROTECTION ALSO COULD BE HERE

SECONDARY CONDUCTORS RECEIVE SUPPLY HERE

CONDUCTOR OVERCURRENT PROTECTION IF PER 240-21(b)(3) COULD ALSO PROTECT TRANSFORMERS IF OVERCURRENT DEVICE DOES NOT EXCEED VALUES REQUIRED PER 450-3(b). SEE 240-3(e)

CONDUCTORS TERMINATE IN SINGLE SET OF FUSES OR CB

277/480 V, 3φ, 4-WIRE WYE PRIMARY

120/208 V, 3φ, 4-WIRE SECONDARY

LIGHTING AND APPLIANCE PANELBOARD

PRIMARY CONDUCTORS RECEIVE SUPPLY HERE

TO POWER SOURCE

120/240 V, 1 φ, 2-WIRE (SINGLE VOLTAGE) SECONDARY

15 A 240 V   NO. 14 Cu CONDUCTOR   15 A OVERCURRENT PROTECTION   15 A PRIMARY FUSE PROTECTS 30 A SECONDARY CONDUCTORS   NO. 10 Cu CONDUCTOR   30 A 120 V

PRIMARY OCPD

25' MAXIMUM

TOTAL LENGTH OF PRIMARY PLUS SECONDARY

PRIMARY CONDUCTORS

SECONDARY CONDUCTORS

TRANSFORMER

| PRIMARY | 15 A; 240 V |
| SECONDARY | 30 A; 120 V |

OVERCURRENT PROTECTION FOR PANELBOARD

LIGHTING AND APPLIANCE PANELBOARD

SUMMARY

The drawing and notes deal with two NEC® requirements. One is that overcurrent protection shall be provided for all conductors where they receive their supply (240-21) and the other is that proper transformer overcurrent protection per 450-3(b) shall be provided. Because of the two different and complex requirements, it is not unusual to concentrate on providing overcurrent protection for a transformer and neglect the required overcurrent protection of the conductors or the panelboard. The only time that a secondary conductor of a transformer is protected by the primary overcurrent device is where a 1φ, 2-wire (single-voltage) transformer is used, the primary is protected per 450-3(b), and the voltage ratio of the transformer protects the secondary conductors to their ampacity. See 240-3(i).

240-3(i) The secondary conductors of a 1φ, 2-wire (single-voltage) secondary are considered protected if the primary overcurrent conforms to 450-3 and is not more than the current determined by multiplying the secondary conductor amps by the secondary-to-primary transformer voltage ratio.

240-21 All conductors are required to be protected where they receive their supply.

240-21(b)(3) A transformer feeder tap is permitted if the primary plus secondary conductors are not over 25' long and all of the following conditions are met:

(a) Primary conductors have ampacity of at least one third of the rating of the overcurrent device protecting the feeder conductors.

(b) The transformer secondary conductors have ampacity that when multiplied by the ratio of the secondary to primary voltage is not less than one third the

ampacity of the rating of the overcurrent device protecting the feeder conductors.

(c) The total length of one primary plus one secondary conductor that is protected at its ampacity is not over 25'.

(d) Physical protection is provided for both the primary and the secondary conductors.

(e) Secondary conductors terminate in a single set of fuses or CBs which limits load to ampacity per 310-15.

384-16(a) Lighting and appliance panelboards are required to be individually protected on their supply side by an overcurrent device not exceeding the rating of the panelboard. See Exceptions.

SERVICES OVER 600 V

A. 230-202(a) Service-entrance cables shall not be smaller than No. 6 unless in multiconductor cable, which shall not be smaller than No. 8.

B. 230-202(b) Wiring methods for services over 600 V shall be per 300-37. For aboveground conductors, wiring methods shall be: RMC, IMC, EMT, RNMC, cable tray, busbars, cablebus, other identified raceways, or open runs of MC cable suitable for use and purpose. Underground conductors shall be identified for voltage and conditions of use.

C. 230-203 Signs worded "Danger – High Voltage Keep Out" shall be posted where unauthorized persons might come into contact with energized parts.

D. 230-205(b) A CB or equivalent disconnecting means shall have a fault-closing rating not less than the maximum short-circuit current available at its supply terminals.

E. 230-6 Raceways in 2″ of concrete or brick are considered outside the building.

SERVICES OVER 600 V — OVERCURRENT PROTECTION

230-205(b) The disconnect shall open all ungrounded conductors simultaneously and shall have a fault-closing rating not less than the maximum short-circuit current available at its supply terminals. Fuses are permitted to contribute to the fault-closing rate of disconnect.

230-208(a) Short-circuit protective devices shall be installed on the load side or as an integral part of the service disconnect. The device shall protect all ungrounded conductors and be rated for interrupting values at its location. Equipment used for protection of service-entrance conductors shall be per 710. See 710-20 and 710-21.

490-21(a)(1 – 9) If the CB is closed on an overload, it shall be free to open the circuit. This may be done by using trip-free CBs. They shall have ratings not less than the maximum available fault current at their terminals and clearly indicate if open or closed.

SERVICES OVER 600 V — ISOLATING SWITCHES

A. 230-204(a) Where the service disconnect is an oil switch or an air or oil CB, an air-break isolating switch is required on the supply side of the disconnect and all service equipment.

230-204(c)(d) Air-break isolators shall be accessible to qualified persons only and arranged for grounding on the load side.

SERVICES OVER 600 V — FUSED ISOLATING SWITCH

A. 230-204(b) Fuses shall be permitted to serve as an isolating switch if they can be operated as a disconnecting switch.

SERVICES OVER 600 V — DISCONNECTING MEANS

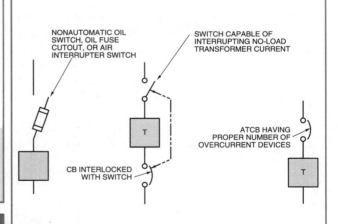

110-31(b)(1) Outdoor electrical installations shall comply with 225 if open to unqualified persons.

110-31(b)(2) Outdoor installations with exposed live parts shall be accessible to qualified persons only and shall also comply with 110-34, 110-35, and 110-36.

230-208(a) Disconnecting means for over 600 V services shall open all ungrounded conductors of the circuit simultaneously and shall have a sufficient fault-closing rating for the maximum available short-circuit current. See 230-205. Some acceptable forms of disconnecting means and fault protection are: (1) oil fuse cutout; (2) nonautomatic oil switch; (3) air load interrupter switch; and (4) automatic trip CB.

NUMBER OF OVERCURRENT DEVICES TO PROTECT CIRCUIT CONDUCTORS – 600 V OR LESS

240-20(a) For circuits of 600 V or less, overcurrent protection is required in series with each ungrounded conductor.

240-20(b) For circuits of 600 V or less, CBs are required to open all conductors of the circuit. Single-pole breakers with handle ties are permitted for 3-wire DC or 1φ circuits, and lighting and appliance branch circuits connected to 4-wire, 3φ systems or 5-wire, 2φ systems if the circuits are supplied from a system having a grounded neutral, and the voltage does not exceed that permitted by 210-6.

240-100(a) For circuits over 600 V, overcurrent protection is required for each ungrounded conductor.

240-100(a)(1) (General Rule) On circuits over 600 V, where CBs are used for AC 3φ circuits, they shall have at least 3 overcurrent relays operated from 3 current transformers. See Ex. 1, 2.

240-100(a)(2) For circuits over 600 V, where fuses are used for overcurrent protection, a fuse is required in each ungrounded conductor.

CAPACITORS – MOTOR CIRCUITS

460-8(a) The rating of capacitor circuit conductors shall be at least ⅓ that of the motor circuit conductors and 135% of the capacitor current rating.

460-8(b)(1), Ex. An overcurrent device is not required in each conductor if connected on the load side of the motor controller.

A. 460-8(c)(1), Ex. No separate disconnect is required for a capacitor.

B. 460-9 If a capacitor is connected to the load side of a motor overload device, the rating or setting of the device is computed from the improved power factor circuit. For the purposes of 430-22, the effect of the capacitor shall be disregarded.

OVER 600 V – GROUNDING OR SHIELDING TAPE

300-40 Metallic shielding and all conducting or semiconducting static shielding parts of shielded cable shall be stripped back at terminations a safe distance as determined by the circuit voltage. All metallic shielding components shall be grounded.

CAPACITORS – OVER 600 V

460-2(a) Vaults or outdoor enclosures per 300, Part B are required for capacitors containing over 3 gal. of flammable liquid. This limit shall apply to any single unit in an installation of capacitors.

460-24(a)(1 – 4) A group-operated switch is rated to carry continuously at least 135% of the rated capacitor bank current, rated to interrupt load of bank, rated for minimum inrush currents, and rated to carry capacitor fault currents.

460-25 Overcurrent protection is required to detect and interrupt dangerous fault current. A 3φ device is permitted to protect a 3φ bank.

460-26 The capacitor's nameplate shall list the maker's name, rated voltage, kVar or A, frequency, phases, and type and amount of liquid.

460-27 If grounded, the capacitor enclosure and neutrals shall comply with 250, except if on a structure operating at other than ground potential.

460-28(a)(b) A discharge means is required to reduce the residual voltage to 50 V or less in 5 min or less. The discharge circuit shall be permanently connected or have an automatic connection after the capacitor bank is disconnected from the source.

CAPACITORS – 600 V OR LESS

460-2(b) Live parts shall be enclosed or guarded except when enclosed and accessible only to qualified personnel.

A. 460-8(a) The ampacity of conductors shall be at least 135% of the capacitor current rating.

B. 460-8(b)(2) Overcurrent protection shall be as low as practicable.

C. 460-8(c) The disconnecting means shall open all ungrounded conductors simultaneously. Its continuous ampacity shall not be less than 135% of the rated capacity current.

460-10 Capacitor cases shall be grounded per 250, except if on a structure operating at other than ground potential.

460-28(a) For over 600 V, a discharge means is required to reduce the residual voltage to 50 V or less in 5 min or less.

OVER 600 V – BRAID-COVERED INSULATED CONDUCTORS, OPEN RUNS

A. 300-39 Where braid-covered insulated conductors are used for open runs for circuits over 600 V, the braid shall be flame retardant.

B. 300-39 Braid shall be stripped back a safe distance and shall not be less than 1″ for each kilovolt to ground in circuit.

RESISTORS AND REACTORS (600 V AND UNDER)

A. **470-2** Resistors and reactors are not to be installed where exposed to physical damage.

B. **470-3** A space of 12″ is required between resistors and reactors and any combustible material. If the space is less than 12″, a thermal barrier is required.

470-4 Insulated conductors shall be rated for 90°C when used for connections between resistance elements and controllers, except for motor starting service, other insulations permitted.

MOTOR-GENERATOR WELDER

A. **630-11(a)** The supply conductor rating for an M-G arc welder having 50% duty cycle is .75 of the rated primary amps per the welder nameplate.

B. **630-12(a)** The overcurrent protection of the welder shall be not over 200% of the rated primary current. Protection may also protect the supply conductors.

C. **630-12(b)** The welder supply conductors shall be protected at not over 200% of the conductor rating.

D. **630-13** Each welder shall have a disconnect. The rating shall not be less than required to accommodate the overcurrent protection required by 630-22.

TRANSFORMER ARC WELDER

630-11(a) The ampacity of supply conductors shall not be less than the rated primary current times the duty-cycle factor stated in the table.

A. **630-12(a)** The welder overcurrent protection shall not be over 200% of the primary current.

B. **630-12(b)** The conductors shall be protected at not over 200% of their rating.

C. **630-13** The disconnect switch or CB shall have an ampere rating sufficient to accommodate the required overcurrent protection per 630-12.

RESISTANCE WELDER

A. **630-31(a)(1)** The minimum rating of the supply conductors for a welder operating at varying duty is 70% of the rated primary current and 50% for manual welders.

B. **630-31(a)(2)** The rating of supply conductors for specific operation per the duty cycle table is 0.71 for 50% duty and 0.45 for 20% duty.

C. **630-32(a)** The maximum rating of the overcurrent device is 300% of the rated primary current.

D. **630-33** The rating of the disconnect shall not be less than that of the supply conductors as required per 630-31.

GROUP OF TRANSFORMER ARC WELDERS

A. **630-11(b)** The conductor rating for a group of welders shall be determined in each case by the nature of use. Adequate conductor size should be obtained if the following percentages of current values determined per 630-11(a) are applied: 100% for the two largest welders, plus 85% for the third largest, plus 70% for the fourth largest, plus 60% for all others. See 630-11, FPN.

B. **630-12(b)** The overcurrent device shall be set at not more than 200% of the conductor rating.

GROUP OF RESISTANCE WELDERS

A. **630-31(a)(2)(b)** The conductor rating is based on 630-31(a)(2) value obtained for largest welder plus 60% of that for the others.

B. **630-32(b)** The overcurrent setting shall not be over 300% of the conductor rating.

C. **630-33** The rating of the disconnect shall not be less than that of the supply conductors as required by 630-31.

MOBILE HOMES – 550

A. 550-2: Mobile Home – A mobile home is a transportable factory-assembled structure or structures, constructed on a permanent chassis, for use as a dwelling. A mobile home is not constructed on a permanent foundation but is connected to the required utilities. The term "mobile home" includes manufactured homes.

FOR USE WITH MOBILE HOMES – 50 AMPERES

E CORD MARKING

G MINIMUM LIGHTING LOAD = L x W x 3 VA

MOBILE HOME

B 50 A POWER SUPPLY CORD

D 21′ MINIMUM 36½′ MAXIMUM

A PERMANENT CHASSIS

K OUTDOOR RECEPTACLE

C 125/250 V, 50 A, 3-POLE, 4-WIRE GROUNDING TYPE

G Y X W — RECEPTACLE

X Y G W — CAP

J 6′ MAXIMUM

LESS THAN 12″ – NO RECEPTACLE REQUIRED

SINK

GFCI

H RECEPTACLE REQUIRED WITHIN 6′ OF SINK

12″ MINIMUM REQUIRES RECEPTACLE

6′ MAXIMUM 6′ MAXIMUM

6′ MAXIMUM

FLOOR LINE

J

I GFCI NOT REQUIRED   REF

F 30″x 30″ WORKING SPACE

6′ MAXIMUM

6′ MAXIMUM

550-4(a) Mobile homes not intended for use as dwelling units (e.g.: on-site construction office) are not required to have the number or capacity of circuits required by this article. All applicable provisions of 550 apply, however, when the service is 120/240 V, 3-wire.

B. **550-5(a)** The power supply shall be a feeder assembly of not more than one listed 50 A power-supply cord.

550-5(a), Ex. 1 A 40 A power-supply cord is permitted for mobile homes factory-equipped with gas or oil-fired central heating equipment.

550-5(b) The power-supply cord shall be permanently connected to the distribution panelboard with adequate strain relief. The free end shall have an attachment plug cap. The listed cord shall have four conductors with the grounding conductor identified by a continuous green color or continuous green color with one or more yellow stripes.

C. **550-5(c)** The attachment plug shall be a 125/250 V, 50 A, 3-pole, 4-wire, grounding type.

D. **550-5(d)** The overall length of the power-supply cord shall be not less than 21′ and not more than 36½′. The cord shall be not less than 20′ in length from the face of the attachment plug cap to the point where the cord enters the mobile home.

E. **550-5(e)** The power-supply cord shall be marked "For use with mobile homes – 40 amperes" or "For use with mobile homes – 50 amperes".

550-5(f) The point of entrance for the feeder assembly shall be in the exterior wall, floor, or roof.

550-5(g) The cord shall be protected at the point of entry by conduit, bushings, etc.

550-5(h) The attachment plug cap shall be protected against corrosion and mechanical damage.

550-5(i) For calculated loads over 50 A, the supply shall be by a mast weatherhead or either a metal or nonmetallic rigid raceway.

F. **550-6(a)** A single disconnecting means is required inside a mobile home. The main CBs or fuses shall be clearly marked "Main". The neutral bar shall be insulated per 550-11(a). The distribution panelboard shall be located in an accessible location, but not in a bathroom or closet. A 30″ wide × 30″ deep clear working space shall be maintained in front of the panelboard from the floor to the top of the panelboard.

G. **550-7(a)** The minimum lighting load shall be calculated by multiplying the area of the mobile home (using outside dimensions) by 3 VA per sq ft.

550-8(a) Plug receptacles shall be the grounding type and installed per 210-7.

H. **550-8(b)** All 120 V, 1φ, 15 A and 20 A receptacle outlets installed outdoors and in bathrooms shall have GFCI protection. All receptacle outlets within 6′ of a lavatory or sink shall have GFCI protection.

I. **550-8(b), Ex.** Receptacle outlets for dedicated appliances do not require GFCI protection.

J. **550-8(d)** Receptacles shall be installed in all rooms other than the bath, closet, and hall areas. No point along the floor line shall be more than 6′ from a receptacle outlet. Receptacles serving countertops shall be no more than 6′ apart.

550-8(d), Ex. 1 Interrupted spaces at least 2′ along the floor line and 12″ for countertops shall be provided with receptacle outlets.

K. **550-8(e)** At least one receptacle outlet shall be installed outdoors.

RECREATIONAL VEHICLES

A. **551-2: Recreational Vehicle** – A vehicular type unit which is self-propelled or is mounted on or pulled by another vehicle and is primarily designed as temporary living quarters for camping, travel, or recreational use. Normally identified as Travel Trailer, Camping Trailer, Truck Trailer, Truck Camper, or Motor Home.

551-2: Travel Trailer – A vehicle that is mounted on wheels, has a trailer area less than 320 sq ft (excluding wardrobes, closets, cabinets, kitchen units, fixtures, etc.), is of such size and weight that a special highway use permit is not required, and is designed as temporary living quarters while camping or traveling.

RECEPTACLES

CAPS

15 A   20 A   30 A   50 A

E 125 V, 2 POLE, 3-WIRE GROUNDING TYPE   125/250 V, 3 POLE, 4-WIRE GROUNDING TYPE

BRANCH CIRCUITS

TEMPORARY LIVING QUARTERS   LIVING AREA LESS THAN 320 SQ FT

RECREATIONAL VEHICLE

15' MAXIMUM

THIS CONNECTION IS FOR _____ VOLT AC 60 HZ _____ AMPERE SUPPLY

F LABEL

G POINT OF ENTRANCE FOR POWER SUPPLY

INSIDE WALL

B   **B**

6' MAXIMUM   12' MAXIMUM

FLOOR   WALL RECEPTACLES

RECEPTACLES

GFCI   GFCI   WITHIN 6' OF SINK **D**

SINK   12" OR WIDER **C**

551-3 When requirements for travel trailers differ from other articles of the NEC®, the requirements of 551 shall prevail.

551-10(a) Low-voltage circuits installed by the manufacturer are subject to the NEC® except automotive vehicle circuits. Circuits supplying lights subject to federal and state regulations shall comply with government regulations and NEC®.

551-10(b) Low-voltage wiring shall be: (1) Cu except the metal chassis or frame is permitted as the return path; (2) Type HDT, SGT, SGR, SLX, or insulated per Table 310-13 or equal. Conductors No. 6 through No. 18 shall be listed; (3) single wires to be stranded; (4) wires to be marked on surface every 4′ and listed; and (5) conductors shall have a minimum insulation rating of 90°C.

551-40(a) The electrical equipment and material indicated for connection to a wiring system rated 120 V, 2-wire with ground, or 120/240 V, 3-wire with ground, shall be approved per 551, Parts A through F.

551-40(b) All electrical equipment and materials for the recreational vehicle shall be listed.

B. **551-41(a)** Receptacles shall be installed at wall spaces 2′ long or longer. No point along the floor line shall be more than 6′ from a receptacle. See Exceptions 1 and 2 for bath and hall areas, occupied wall spaces, and wall spaces behind doors.

551-41(b) Receptacle outlets shall be installed in kitchens of recreational vehicles as follows:

C. **551-41(b)(1)** Adjacent to countertops in kitchens with at least one on each side of the sink if there are 12″ or wider countertops on each side of the sink.

551-41(b)(2) Adjacent to the refrigerator and gas range space except if the appliances are gas-fired and require no electrical connection.

551-41(b)(3) Adjacent to countertop spaces 12″ or wider that cannot be reached by a 6′ cord without crossing a traffic area, sink, etc.

551-41(c) Each 125 V, 1φ, 15 A or 20 A receptacle outlet shall have GFCI protection where installed in the following locations:

551-41(c)(1) Adjacent to a bathroom lavatory.

D. **551-41(c)(2)** Within 6′ of any lavatory or sink. See Exceptions.

551-41(c)(3) In the area occupied by a toilet, shower, tub, or any combination of these.

551-41(c)(4) On the exterior of the vehicle except if located in the access panel. See Exception.

E. **551-42** Each recreational vehicle with a 120 V electrical system shall have one of the following:

551-42(a) A 15 A supply if one 15 A circuit.

551-42(b) A 20 A supply if one 20 A circuit.

551-42(c) A 30 A supply if two or more 15 A or 20 A circuits. See Exceptions.

551-42(d) A 50 A, 120/240 V supply if 6 or more circuits.

551-46(b) The power-supply cord shall be a minimum of 23′ when the point of entrance is on the side of the vehicle. It shall be a minimum of 28′ when the point of entrance is on the rear of the vehicle.

F. **551-46(d)** A label at least 1¾″ × 3″ shall be permanently affixed near the point of entrance. The label shall read, as appropriate, "THIS CONNECTION IS FOR _____ VOLT AC, 60 HZ _____ AMPERE SUPPLY." The correct voltage and ampere ratings shall be marked.

G. **551-46(e)** The point of entrance for the power-supply assembly shall be located within 15′ of the rear, on the left (road) side, or at rear, and within 18″ of the outside wall. See Ex. 1 and 2.

551-54(a) The grounding conductor in the power-supply cord shall be connected to the grounding bus or approved grounding means in the distribution panelboard.

551-55(a) All exposed metal parts, enclosures, frames, light fixture canopies, etc. shall be bonded to grounding terminals or enclosure of the distribution panelboard.

X-RAY EQUIPMENT REQUIREMENTS

660-4(a) Hard-service cord and plug may be used to supply fixed and stationary unit not over 30 A. Over 30 A requires wiring which meets general NEC® requirements.

660-4(b) Individual branch circuits are not required for portable units with not over 60 A rating.

A. **660-5** A switch or CB at least 50% of the capacity of the momentary rating or 100% of the long time rating, whichever is greater, shall be readily accessible from the X-ray control station. Where 120 V, 30 A or less, a plug and receptacle may serve as the disconnect.

660-6(a) The ampacity of conductors and overcurrent devices shall be 50% of the momentary rating or 100% of the long time rating, whichever is greater.

B. **660-10** All new, reconditioned, or used equipment moved to a new location shall be of the approved type.

660-21 A manually-controlled device shall be located in or on equipment required for portable X-rays and shall comply with 660-20.

660-24 Where two or more pieces of equipment use the same high-voltage devices, each shall have a separate disconnect.

C. **660-35** Transformers and capacitors (part of X-ray) need not conform to 450 and 460.

660-36 Enclosures of grounded metal or insulating material shall be provided for capacitors.

D. **660-47(a)** The connection from high-voltage equipment to X-ray tubes shall be made with high-voltage shielded cables.

660-48 Noncurrent-carrying metal parts shall be grounded. Portable equipment shall have an approved grounding-type plug.

MOTOR-GENERATOR EQUIPMENT

517-71(a)(b) Equipment shall be permanently installed unless portability is essential. Where portable, a 3-wire hard-service type cord with insulated, green, or green with yellow stripe(s) is required for the EGC.

665-20 The converting apparatus shall be enclosed with noncombustible material.

A. **665-21** The control panels shall be of dead-front construction.

B. **665-22** Doors or panels are required for internal access. Detachable panels (not normally used for access) shall be difficult to remove. Doors and detachable panels giving access to more than 500 V shall be interlocked with a circuit-interrupting device.

C. **665-23** Warning labels where necessary shall read: "DANGER – HIGH VOLTAGE – KEEP OUT".

665-24 Capacitors shall have bleeder resistors per 460-6(a).

665-25 The work applicator shall be guarded.

D. **665-26** Grounds or bonds shall limit stray R-F potentials.

665-28 Low-frequency AC or DC control is limited to 150 V. Conductors shall not be smaller than No. 18.

665-44(a) The output circuit shall be isolated from ground. If over 500 V, a DC ground protector is required. DC impressed on the output circuit shall not exceed 30 V and 5 mA.

665-44(b) Components shall be connected by multi-conductor or coaxial cable or busbar. Conduit shall be nonferrous.

665-47(a) A local remote switch is required if remote application controls are employed.

665-47(b) Foot switches shall be shielded from accidental closing.

665-60 Equipment other than motor generators includes all static multipliers and oscillator-type units utilizing vacuum tubes and/or solid-state devices and shall be capable of converting AC or DC to AC frequency for induction and/or dielectric heating use.

E. **665-61(a)** For a single unit, the ampacity of supply conductors shall be according to the unit nameplate.

F. **665-61(b)** For two or more pieces of equipment, the feeder ampacity shall not be less than the sum of the nameplate currents, except where simultaneous operation is not possible.

665-64 The output circuit includes all output parts external to the converting device, including transformers, contactors, busbars, and other conductors, and shall comply with 665-64(a)(b).

665-64(a) The converter output circuit shall be isolated except where DC could exist on failure, then the DC is grounded.

665-64(b) Where connections between the converter and the work applicator exceed 2′, the connections shall be enclosed or guarded with nonferrous, noncombustible material.

665-68(a) "Local-Remote" switches shall be interlocked.

665-68(b) Foot switches shall be shielded from accidental contact.

SWIMMING POOLS – 680 . . .

680-4: Dry-Niche Lighting Fixture – A lighting fixture installed in the pool wall in a niche sealed against pool water entry.

Forming Shell – A structure mounted in a pool and designed to support a wet-niche lighting fixture assembly.

Hydromassage Bathtub – A permanently installed bathtub with a recirculating pump designed to accept, circulate, and discharge water for each use.

No-Niche Lighting Fixture – A lighting fixture installed above or below the water without a niche.

A. Permanently Installed Swimming Pool – A pool constructed in ground or partially above ground and designed to hold over 42″ of water, and all indoor pools regardless of depth.

Spa (Hot Tub) – An indoor or outdoor hydromassage pool or tub that is not designed to have the water discharged after each use.

Storable Swimming Pool – A pool constructed on or above ground and designed to hold less than 42″ of water, or a pool with nonmetallic, molded polymeric walls or inflatable fabric walls regardless of size.

Wet-Niche Lighting Fixture – A lighting fixture installed in a forming shell and completely surrounded by water.

680-2 All electric equipment installed in, on, or associated with pools, fountains, and similar installations shall conform to 680.

680-5(a) Transformers that supply underwater fixtures shall be the two-winding type, with a grounded metal barrier between the primary and secondary windings.

B. 680-6(a)(1) General-use receptacles shall be located at least 10′ from the inside pool walls.

C. 680-6(a)(2) Where a pool is installed for a dwelling, at least one receptacle is required between 10′ and 20′ from the pool.

D. 680-6(a)(3) All receptacles located within 20′ of the pool shall have GFCI protection.

E. 680-6(b)(1) Lighting outlets, fixtures, and ceiling fans shall not be installed over the pool or over an area 5′ horizontally from the inside wall of the pool, unless the fixture is at least 12′ above the maximum water level.

F. 680-6(b)(2) Existing lighting fixtures and lighting outlets located less than 5′ horizontally from the inside wall of a pool and at least 5′ above the maximum water level shall be rigidly mounted and GFCI-protected.

680-6(b)(4) Lighting fixtures and lighting outlets installed between 5′ and 10′ horizontally from the inside wall of a pool shall be GFCI-protected unless at least 5′ above the maximum water level and rigidly mounted.

G. 680-6(b)(5) For cord-connected lighting fixtures installed within 16′ of the nearest water surface, the flexible cord shall be no more than 3′ in length, have a Cu EGC, and be no smaller than No. 12 with a grounding-type attachment plug. See 680-7.

H. 680-8 The following parts of swimming or wading pools are not permitted under existing service drops or other overhead wiring, and overhead wiring is not permitted to be installed above: 1. swimming or wading pools, including the area extending 10′ horizontally from the walls of the pool; 2. diving structures; or 3. towers, platforms, or observation stands.

I. 680-8 3. Structures listed in 680-8 are permitted under supply lines or service drops with clearances per Table 680-8.

J. 680-20(a)(1) Underwater fixtures, during normal use (not relamping), shall be free from shock hazard by design. If operating at over 15 V, GFCI protection is required to eliminate shock hazard during relamping.

K. 680-20(a)(2) No underwater lighting fixture shall operate at over 150 V between conductors.

L. 680-20(a)(3) Lighting fixtures shall be installed in pool walls so that the top of the lens is at least 18″ below the normal water level. Lighting fixtures identified for the purpose are permitted 4″ below the normal water level.

M. 680-20(b)(1) Wet-niche fixtures shall have approved forming shells with threaded conduit entries. Metal conduit for wet-niche fixtures shall be of brass or other corrosion-resistant metal. Conduit shall be RMC, IMC, LTFNMC, or RNMC. Where PVC is used, a No. 8 insulated Cu wire is required in the PVC with provisions to terminate in the forming shell, a junction box or transformer enclosure, or a GFCI enclosure. Termination of the No. 8 conductor in the forming shell shall be protected by covering or encapsulating with listed potting compound.

680-20(b)(2) Where flexible cord is used, the end of the cord jacket and wire terminals in the forming shell shall be similarly protected.

680-20(b)(3) Noncurrent parts of wet-niche fixtures shall be grounded to the forming shell by a positive locking device. A tool shall be required to remove the fixture from the forming shell.

680-21(a)(3) Junction boxes connected to conduit that extends to a forming shell shall be of copper, brass, suitable plastic, or other approved material.

680-21(a)(4) Junction boxes connected to conduit that extends to a forming shell shall provide continuity by means of copper, brass, or other approved corrosion-resistant metal that is integral with the box.

680-21(b)(5) Junction boxes connected to conduit that extends to a forming shell shall be not less than 4′ from the inside wall of the pool, unless separated by barrier and not less than 4″ from the inside bottom of the box above grade level or 8″ above the maximum water level, whichever is greater.

680-21(d) Junction boxes shall have one more grounding terminal than conduit entries.

N. 680-22(a) The following parts shall be bonded: All metal parts of electric pool water handling equipment and other metallic elements within or attached to pool structure, including reinforcing rods, coping stones, deck, piping systems, ladders, rails, diving board platforms, drains, skimmers, refill pipes, forming shells and all other metal piping or fixed metal parts within 5′ of the inside walls of the pool or within 12′ above the maximum water level of the pool or any observation stands, towers, or platforms (unless separated from pool by permanent barrier).

680-22(a)(1) Steel tie wires do not require bonding.

680-22(a)(3) Isolated metal parts, no more than 4″ in any dimension and which do not penetrate into the pool structure more than 1″ do not require bonding.

680-22(b)(1) Structural reinforcing steel is permitted as a common bonding grid for nonelectrical parts.

680-22(a)(4) Metal parts of double-insulated equipment do not require bonding.

680-22(b) The parts required to be bonded by 680-22(a) shall be connected to a common bonding grid with not less than No. 8 Cu wire, insulated, covered or bare, and connected by pressure connectors or clamps of brass, copper, or copper alloy. The common bonding grid is permitted to be: (1) reinforcing steel with usual tie wires; (2) the wall of a bolted or welded metal pool; (3) No. 8 or larger solid Cu wire, insulated, covered or bare; or (4) RMC or IMC of brass or other corrosion-resistant material.

680-22(c) Pool heaters over 50 A shall be grounded and bonded per specific instructions.

680-24 The following electrical equipment shall be grounded: (1) Wet-niche and no-niche fixtures, (2) dry-niche fixtures, (3) equipment within 5′ of the pool, (4) all associated recirculating equipment, (5) junction boxes, (6) transformer enclosures, (7) GFCIs, and (8) panelboards that are not part of the service equipment.

680-25(b)(1) Wet-, dry-, or no-niche fixtures shall be connected to an insulated, Cu EGC sized per Table 250-122, but no smaller than No. 12. The insulated EGC shall be run in RMC, IMC, or RNMC.

680-25(b)(5) Flexible cords to wet-niche lighting fixtures shall contain an insulated, Cu EGC not smaller than No. 16.

680-25(d) An EGC connected to the subpanel shall continue to the service ground.

680-25(f) Other electrical equipment shall be grounded per 250 and wiring shall be per Ch 3.

... SWIMMING POOLS — 680

Ⓒ Ⓓ
AT LEAST ONE GFCI
RECEPTACLE REQUIRED

Ⓑ RECEPTACLES
NOT ALLOWED

INSIDE WALL
OF POOL

10' 10'

DWELLING

OVER
42"

PERMANENTLY
INSTALLED
SWIMMING POOL Ⓐ

Ⓘ HORIZONTAL CLEARANCES
ARE 10' MINIMUM FROM
INSIDE WALL OF POOL

OVERHEAD
CONDUCTORS

15 - 50 kV
0 - 15 kV
0 - 750 V

14' 16' 18' DIVING
BOARD 18' 25' 27' 14' 16' 18'

Ⓔ EXTERIOR LIGHTING
FIXTURE OR CEILING FAN

GFCI-PROTECTED
LIGHTING FIXTURE Ⓕ

20 A RECEPTACLE
W/GROUNDING PLUG

20 A EQUIPMENT Ⓖ

16'

MAXIMUM
WATER LEVEL

12'

5'

5'

3' OR
LESS CORD

DIVING
BOARD

DECK
BOX

Ⓛ NORMAL
WATER
LEVEL

18" MINIMUM

15 V CIRCUIT

GFCI PROTECTION
REQUIRED
Ⓙ

APPROVED
SHELL
Ⓜ

UNDERWATER
LIGHTING
FIXTURE Ⓚ

UTILITY
POLE

DIVING
BOARD

SERVICE
DROP

SWIMMING
POOL

10'
MINIMUM Ⓗ

POINT OF
ATTACHMENT

DWELLING

DECK BOX

DIVING BOARD

CONDUIT

SKIMMER

PUMP MOTOR

DECK STEEL

LADDER

LIGHT
FIXTURE

COMMON STEEL
(COMMON BONDING GRID)

BONDING REQUIRED

Ⓝ

OFFICE FURNISHINGS — 605

A. **605-1** This article deals with lighting accessories, electrical equipment, and wiring systems that are contained within or installed on relocatable wired partitions.

B. **605-2** Partitions are not permitted to extend from floor to ceiling. Wiring systems providing power for appliances and lighting accessories in wired partitions shall be identified for such use.

605-2, Ex. The AHJ can permit relocatable wired partitions to extend to the ceiling but not penetrate the ceiling.

C. **605-3** Wireways in partitions containing conductors and connections shall be identified as suitable for the use.

D. **605-4** A flexible assembly, identified for use with wired partitions, shall be used for the electrical connection between partitions.

605-4, Ex. Flexible cord is permitted to interconnect partitions when partitions are mechanically contiguous, cord is extra-hard usage type, cord is as short as possible, but no longer than 2′, and cord terminates in an attachment plug and cord-connector with strain relief.

605-5 Lighting equipment shall be identified for use with wired partitions and shall: (a) have adequate support; (b) where cord-and-plug connected shall not exceed 9′, be not smaller than No. 18, contain an EGC, and shall be of the hard usage type; and (c) lighting accessories shall not contain receptacle outlets.

605-6 Partitions secured to the building shall be permanently wired by wiring methods in 300.

605-7 Freestanding partitions shall be permitted to be permanently connected to the building electrical system by wiring methods in 300.

E. **605-8** Free standing individual partitions that are not over 30′ long and are mechanically contiguous are permitted to be cord-and-plug connected providing: (a) cord is at least No. 12 AWG with insulated grounding conductor of extra-hard usage type and not over 2′ long; (b) permanent building wiring connection is permitted by cord if the receptacle is on a separate circuit and located not more than 12″ from the partition; (c) not more than thirteen 15 A, 125 V receptacles are permitted in individual or group of interconnected partitions; and (d) multi-wire circuits not permitted.

605-8(a) The flexible power supply cord shall be a No. 12 or larger extra-hard usage cord with insulated grounding conductor not over 2′ long.

Trade Competency Test 1

NEC®		Answer	

_____   T   F   **1.** The secondary circuits of 460 V wound-rotor induction motors are permitted to be protected against overload by the motor-overload device.

_____   T   F   **2.** The primary windings of voltage transformers installed indoors shall normally be protected with fuses.

_____   **3.** The conductors in a pull box having any dimension over 6′ shall be cabled or _____ in an approved manner.

_____   T   F   **4.** A transformer is a polyphase or single transformer, identified by a single nameplate.

_____   _____   **5.** High fire point, liquid-insulated transformers installed indoors shall be installed in vaults if rated over _____ V.

     A. 5000   C. 25,000
     B. 15,000   D. 35,000

_____   _____   **6.** In 430, the term "in sight" refers to a distance of not more than _____′.

_____   _____   **7.** The motor disconnecting means shall disconnect both the motor and the _____.

_____   _____   **8.** Enclosures for CBs, switches, and fuses that are located in agricultural buildings shall be dustproof and _____.

_____   _____   **9.** Except for polyphase wound-rotor units, AC motors _____ HP and larger shall have code letters or locked-rotor amps marked on their nameplates.

_____   T   F   **10.** Capacitors used with motors increase motor current.

_____   _____   **11.** The ampacity of supply conductors for a transformer arc welder with a 100% duty cycle and 30 A primary current is _____ A.

     A. 30   C. 60
     B. 37.5   D. 90

_____   T   F   **12.** A ¾ HP, 115 V motor on an automatic air compressor does not need overload protection.

_____   _____   **13.** Separate running overcurrent protection for continuous-duty motors more than 1 HP shall be based on the FLC, which is listed on the _____ of the motor.

_____   _____   **14.** Wired office partitions may extend from floor to _____ with permission of the AHJ.

_____   T   F   **15.** The rated load current of a sealed (hermetic) motor-compressor shall be marked on either the equipment or the nameplate.

16. The controller may be an attachment plug and receptacle for portable motors rated at _____ HP or less.

17. On an ungrounded circuit, _____ thermal overcurrent units are needed to protect a 3φ induction motor from overload.

18. The running overcurrent protection for a DC motor with a 1.5 service factor and a 40°C temperature rise is _____%.

19. A suitable plug and hard-usage cord can supply a fixed or stationary nonmedical X-ray unit if the branch circuit is rated at no more than _____ A.

20. The maximum size of the overload relay used to protect a 50 A, 230 V, 40°C motor is _____ A.

A. 50   C. 70
B. 75   D. 100

21. Fuses used as motor overload protection devices shall be inserted in each _____ conductor.

22. The conductors of mineral-insulated, metal-sheathed cable (Type MI) are made of solid _____ or nickel-clad copper.

23. FMC used in a manufactured wiring system shall contain at least a No. _____ AWG Cu grounding conductor.

24. The minimum clearance from the 0 – 750 V service conductors to the high-water line of a swimming pool is _____′.

25. Exposed live parts of motors and controllers operating at _____ V or more between terminals shall be guarded.

26. A general-use snap switch suitable only for use on AC may be used to disconnect a 2 HP, 240 V motor if the FLC rating of the motor does not exceed _____% of the ampere rating of the switch.

27. A 3 HP, 240 V, 3φ squirrel-cage motor with no code letter is started at full voltage. The size of a TW branch-circuit conductor shall be not less than No. _____.

A. 8   C. 12
B. 10   D. 14

28. Unless otherwise designed and marked, busways shall be supported at least every _____′.

29. The minimum length of a pull box for a straight-through pull of three No. 4/0 conductors in 2½″ conduit is _____″.

30. For stationary motors of 2 HP or less and 300 V or less, a general-use switch may be used as a controller if the rating is at least _____ times the full-load motor current.

T   F   **31.** Transformers may have overcurrent protection devices in both primary and secondary circuits.

T   F   **32.** Voltage transformers shall be fused on the primary side when installed indoors.

Trade Competency Test 2

Name _____

Date _____

NEC®	Answer	

_____ T F **1.** An overload in electrical apparatus includes short circuits or ground faults.

_____ _____ **2.** Control circuit transformers rated less than _____ VA require no additional protection.

_____ _____ **3.** A 30 HP, 230 V, 3φ squirrel-cage motor with code letter B is started with an autotransformer compensator. The initial rating of a dual element fuse for branch-circuit protection shall not exceed _____ A.
A. 100   C. 150
B. 125   D. 175

_____ _____ **4.** No underground lighting fixture shall operate at a voltage greater than _____ V between conductors.

_____ _____ **5.** A(n) _____ building is of closed construction and is made or assembled in manufacturing facilities on or off the building site for installation or assembly and installation on the building site, other than manufactured homes, mobile homes, park trailers, or recreational vehicles.

_____ _____ **6.** A remote No. 14 control circuit tapped from the line side of a magnetic starter requires no additional protection if the branch circuit is protected by a(n) _____ A device or less.

_____ _____ **7.** Sealed refrigeration units have _____ refrigerant motor-compressors.

_____ _____ **8.** In general, a controller used to stop and start a motor shall have a rating equivalent to the _____ of the motor.

_____ _____ **9.** The maximum overcurrent protection of a No. 6 THHN feeder that supplies two 30 A transformer arc welders rated for continuous duty shall be _____ A.

_____ T F **10.** A nonautomatically started, not permanently installed, 1 HP motor within sight of the controller location shall be permitted to be protected against overload by the branch-circuit, short-circuit, and ground-fault protective device.

_____ _____ **11.** Overhead 120/240 V conductors crossing a swimming pool shall have a clearance of _____ ′ above water level.

_____ T F **12.** The overcurrent protection device for a 30 A transformer arc welder shall not exceed 30 A.

_____ _____ **13.** Conductors supplying a capacitor with a nameplate current of 50 A shall be rated not less than _____ A.
A. 57.5   C. 67.5
B. 62.5   D. 72.5

_____ T F **14.** A 277 V-to-ground circuit is permitted to supply lighting fixtures incorporating mogul-base screw-shell lampholders.

_____ _____ **15.** The basic rule requires an individual disconnecting means for _____ motor.

_____ _____ **16.** A manually-operated resistance welder draws 50 A according to the nameplate. Circuit conductors are permitted to be chosen on the basis of _____ A.

    A. 25   C. 75

    B. 50   D. 100

_____ _____ **17.** A busway may be reduced in size provided the smaller busway has a current rating of at least one-third of the rating of the main busway overcurrent protection device and does not extend more than _____′.

_____ _____ **18.** Pool heaters rated over _____ A are to be grounded and bonded.

_____ T F **19.** An adjustable overload protection device protecting a 50°C motor with an FLC rating of 20 A shall be selected to trip at not more than 23 A.

_____ _____ **20.** Junction boxes housing conductors for swimming pool equipment shall have _____ more grounding terminal(s) than the number of conduit entries.

_____ T F **21.** The disconnecting means for 460 V capacitors need not open all ungrounded conductors simultaneously.

_____ _____ **22.** The conductors to an adjustable speed drive shall be based on the rated _____ of the power conversion equipment.

_____ _____ **23.** The three types of listed optical fiber cables are the nonconductive, the conductive, and the _____.

_____ _____ **24.** A No. 14 motor-control circuit conductor requires separate overcurrent protection when the control circuit extends beyond the control equipment enclosure and the motor branch-circuit protective device exceeds _____ A.

    A. 30   C. 60

    B. 45   D. 75

_____ _____ **25.** The disconnecting means for a motor shall have an ampacity of at least _____% of the FLC rating of the motor.

_____ _____ **26.** The minimum height of the door sill or curb in an approved transformer vault is _____″.

_____ _____ **27.** The controller can be an attachment plug and receptacle for a(n) _____ motor rated at $\frac{1}{3}$ HP or less.

_____ _____ **28.** The FLC rating for a 10 HP, 230 V, 2ϕ AC motor is _____ A.

_____ _____ **29.** The minimum size of conductors permitted in a cablebus is No. _____.

    A. 1   C. 1/0

    B. 2   D. 2/0

_____ _____ **30.** It is mandatory that one of the disconnecting means to a motor be _____ accessible.

_____ _____ **31.** Transformers used near swimming pools shall be provided with a metallic _____ between primary and secondary windings.

_____ _____ **32.** Overcurrent protection for a manually operated resistance welder that draws 50 A shall not exceed a maximum value of _____ A.

    A. 150   C. 200

    B. 175   D. 250

_____ _____ **33.** The allowable ampacities of wires used in crane circuits is _____ than for general electrical construction work.

_____ _____ **34.** The metal shell of wet niche fixtures shall be _____.

Name _____

Date _____

NEC®	Answer	
_____	_____	**1.** Flexible cords used to wire wet-niche lighting fixtures shall be equipped with an EGC not smaller than No. _____.
_____	T F	**2.** A stop-and-start station that can be locked open shall not be used as a disconnecting means.
_____	T F	**3.** A $\frac{1}{40}$ HP, split-phase, automatically started motor does not require running overcurrent protection if protected by the branch-circuit, short-circuit, and ground-fault protection device.
_____	_____	**4.** Conductors rated over 15,000 V shall enter metal-enclosed switchgear or be installed in a transformer _____.
_____	T F	**5.** Motor overload protection shall never be shunted during the starting period.
_____	_____	**6.** Motors with no starting current problems and with a temperature rise of 50°C shall have overload protection provided at _____% of nameplate current.
_____	T F	**7.** Replacement parts may be stored in transformer vaults if at least 3′ from operating equipment.
_____	_____	**8.** Surge arresters shall be located so they are _____ only to qualified personnel, unless listed for the installation.
_____	T F	**9.** The second largest X-ray unit on a feeder, in a health care facility, has a demand factor of 150% of the momentary rating.

10. The walls and roofs of transformer vaults shall have a minimum fire resistance of _____ hour(s).

A. 1   C. 3
B. 2   D. neither A, B, nor C

_____	_____	**11.** For other than industrial establishments, overcurrent protection shall be required where _____ are reduced in ampacity.
_____	T F	**12.** An air-break isolating switch is required on the supply side of the disconnect and all service equipment for services over 600 V.
_____	T F	**13.** A motor branch-circuit, short-circuit protection device consisting of an NTDF shall never have a rating in excess of 400% of the full-load motor current.
_____	_____	**14.** A travel trailer has a maximum gross trailer area less than _____ sq ft.
_____	T F	**15.** Foreign pipes or ducts shall not pass through a transformer vault.

_____ T F **16.** Infrared heating units may be supplied from a 50 A branch circuit in an industrial occupancy.

_____ _____ **17.** Raceways routed in _____″ of concrete or brick are considered to be outside the building.

_____ _____ **18.** The secondary conductors of a 2-wire to _____ transformer can be protected by the primary overcurrent protection device.

_____ T F **19.** Single-phase conductors supplying a phase converter shall have an ampacity of not less than 125% of the 1ϕ input FLC rating.

_____ _____ **20.** A feeder circuit supplying power to a mobile home shall be provided with a(n) _____ conductor.

_____ T F **21.** An inherently short-time intermittent duty motor usually does not require separate running overload protection.

_____ _____ **22.** A ceiling fan shall be hung no less than _____′ above the maximum water level of a swimming pool.

_____ _____ **23.** The feeder circuit to a number of resistance welders shall not exceed _____% of the conductors' rating.

_____ T F **24.** In some cases, overload protection for a 50 A continuous duty motor may be set at 70 A.

_____ _____ **25.** The ampacity of branch-circuit conductors supplying power to X-ray equipment shall be _____% of the momentary rating or 100% of the long time rating, whichever is greater.

_____ _____ **26.** Enclosed overcurrent protection devices for services rated over 600 V are not required to be derated _____% for continuous operation.

_____ _____ **27.** Secondary open circuit voltage of 7500 V or less shall have an output not exceeding _____ mA.

A. 240 C. 280
B. 260 D. 320

_____ _____ **28.** All metal enclosures within the walls of a swimming pool shall be bonded to the grid with a common conductor No. _____ or larger in size.

_____ _____ **29.** A resistance welder with a 50% duty cycle has a multiplier of _____%.

_____ _____ **30.** Each welder is required to have a disconnect either at the welder or in the _____.

_____ _____ **31.** Dry-type transformers rated over _____ V, primary or secondary, are required to be installed in a vault.

_____ _____ **32.** For capacitors rated over 600 V, the discharge means is required to be reduced to 50 V or less in _____ minutes or less.

_____ _____ **33.** The minimum rating of supply conductors for an automatically fed resistance welder operating at varying duty is _____% of the rated primary current.

_____ _____ **34.** Control conductors shall be used with ampacities rated for _____°C.

A. 60 C. 90
B. 75 D. 120

_____ _____ **35.** Conductors from surge arresters to line and ground should be kept as _____ as possible.

A *hazardous location* is a location where fire or explosion hazards may exist because of flammable gases and liquids, combustible dust, or ignitable fibers or flyings.

HAZARDOUS LOCATIONS 6

CLASSES

Some locations, because of their nature, call for higher standards with respect to materials or for methods of installation than are thought necessary in the general run of electrical wiring. The NEC® classifies hazardous locations under three headings: Class I, Class II, and Class III. Class I concerns activities that may cause fire or explosion from gases or vapors. Class II concerns the fire or explosion risk from dusts. Class III concerns the dangers of combustible fibers.

Each class has two divisions. In Division 1 locations, the particular danger is imminent at any or all times during the normal course of operation. In Division 2 locations, the danger is not present under normal conditions, but is likely to arise from a reasonable, foreseeable, accidental occurrence. For example, a building that houses a machine for compressing acetylene gas falls under Class I, Division 1. If high-pressure mains from this machine pass into another building where shut-off valves and pressure-reducing instruments are inserted in the lines, the adjacent building falls under Class I, Division 2.

Plants where combustible or electrically conductive dusts are present are in Class II, Division 1. Locations where readily-ignitable fibers are merely stored or handled, other than in the actual process of manufacture, are in Class III, Division 2. This classification applies mostly to warehouses where manufactured goods, in bales, are held for shipment to consumers.

Methods for Reducing Hazards

The NEC® requirements for proper seals in Class I, Division 1 and 2 locations are of utmost importance. Seals, when required, are installed within 18″ of an enclosure for switches, relays, CBs, fuses, or other equipment that could produce arcs, sparks, or high temperatures. The purpose of these seals is to restrict an explosion to one enclosure. Seals are also required in Class I, Division 1 and 2 areas when a conduit run of 2″ or larger enters an enclosure or fitting that contains terminals, splices, or taps. This seal is provided to limit the total volume that could be exposed to an explosion.

Seals are also required in conduits passing from a hazardous to a nonhazardous location. The purpose of this seal is to prevent an explosive mixture from being communicated through the conduit from a hazardous to a nonhazardous area.

In Class II areas, an alternate method of sealing is provided that permits a horizontal raceway not less than 10′ long or a vertical raceway extending not less than 5′ downward from a dust-ignition proof enclosure. All sealing fittings and compound shall be approved for the purpose. At the present time, Underwriters Laboratories Inc. lists a sealing fitting, but does not list a compound except when a compound is used with the explosion-proof seal fittings of a specific manufacturer.

Intrinsically safe equipment and associated wiring that is approved for a specific location is permitted in hazardous areas. This equipment is such that under normal or abnormal conditions, it is not capable of releasing sufficient electrical or thermal energy to cause ignition of a specific explosive mixture.

The NEC® recommends that wherever possible, electrical equipment for hazardous locations be located in less hazardous areas. It also suggests that by adequate, positive-pressure ventilation from a clean source of outdoor air, the hazards may be reduced, or hazardous locations may be limited or eliminated. In many cases, the installation of dust collection systems can greatly reduce the hazards in a Class II area.

RMC, IMC, or Type MI cable are prescribed as the only acceptable wiring methods in the majority of hazardous locations. Grounding continuity shall be assured by means of bonding and not by double lockouts or locknut and bushing. Fittings shall be explosion-proof, dust-ignition proof, or dusttight, depending on the nature of the hazard.

Motors in the scope of Class I, Division 1 shall be explosion-proof. Under Class I, Division 2,

standard motors can be used if they have no sliding contacts. Motors in Classes II and III shall be totally enclosed or pipe-ventilated. Flexible cords of the three class areas shall be approved for extra hard use.

Specially Listed Occupancies

The NEC® considers all areas up to a level of 18″ in a commercial garage as Class I, Division 2 locations. A commercial garage where only exchange of parts and routine maintenance work are performed is considered a nonhazardous location. Aircraft hangars are given stricter, but similar protection. Space in the immediate vicinity of a gasoline dispensing island is Class I, Division 2 to the top of the dispenser enclosure extending 18″ horizontally to finished grade. Surrounding territory within a radius of 20′ of the island is Class I, Division 2 to a height of 18″ above grade. Bulk storage plants for gasoline are subject to comparable restrictions.

Finishing processes where paint, lacquer, or other flammable coatings are applied by spraying, dipping, brushing, or similar means, shall be carried out under prescribed safeguards for electrical wiring and equipment. Of particular interest here are electrostatic units that incorporate high-voltage transformers.

Anesthetizing locations of hospitals are Class I, Division 1 to a height of 5′ above floor level. Special rules apply to lighting and surgical equipment in the operating room, isolating transformers being generally required. Gas storage rooms are designated as Class I, Division 1 throughout.

Theaters and movie studios are given adequate consideration in the NEC®. Unless theaters are provided with sufficient emergency lighting, there is a likelihood of panic in case of fire. Fires may occur suddenly from temporary wiring used in connection with stage scenery or from projection room equipment. Movie studios also have a threat of fire from temporary wiring, arc lamps, and portable devices. Radio or television studios and receiving stations shall be guarded from the danger of high-voltage electronic wiring systems or equipment.

MATERIAL	CLASS I		CLASS II			CLASS III	
	Division 1	Division 2	Division 1	Division 2	Metal dust	Division 1	Division 2
RACEWAY							
Rigid	✓	✓	✓	✓	✓	✓	✓
IMC	✓	✓	✓	✓	✓	✓	✓
EMT				✓			
Open wiring							✓e
BOXES – Fittings							
Threaded	✓	✓	✓	✓	✓	✓	✓
Explosion-proof	✓						
Dust-ignition proof			✓c		✓		
Dusttight			✓d				
Dust-minimal				✓		✓	✓
FLEXIBLE CONDUIT							
Explosion-proof	✓						
Standard							
Extra-hard-usage cord	✓	✓	✓	✓	✓	✓	✓
SWITCHES							
Explosion-proof	✓	✓					
General purpose – hermetica		✓					
General purpose – oilb		✓					
Dust-ignition-proof			✓		✓		
Dust-minimal				✓		✓	✓

aGeneral purpose enclosure – contacts hermetically sealed
bGeneral purpose enclosure – contacts immersed in oil
cWhere joints or terminals
dWhere no joints or terminals
eStorage areas only

HAZARDOUS LOCATIONS – 500

Hazardous Location – A location where there is an increased risk of fire or explosion due to the presence of flammable gases, vapors, liquids, combustible dusts, or easily-ignitable fibers or flyings.

Location – A position or site.

Flammable – Capable of being easily ignited and of burning quickly.

Gas – A fluid (such as air) that has no independent shape or volume but tends to expand indefinitely.

Vapor – A substance in the gaseous state as distinguished from the solid or liquid state.

Liquid – A fluid (such as water) that has no independent shape but has a definite volume. A liquid does not expand indefinitely and is only slightly compressible.

Combustible – Capable of burning.

Ignitable – Capable of being set on fire.

Fiber – A thread or piece of material.

Flyings – Small particles of a material.

Dust – Fine particles of matter.

Classes	Likelihood that a flammable or combustible concentration is present
I	Sufficient quantities of flammable gases and vapors present in air to cause an explosion or ignite hazardous materials
II	Sufficient quantities of combustible dust are present in air to cause an explosion or ignite hazardous materials
III	Easily-ignitable fibers or flyings are present in air, but not in a sufficient quantity to cause an explosion or ignite hazardous materials

Divisions	Location containing hazardous substances
1	Hazardous location in which hazardous substance is normally present in air in sufficient quantities to cause an explosion or ignite hazardous materials
2	Hazardous location in which hazardous substance is not normally present in air in sufficient quantities to cause an explosion or ignite hazardous materials

Groups	Atmosphere containing flammable gases or vapors or combustible dust	
Class I	**Class II**	**Class III**
A B C D	E F G	none

DIVISION I EXAMPLES

Class I:

- Spray booth interiors
- Areas adjacent to spraying or painting operations using volatile flammable solvents
- Open tanks or vats of volatile flammable liquids
- Drying or evaporation rooms for flammable solvents
- Areas where fats and oil extraction equipment using flammable solvents are operated
- Cleaning and dyeing plant rooms that use flammable liquids
- Gas generator rooms
- Pump rooms for flammable gases or volatile flammable liquids that do not contain adequate ventilation
- Refrigeration or freezer interiors that store flammable materials
- All other locations where sufficient ignitable quantities of flammable gases or vapors are likely to occur during routine operations

Class II:

- Grain and grain products
- Pulverized sugar and cocoa
- Dried egg and milk powders
- Pulverized spices
- Starch and pastes
- Potato and woodflour
- Oil meal from beans and seeds
- Dried hay
- Any other organic materials that may produce combustible dusts during their use or handling

Class III:

- Portions of rayon, cotton, or other textile mills
- Manufacturing and processing plants for combustible fibers, cotton gins, and cotton seed mills
- Flax processing plants
- Clothing manufacturing plants
- Woodworking plants
- Other establishments involving similar hazardous processes or conditions

HAZARDOUS (CLASSIFIED) LOCATIONS

STANDARD SYSTEM
500; 501; 502; 503

CLASS I FLAMMABLE GASES OR VAPORS
DIVISION 1 NORMALLY PRESENT IN AIR
DIVISION 2 NOT NORMALLY PRESENT IN AIR

CLASS II COMBUSTIBLE DUST
DIVISION 1 NORMALLY PRESENT IN AIR
DIVISION 2 NOT NORMALLY PRESENT IN AIR

CLASS III IGNITIBLE FIBERS OR FLYINGS
DIVISION 1 NORMALLY PRESENT IN AIR
DIVISION 2 NOT NORMALLY PRESENT IN AIR

OPTIONAL SYSTEM
500; 501; 502; 503; 505

ZONE 0 FLAMMABLE GASES OR VAPORS ARE PRESENT CONTINUOUSLY OR FOR LONG PERIODS OF TIME

ZONE 1 FLAMMABLE GASES OR VAPORS ARE LIKELY TO EXIST UNDER NORMAL OPERATING CONDITIONS OR EXIST FREQUENTLY

ZONE 2 FLAMMABLE GASES OR VAPORS ARE NOT NORMALLY PRESENT OR ARE PRESENT FOR SHORT PERIODS OF TIME

CLASS I, DIVISION 1

VESSELS CONTAINING FLAMMABLE GASES

CLASS I, DIVISION 2

ZONE 0

ZONE 1

ZONE 1

INTERCONNECTING PIPING

AREA CLASSIFICATIONS, WIRING METHODS, AND EQUIPMENT SELECTION BY REGISTERED P.E.

Hazardous Location – Any location where fire or explosion hazards may exist due to flammable gases or vapors, flammable liquids, combustible dust, or ignitable fibers or flyings.

505-6(a) The optional classification system can only be used under supervision of a qualified Registered Professional Engineer.

INTRINSICALLY SAFE SYSTEMS – 504

NONHAZARDOUS LOCATION ◄——————► HAZARDOUS LOCATION

CONDUIT OR OTHER ENCLOSURE

SEAL

INTRINSICALLY SAFE TRANSMITTER

CONTROL ROOM EQUIPMENT

INTRINSICALLY SAFE BARRIERS

INTRINSICALLY SAFE SENSOR

504-2: Intrinsically Safe System – An assembly of interconnected apparatus and cables that may be used in hazardous locations making a spark incapable of causing ignition in circuits.

504-20 Any wiring method suitable for unclassified locations, including Ch 7 and 9 methods, is permitted. Seals shall be provided per 504-70.

504-30(a)(1) Open wiring methods, including conductors and cables, shall be separated and secured from other conductors of power and lighting circuits by at least 1.97″.

504-30(a)(2) Conductors shall not be placed in any raceway, cable tray, or cable with conductors that is not intrinsically safe.

504-30(a)(3) Intrinsically safe systems installed within enclosures shall have conductors separated at least 1.97″ from other conductors.

504-50(a) Associated metal apparatus such as cable shields, enclosures, and raceways shall be grounded.

504-50(b) Intrinsically safe systems shall be grounded to the grounding electrode per 250-50(a – d) where available.

504-50(c) Shields for shielded conductors or cables shall be grounded unless they are part of an intrinsically safe circuit. See Exception.

504-60(a) Bonding in hazardous locations shall comply with 250-100.

504-70 Conduits and cables shall be sealed to minimize passage of gases, vapors, or dust per 501-5 and 502-5.

504-80(a) Intrinsically safe circuits shall be suitably identified at all terminals and junction locations to avoid unintentional interference.

504-80(b) Labels, worded "Intrinsic Safety Wiring," shall be permanently affixed to all wiring so as to be visible after installation and at intervals not exceeding 25′ in length.

504-80(c) Color coding of intrinsically safe conductors and raceways, boxes, and cable trays that contain only intrinsically safe wiring shall be permitted to be light blue provided no other light blue conductors are used.

CLASS I, DIVISION 1 LOCATIONS – 501

500-7(a): Class I, Division 1 Locations – Areas where flammable gases or vapors are normally present in air. See 500-5 for special precautions.

501-2(a)(1)(2) Transformers or capacitors containing a liquid that will burn shall have no openings into the Division 1 area.

501-3(a) Meters, instruments, and relays shall be in enclosures approved for Class I, Division 1 locations.

A. **501-4(a)** Wiring methods permitted for Class I, Division 1 locations are RMC, IMC, or Type MI cable with approved termination fittings. See Ex. 1 for RNMC and Ex. 2 for Type MC cable. Explosion-proof flexible connection with fittings shall be listed for Class I, Division 1 locations.

B. **501-5(a)(1)(2)** A seal shall be installed within 18″ of enclosures for spark- or heat-producing apparatus – also in each run of 2″ or larger conduit where splices are made, and within 18″ thereof.

501-5(a)(4) Seals are required where a conduit leaves a Class I, Division 1 location. Seals may be installed on either side and within 10′ of the boundary.

C. **501-5(c)(1)** Seal fittings shall be accessible.

501-5(c)(2)(3) Specifications for sealing compound are detailed in the NEC®.

501-5(f)(3) Canned pumps, process connections for flow, pressure, or analysis measurement, etc. that depend upon a single seal, diaphragm, or tube to keep fluid from entering the electrical raceway or conduit system require an additional approved seal, barrier, or other means to keep fluids from entering the conduit system.

D. **501-6(a)** Switches, CBs, motor controllers, and fuses shall be in enclosures approved for Class I, Division 1 locations.

E. **501-7(a)** Switching mechanisms associated with transformers, impedance coils, and resistors shall be in enclosures approved for Class I, Division 1 locations.

F. **501-8(a)(1)** Motors, generators, and other rotating machinery shall be approved for use in Class I, Division 1 locations.

G. **501-8(a)(2)** Motors and generators in Class I, Division 1 locations shall be approved for location. One alternative is to use totally enclosed types of equipment with positive-pressure ventilation that ensures enclosures are purged with ten volumes of air before equipment is energized.

H. **501-9(a)(1)** Portable light fixtures used as a complete assembly shall be approved for use in Class I, Division 1 locations.

I. **501-9(a)(1 – 4)** Lighting fixtures shall be approved as complete units for use in Class I, Division 1 locations. Stems over 12″ shall be braced.

J. **501-10(a)** Utilization equipment shall be approved for use for Class I, Division 1 locations.

K. **501-11** Flexible cord, extra-hard usage type, with grounding conductor shall be terminated in approved manner, suitably supported, and with suitable seals.

L. **501-12** Receptacles and plugs shall be approved for use in Class I, Division 1 locations.

M. **501-14(a)** Signal devices shall be approved for use in Class I, Division 1 locations.

501-15 No exposed live parts are permitted in Class I, Division 1 locations.

501-17(a) Surge arrestors shall be in an enclosure approved for use in Class I, Division 1 locations and shall comply with 280.

CLASS I, DIVISION 2 LOCATIONS — 501

500-7(b): Class I, Division 2 Locations – Areas where flammable mixtures are not normally present, but may appear through mechanical failure or accident. Areas where flammable liquids or gases are handled, processed, or used, but are normally contained, and where positive mechanical ventilation is provided or areas where a barrier is provided, but which could become hazardous upon a malfunction shall be classed as a Class I, Division 2 area.

501-2(b) Transformers are permitted to be installed as in nonhazardous locations and shall comply with 450-21 – 450-27.

A. 501-3(b)(1) Meters, relays, resistors, etc., without Class I enclosures shall have make-and-break contacts immersed in oil or in hermetically sealed chambers unless on circuits of negligible energy (nonincendive circuits).

501-3(b)(5) Instrument fuses are permitted in general-purpose enclosures per 501-3(b)(1 – 4) if not subject to overloading and preceded by a switch conforming to 501-3(b)(1).

B. 501-4(b) FMC, LTFMC, LTFNMC, or extra-hard usage cord with approved fittings are permitted. Wiring methods permitted in Class I, Division 2 locations include: threaded RMC, IMC, enclosed gasketed wireways or busways, Type PLTC cable per 725, Type ITC, MI, MC, MV, or TC cable (with approved fittings). Boxes and fittings generally need not be explosion-proof, except as required by 501-3(b)(1), 501-6(b)(1), and 501-14(b)(1).

C. 501-5(b)(2) A sealing fitting is required in the conduit run passing to the nonhazardous area.

501-6(b)(1) Motor controllers, switches, and CBs shall be listed as Class I, Division 1 equipment per 501-3(a) unless make-and-break contacts are immersed in oil or in a chamber hermetically sealed against the entrance of gas or vapors, in a factory-sealed explosion-proof chamber, or the device is solid state.

D. 501-6(b)(4) Approved cartridge fuses are permitted for supplementary protection with lighting fixtures.

E. 501-7(b)(2) General-purpose enclosures are permitted for transformers, solenoids, or impedance coils in Class I, Division 2 locations.

F. 501-8(b) Standard squirrel cage induction motors may be used in Class I, Division 2 locations. If the motor has sliding contacts, it must be approved for Class I, Division 2 locations, unless the contact arrangements are especially approved for the location.

G. 501-9(b)(1 – 5) Fixed fixtures shall have protection from physical damage. Note surface temperature limitation. Portable lamps shall be specifically approved for the purpose. Pendants to have reinforcing supports if over 12″ long. Fixture switches shall comply with 501-6(b)(1). Electric discharge starting and control equipment shall be per 501-7(b), except the potted thermal protector in the ballast of fluorescent lighting shall be approved for hazardous locations.

H. 501-10(b)(1) Electrically-heated appliances shall be approved for Class I, Division 2 locations.

I. 501-12 Plug receptacles shall be approved for the location.

J. 501-14(b)(1) Signal unit shall be explosion-proof, or contacts immersed in oil, or hermetically sealed unless on circuits of negligible energy. See Exception.

501-15 No exposed live parts are permitted in Class I, Division 2 locations.

501-16(b) Locknut-bushing type contacts are not approved for general bonding purposes in hazardous locations. Bonding jumpers are required for FMC. External bonding jumpers shall be per 250-100. See Exception.

CLASS II, DIVISION 1 LOCATIONS – 502

500-8(a): Class II, Division 1 Locations – Areas where combustible dust in dangerous concentrations normally suspended in the air or where metallic dust is present.

CLASS II, DIVISION 1 LOCATION

500-5(b)(1 – 3) Classification is by group of specified dust: Group E, metal dusts: aluminum, magnesium; Group F: carbon black, charcoal, coal dust; Group G: flour, starch, or grain dust.

A. **502-2(a)(1)** Transformers and capacitors having liquid that will burn shall be in an approved vault with self-closing fire doors inside and out per 450-41 – 450-48.

502-2(a)(2) If nonburning liquid, transformers shall be in an approved vault or specially approved complete assembly.

502-2(a)(3) No transformer or capacitor shall be installed in a location where metal dusts may be present.

502-4(a) The wiring method shall be RMC, threaded steel IMC, or Type MI cable. See Exception.

B. **502-4(a)(1)** Boxes and fittings shall have threaded bosses and close-fitting covers and no holes for attachment screws, except in locations where dusts are of a combustible electrically-conductive nature, and they contain taps, joints, or terminal connections, then box or fitting shall be approved for Class II locations.

C. **502-4(a)(2)** For flexible connections, extra-hard usage cord, LTFMC, LTFNMC, or dusttight flexible connectors are permitted.

D. **502-5** Ten foot horizontal raceway acts as seal. Five foot vertical raceway may act as a dust seal if it extends down from a dust-ignition proof enclosure.

E. **502-6(a)(1)** Dust-ignition proof switch and controller shall be approved for use in Class II, Division 1 locations.

F. **502-7(a)** Dust-ignition proof control transformer shall be approved for use in Class II, Division 1 locations.

G. **502-8(a)** Motors and generators shall be approved for Class II, Division 1 locations or be totally enclosed pipe-ventilated and meet temperature limitations of 502-1.

H. **502-10** Appliances shall be approved for use in Class II, Division 1 locations.

I. **502-11(a)(1)** Lighting fixtures shall be marked for maximum lamp wattage and approved for use in Class II, Division 1 locations.

J. **502-13(a)** Receptacles and plugs shall be approved for use in Class II, Division 1 locations.

K. **502-14(a)(1 – 6)** Signaling, alarm, remote-control, and communications systems wiring method shall be RMC, threaded steel IMC, or MI cable. Any breaking contacts in the circuit shall be approved for Class II, Division 1 locations.

502-18 Multiwire branch circuits are not permitted in Class II, Division 1 locations. See Exception.

CLASS II, DIVISION 2 LOCATIONS – 502

CLASS II, DIVISION 2 LOCATION

500-8(b): Class II, Division 2 Locations – Areas where combustible dust is not usually present in dangerous mixture but likely to accumulate.

502-2(b)(1) Transformers containing liquid that will burn shall be in vaults per 450-41 – 450-48.

A. **502-2(b)(2)** Transformers containing askarel and rated over 25 kVA shall have means of disposing of gas and shall be at least 6″ from combustibles.

502-2(b)(3) Dry-type transformers shall be in vaults or have terminals fully enclosed and shall operate at voltage not exceeding 600 V.

B. **502-4(b)** Wiring methods for Class II, Division 2 locations shall be: (1) RMC; (2) IMC; (3) EMT; (4) dusttight wireways; (5) Type MC or MI cable with approved fittings; (6) Type PLTC in cable tray; (7) Type MC, ITC, or TC in specific type of cable tray. See Exceptions.

502-4(b)(1) Boxes and fittings to resist entrance of dust shall have telescoping or close-fitting covers to contain sparks. No holes in the box shall be left exposed after installation.

502-5 Sealing is the same as for Class II, Division 1 areas.

C. **502-6(b)** Dusttight metal enclosures are required in Class II, Division 2 locations.

D. **502-8(b)** Motors shall be totally-enclosed nonventilated, totally-enclosed pipe-ventilated, totally-enclosed water-air cooled, totally-enclosed fan-cooled, or dust-ignition proof with no external openings and maximum full load external temperature per 500-5(f).

E. **502-10(b)(2)** Motor-driven utilization equipment shall be dust-ignition proof per 502-8(b).

F. **502-11(b)(1)** Portable lamps shall be approved for Class II, Division 2 locations and marked for maximum lamp wattage.

G. **502-11(b)(2)** Fixed lighting shall be approved for Class II, Division 2 or effectually dusttight enclosures marked for maximum lamp wattage and shall comply with 500-5(f).

H. **502-13(b)** Receptacle and plug shall provide connection to grounding conductor and be of type that supply circuit cannot be broken while live parts are exposed.

I. **502-14(b)(1)** Dusttight enclosures for signaling apparatus are required for Class II, Division 2 locations.

502-16 Equipment in Class II, Division 1 and 2 locations shall be grounded per 250. For bonding and types of EGCs, see 502-16(a)(b).

CLASS III, DIVISION 1 LOCATIONS – 503

500-9(a): Class III, Division 1 Locations – Areas where easily-ignited fibers or materials producing combustible flyings are handled, manufactured, or used.

A. 503-2 Transformers and capacitors containing liquid that will burn shall be in a vault per 450-41 – 450-48. They shall comply with 502-2(b).

B. 503-3(a) Wiring methods shall be RMC, IMC, approved Types MI or MC cables, or RNMC, EMT, or dusttight wireways.

C. 503-3(a)(1) All boxes and fittings shall be dusttight.

D. 503-4 Switches and controllers shall be in dusttight enclosures.

E. 503-6, Ex. a, b, c Motors shall be totally-enclosed or pipe-ventilated. Self-cleaning squirrel cage textile motors may be locally acceptable. Partially-enclosed or splash-proof motors shall not be installed in Class III, Division 1 locations.

F. 503-8(a) Utilization equipment shall be approved for Class III locations.

G. 503-9(a) Lamps shall be protected against entrance of lint and fibers. They shall also be marked for their maximum wattage. Their surface temperature shall not exceed 165°C.

H. 503-9(d) Portable light fixtures shall be unswitched. Exposed metal parts shall be grounded.

I. 503-11 Receptacles and plugs shall be the grounding type and designed to minimize entry or accumulation of fibers or filings and prevent escape of sparks. See Exception.

J. 503-12 Signals, remote-control, and intercommunications equipment shall comply with the requirements of 503.

CLASS III, DIVISION 2 LOCATIONS – 503

500-9(b): Class III, Division 2 Locations – Areas where easily-ignited fibers are stored or handled.

A. 503-3(b) The same wiring methods as required for Class II, Division 1 are acceptable, except open wiring is permissible in storage areas that have no machinery. See Exception.

B. 503-9(a) Lamps shall be protected against entrance of lint and fibers. They shall also be marked for their maximum wattage. Their surface temperature shall not exceed 165°C.

C. 503-13 Cranes operating over combustible fibers, etc. shall be specially wired: (a) power supply to contact conductors isolated and ungrounded; (b) contact conductors to be guarded against accidental contact; (c) current collectors to be guarded against escape of hot particles or sparks; and (d) control equipment per 503-4 and 503-5.

503-15 No exposed live parts are permitted in Class III, Division 2 locations.

503-16 Equipment and wiring shall be grounded per 250. Bonding and EGCs shall conform to 503-16(a)(b).

Note: In general, provisions for wiring in Class III, Division 1 locations apply in Class III, Division 2 locations.

COMMERCIAL GARAGES – 511

A. 511-2 Where the garage is used for parking and storage and only routine maintenance requiring no electrical tools, or where exchange of parts is performed in adequate ventilation, it is not classified as a hazardous area. A residential or commercial parking garage could include large apartment building garages. If the area is not classified as hazardous, the door to the adjacent area is no problem, except fire rating per the building code. There is no restriction on the number of cars using volatile flammable liquids for fuel or power.

B. 511-3(a) The area to 18″ above the floor is a Class I, Division 2 area except when proper mechanical ventilation and the AHJ determines otherwise. See Exception.

C. 511-3(b) A pit is a Class I, Division 1 area unless ventilated and approved by the AHJ, then it may be a Class I, Division 2 area.

511-3(b), Ex. The pit is unclassified if there is no dispenser unit. See Table 514-2.

511-3(d) Adjacent areas may not be considered to be hazardous areas per the AHJ.

D. 511-3(f) If a lamp and cord can be used in a hazardous area, it shall be approved for a Class I, Division 1 area. The outlet shell shall be molded composition or other suitable material.

E. 511-4 A raceway under the floor is considered in a hazardous area if any connections or extensions lead into or through this area. See Exception.

F. 511-6(a) Fixed wiring above hazardous areas shall be in metallic raceways, RNMC, or MI, TC, FMC, LTFMC, LTFNMC, or MC cable. Conductors are not permitted in any cell, header, or duct containing foreign pipes.

511-7(a) Arc or spark producing equipment, such as switches or motors (excluding receptacles, lamps, and lampholders), mounted less than 12′ from the floor, shall be the totally-enclosed type.

G. 511-7(b) Fixtures shall be at least 12′ above the driveway unless they are the totally-enclosed type.

H. 511-8 Battery charger shall be placed outside hazardous area.

511-9(a) Electric vehicle charging equipment shall be installed per 625.

I. 511-10 GFCIs are required in commercial repair garages for all 125 V, 1φ, 15 A and 20 A receptacles where electrical automotive diagnostic equipment, electrical hand tools, and portable lighting are used.

AIRCRAFT HANGARS – 513

A. 513-3(a) Pit shall be designated as a Class I, Division 1 location.

B. 513-3(b) The general hazardous area throughout the hangar is 18″ above the floor and classified as a Class I, Division 2 location.

C. 513-3(c) The area within 5′ horizontally from aircraft fuel tanks and from the floor to 5′ above aircraft wings is a Class I, Division 2 area.

513-3(d) Adjacent areas are nonhazardous areas where adequately ventilated and physically isolated by walls or partitions.

D. 513-4 Wiring under a hangar floor is a Class I, Division 2 location. Plugs and receptacles in hazardous area shall be explosion-proof or designed so they cannot be energized while connections are made or broken.

513-5(a) Fixed wiring not within the hazardous area shall be in metallic raceway, Type MI, TC, or Type MC cable. See Exception.

513-5(c) Extra-hard usage cord with a separate EGC shall be used for portable equipment.

E. 513-6(a) Equipment less than 10′ above wings and engine enclosures shall be of the totally-enclosed type. See Exception.

F. 513-6(b) No metal-shell, fiber-lined lampholders are required in Class I locations.

513-6(c) Portable lamps shall be approved for the locations in which they are used.

513-6(d) Portable utilization equipment shall be approved for Class I, Division 2 locations.

513-7(a) If used in hazardous area, wiring shall be approved for use in Class I, Division 2 locations.

513-7(b)(c) If not for use in hazardous area, and wired per 513-5 and 513-6, a warning sign shall be used ("WARNING – KEEP 5′ CLEAR OF AIRCRAFT ENGINES AND FUEL TANK AREAS").

513-10 Battery charging equipment not in hazardous areas shall be in a separate building or area. Mobile units shall have a sign.

513-11, 513-12 Equipment not suitable for Class I, Division 2 shall have all electrical devices and wiring at least 18″ off floor and shall have a warning sign.

G. 513-16 Metal raceways, metal-jacketed cables, and noncurrent-carrying metal parts of equipment shall be grounded per 250.

GASOLINE DISPENSING AND SERVICE STATIONS – 514

514-1 If the flash point of the liquid is below 100°F, the AHJ may class the area as a nonhazardous area.

A. 514-2 The hazardous area shall not extend beyond an unpierced wall.

B. Table 514-2 The area to a height of 18″ above grade for a distance of 20′ horizontally in every direction from the dispensing pump is a Class I, Division 2 location. Wiring below the surface of Class I, Division 1 and 2 locations shall be Class I, Division 1 locations.

Spherical volume with a 3′ radius at the point of discharge is considered a Class I, Division 1 location. Volume between a 3′ and 5′ radius is considered a Class I, Division 2 location.

Within 18″ horizontally in all directions extending to grade from the dispenser enclosure or the portion of the enclosure with liquid handling components is a Class I, Division 2 location.

Area within 10′ radius of tank fill pipe is considered a Class I, Division 2 location to a height of 18″ above ground level.

514-4 Wiring above Class I areas shall be in metallic raceways, RNMC, or Types MI, TC, or MC cables per 511-6.

514-5 Double-pole switching is required on all circuits to or through dispensing pump including the grounded conductor, if any.

514-5(b)(c) Attended or unattended self-service station shall have emergency control disconnects with the required distances.

C. 514-7(a) Seal fittings are required on all conduits entering or leaving the dispensing area. The seal shall be the first fitting after the conduit emerges from the earth or concrete.

514-8 Conduit runs below the surface of hazardous area are considered Class I, Division 1 to point of emergence above grade. Underground wiring is permitted in RMC or threaded steel IMC. Where RNMC is used, it shall be at least 2′ deep and, if emerging from the underground, the last 2′ of underground to emergence above ground shall be RMC or IMC with threaded fittings. Where PVC is used, an EGC is required. All wiring systems shall comply with 300 including Table 300-5, which requires 24″ depth under areas subject to heavy vehicular traffic.

514-16 Metal raceways, metal-jacketed cables, and noncurrent-carrying metal parts of equipment are grounded per 250.

BULK STORAGE PLANTS – 515

A. 515-1: **Bulk Storage Plant** – A location where volatile flammable liquids are stored in tanks having a total capacity of one carload or more from which liquids are distributed by tank truck or other methods.

B. Table 515-2 When loading tank trucks or cars, the area extending 3′ in all directions from open domes or atmospherically-ventilated closed domes is a Class I, Division 1 location and the distance from 3′ to 15′ is a Class I, Division 2 location. The area within 3′ of a fixed connection used in bottom or in dome loading is a Class I, Division 2 location and where such loading is from a bottom connection, the area shall extend to a height of 18″ within a radius of 10′.

Pumps, bleeders, withdrawal fittings, meters, and similar devices located indoors where flammable liquids are handled under pressure shall be considered Class I, Division 2 locations within a distance of 5′ in all directions from any surface of handling equipment and to a height of 3′ for a distance of 25′.

Where Class I liquids are stored, handled, or dispensed, Table 515-2 shall be used to delineate and classify bulk storage plants. A Class I location shall not extend beyond an unpierced wall, roof, or other substantial partition.

For above-ground tanks, the area above the roof and within the shell of a floating roof tank as a Class I, Division 1 location, while the area above the roof of a closed-roof tank is a Class I, Division 2 location. Areas within 10′ of a tank shell and inside a dike are Class I, Division 2 locations. Areas within 5′ of a vent opening are Class I, Division 1 locations and within 5′ to 10′, Class I, Division 2 locations.

Where drum or container filling is done outdoors or indoors with adequate ventilation, such areas are considered Class I, Division 1 locations within a distance of 3′ from the vent or fill opening extending in all directions and a Class I, Division 2 location within a distance of 3′ to 5′ from the vent or fill opening with a further area to a height of 18″ and a distance of 10′.

Garages for tank vehicles are Class I, Division 2 locations to a height of 18″ above the floor unless special hazard exists.

Pits within Division 1 or Division 2 locations are considered Class I, Division 1 locations unless adequately ventilated, then they become Class I, Division 2 locations. Those outside such locations but containing piping or valves are Class I, Division 2 locations.

515-4 Fixed wiring above hazardous area shall be in metallic raceways, PVC, Sch. 80 RNMC, MI, TC, or Type MC cable. Sparking equipment shall be totally enclosed or guarded.

C. 515-5(a)(b)(c) Underground wiring is permitted in RMC, threaded steel IMC, RNMC, or approved MI cable. Where RNMC is used, it shall be at least 2′ underground, and, if emerging from underground in a Class I area, the last 2′ of underground to emergence above ground shall be in RMC or steel IMC with threaded fittings. Where RNMC is used, an EGC is required. Where cable is used, it shall be in RMC or steel IMC from the lowest buried cable to the connection point above ground. All conductor insulation shall comply with 501-13.

SPRAYING, DIPPING, AND COATING PROCESSES – 516

OPEN SPRAYING

FINISHING PROCESSES

SPRAY BOOTH

PAINT DRYING

A. 516-2(a)(3) The area containing paint spray is considered a Class I, Division 1 location.

B. 516-2(b)(1) The area outside of the Class I, Division 1 location per 516-2(a)(3) is classed as a Class I, Division 2 location.

516-2(b)(3) If open top spray booth, the 3′ above the booth and the 3′ from the other openings are considered as Class I, Division 2 locations.

516-2(b)(4) If the booth is enclosed, 3′ in all directions from all openings are Class I, Division 2 locations.

C. 516-2(d) Drying and baking areas are generally considered nonhazardous if adequately ventilated, protective devices of the ventilating system are interlocked with the heating circuit, and acceptable to the AHJ.

516-3(b) Electrical equipment shall not be installed where subject to vapor or deposit of residue unless listed for purpose.

516-3(c) Illumination through glass panels or equivalent permitted if installed per 516-3(c)(1 – 5).

516-4 Fixed electrostatic spray equipment shall be listed or an approved type.

516-4(a) Electrical equipment within the hazardous area shall be approved for the location. See Exception.

516-4(b) Electrodes and electrostatic atomizing heads shall be adequately supported and insulated from ground.

516-4(j)(1) The separation between grid and work shall be at least twice the sparking distance.

516-5 Electrostatic hand-spraying equipment shall be approved for the location. Transformers, etc. shall be located outside the hazardous area. The handles of spray guns and other conductive objects shall be grounded. Target objects shall maintain contact with a grounded support. The area shall be properly ventilated and interlocked with spraying equipment. Hand guns and connections to power supply are permitted in the hazardous area.

516-7(a) Fixed wiring above hazardous areas shall be in metal raceways, ENT, RNMC, or Type MI, TC, or MC cable.

INHALATION ANESTHETIZING LOCATIONS — 517

A. 501-5; 501-5(a)(4) Seals may be located within 10′ of the point where conduit emerges from the boundary wall provided the junction box contains a seal-off device and the conduit between the box and the fitting is one continuous piece.

B. 517-60(a)(1) Anesthetizing locations shall be Class I, Division 1 locations to a height of 5′ above the floor.

C. 517-60(a)(2) A storeroom for combustible gas or other volatile agents shall be a Class I, Division 1 location throughout.

517-60(b) An inhalation anesthetizing location designated only for use of nonflammable anesthetizing agents shall be considered to be a nonhazardous area. In such cases, 517-61(a)(b), as applied to X-ray systems only, does not apply.

D. 517-61(a)(3) Portable equipment shall be approved for use in Class I, Division 1 locations if more than 10 V. A suction unit shall be approved for Class I, Division 1 locations and also specifically approved for hazardous atmospheres involved. Fixed wiring and equipment, and portable equipment over 10 V between conductors in hazardous areas, shall conform to Class I requirements per 501-1 – 501-15 and 501-16(a)(b), and shall be specifically approved for the location.

517-61(a)(4) Boxes, fittings, or enclosures partially within the hazardous area shall be approved for a Class I, Division 1 location.

E. 517-61(a)(5) Receptacles and attachment plugs shall provide connection for a grounding conductor and be listed for Class I, Group C.

F. 517-61(a)(6) Cords supplying equipment of more than 8 V shall be extra-hard usage type and shall have a take-up reel or equivalent for storage.

517-61(b)(1) Wiring above the hazardous area shall be per 517-60 in RMC, EMT, IMC, or Type MI or MC cable with gas/vapor-tight metal sheath.

517-61(b)(2) Spark-producing devices, except wall-mounted receptacles above hazardous areas, shall be the totally-enclosed type or constructed to prevent escape of sparks. See Exception.

G. 517-61(b)(3) Surgical lighting units shall be approved for Class I, Division 2 locations, except there are no surface temperature limitations. See 501-9(b).

H. 517-61(b)(3), Ex. 1 Fixtures above the hazardous area shall conform to the same requirements as fixtures in Class I, Division 2 locations per 501-9(b)(2), except for temperature limitations and integral or pendant switches.

517-61(b)(3), Ex. 2 An integral or pendant surgical switch is not required to be explosion-proof if it is not lowerable into the hazardous area.

517-61(b)(4) Seals are required when any part of the system extends into a hazardous area per 501-5 and 501-5(a)(4).

517-61(b)(5) Single-phase, 120 V plugs and receptacles above the hazardous area shall be listed for hospital use and voltage, frequency, rating, number of conductors, and provide for connection of the grounding conductor.

517-64(a)(1)(2)(3) Equipment in contact with persons shall not be over 10 V or shall be intrinsically safe or double-insulated and be moisture-resistant.

517-64(c)(2) Isolating transformer for low-voltage circuits shall have the core and case grounded.

517-64(f) Receptacles and plugs on low voltage shall be non-interchangeable with higher voltages.

I. 517-75 All new, used, or reconditioned X-ray equipment moved to a new location shall be an approved type.

517-160(a)(1) Circuits to equipment over 10 V shall be isolated and have a switch in each conductor. Isolated power system may be transformer, M-G set, or batteries.

J. 517-160(a)(2) In general, circuits in anesthetizing locations shall be supplied by isolating transformers rated at not more than 600 V primary or secondary. The primary winding shall have overcurrent protection. Secondary circuits shall have overcurrent protection, double-pole switching, and be ungrounded.

K. 517-160(b)(1) The secondary circuit shall have a ground detector, a green light when under 3.7 mA, a red light and audible when 5.0 mA or over. Lights shall be conspicuously visible to persons in the room. See Exception.

THEATERS AND SIMILAR LOCATIONS — 520

A. **520-1** This article includes the building or part of the building used for presentation, musical, dramatic, motion picture projection, and assembly areas of motion picture and TV studios.

B. **520-5(a)** Fixed building wiring shall be metal raceways, PVC encased in 2″ of concrete, or Type MI or MC cable. See Exception.

C. **520-5(b)** Portable equipment is permitted to be wired with approved flexible cords and cables.

520-5(c) Nonrated areas are permitted to be wired with Type NM, AC cable, or RNMC or ENT.

520-7 No exposed live parts are permitted on stage. Live parts shall be enclosed or guarded.

D. **520-21** Stage switchboard shall be dead-front and per 384, Part C unless suitable as a stage switchboard.

520-22 Exposed live parts on back shall be enclosed by walls, grills, or other approved means and the entrances of enclosures shall have self-closing doors.

E. **520-23** Controls and overcurrent devices for portable branch circuits are required at stage switchboards.

520-24 A metal hood is required unless the stage switchboard is dead-front and dead-back (or recessed).

520-41(a)(b) Circuits for footlights, border lights, and side lights shall not be over 20 A unless they have heavy-duty lamps. See 210 for heavy-duty lampholders.

520-42 Conductors shall have at least 125°C insulation.

F. **520-43(b)** Footlights shall be a metal trough or individual outlets with RMC, FMC, IMC, or Type MI or MC cable. Circuit conductors shall be soldered to lampholder terminals.

520-43(c) An automatic disconnect switch is required for disappearing footlights.

520-44(b)(1) Electric cables for border lights are permitted only where flexibility is necessary. They shall be listed for extra-hard usage and suitably supported.

520-45 Stage receptacles for equipment or fixtures shall be rated in amperes. Conductors shall be rated per 310 and 400.

520-46 Receptacles for portable stage lighting are required to be in suitable pockets or enclosures and shall comply with 520-45.

520-47 Lamps in scene locks shall be guarded and 2″ away from combustible material. See Exception.

G. **520-48** Curtain machines shall be listed.

520-49 There shall be two required switches in the closed circuit flue. The damper control shall be set at the electrician's station.

H. **520-51** A fused switch or CB shall control a portable switchboard.

520-53(g) A pilot light shall be provided in the enclosure.

520-53(h)(1) Supply cords or cables shall be listed for extra-hard usage.

I. **520-61** Arc lamps are required to be listed.

J. **520-62(b)** Portable plugging box receptacles shall have overcurrent protection per 520-45 and be totally enclosed.

K. **520-64** Portable strips shall be constructed per the specifications for border lights.

L. **520-66** "Effect units" shall be guarded against contact with combustible materials.

M. **520-68(a)(2)** A reinforced cord is acceptable for a stand lamp if the circuit protection is not over 20 A.

N. **520-81** All metal raceways, metal frames, and enclosures of equipment (including border lights and portable fixtures) shall be grounded per 250.

CABLE TRAYS — 318

A. **318-3(a)** Cable trays shall be permitted to support: MI cable, ENT, AC, MC, NM, SE, UF, power and control tray cable, ITC, power-limited tray cable, and other factory-assembled tray cables, IMC, RMC, RNMC, EMT, FMT, FMC, optical fiber cables, LTFMC, LTFNMC, fire alarm cable, and communication cables.

318-3(b)(1)(2) In industrial establishments where supervision and maintenance is present, other single conductors and multi-conductor cables are permitted.

318-3(b)(1) c. The EGC is permitted to be No. 4 or larger.

318-6 a. Each run shall be a complete system before cables are installed.

B. **318-6(f)** Cables rated 600 V or less are not permitted with cables rated over 600 V except: (1) where separated by a fixed barrier of material; (2) Type MC cable is used.

318-6(g) Cable trays may pass through partitions, walls, floors, or platforms where the possible spread of fire is prevented per 300-21.

318-6(i) Adequate working space is required about cable trays to permit installation and maintenance of cables.

C. **318-7** All metal parts shall be grounded in the same manner as for conductor enclosures per 250.

318-7(a)(b)(1 – 4) Metal trays shall be grounded per 250, and, if used as EGCs, shall comply with 318-7(b)(2). See Table 318-7(b)(2).

318-8(a) Approved cable splices are permitted in the cable tray if they are accessible and not above the top of the side rails.

318-11(a)(1 – 3) Multiconductor cables shall have ampacity per Tables 310-16 and 310-18 except they may be derated to 95% if more than 6′ of trays have an unventilated solid cover. Derating shall be limited to the number of current-carrying conductors in the cable and not to the number of conductors in the cable tray.

BRACKET FIXTURE WIRING – 520-63

520-63(a) Brackets on scenery to be wired internally, or, if hard usage cord terminating in connector within 18″ is used, may be wired externally.

FESTOONS – 520-65

520-65 Joints in festoon lighting shall be staggered. Lamps, such as lanterns, shall have a guard if enclosed in combustible material.

DIMMERS – 520-25

A. **520-25(a)** Where installed in ungrounded conductors, each dimmer shall be protected by an overcurrent device set at a value not exceeding 125% of its rating.

B. **520-25(b)** Resistance or reactor (series) dimmers may be placed in either line wire and shall comply with 380-1. Dimmers in the grounded neutral shall not open the circuit.

C. **520-25(c)** Voltage between conductors shall not exceed 150 V. The grounded conductor shall be common to both the load and supply.

MULTIPOLE BRANCH-CIRCUIT CABLE CONNECTORS – 520-67

520-67 The female half of the connector shall be attached to the load side of the cord or cable. Cable connectors shall be constructed so the tension on the cord or cable will not be transmitted to connections. Connectors to be rated in amps and designed so that different rated males and females cannot be properly connected and polarized. The OCPD shall not be greater than the feeder ampacity.

SMOKE VENTILATOR CONTROL – 520-49

A. **520-49** The circuit operating device shall be normally closed and shall be controlled by at least two externally operable switches.

DRESSING ROOMS – PART F

110-2 Equipment (including dressing room lighting) shall be approved by the AHJ. Also see 90-4 and 100 – Definitions, approved for purpose.

520-71 Pendant lampholders are not permitted in dressing rooms.

520-72 Lamps within 8′ of the floor shall have riveted, open-end guards or otherwise locked in place.

520-73 Lamps and receptacles located adjacent to mirrors and above the dressing table shall be controlled by wall switches. Switch-controlling receptacles shall have pilot lights.

STAGE OR SET – 530, PART B

530-11, 530-12, 530-13 Studio wiring requirements generally follow those for theater stage.

A. **530-13** A single switch may be used to disconnect all contactors on a location board that is not more than 6′ away.

B. **530-18(b)** Overcurrent devices for feeders may be set at not over 400% of the conductor rating.

C. **530-18(e)** Where plugging box does not have overcurrent devices, every cord or cable smaller than No. 8 supplied through it shall have its own overcurrent device.

530-19(a) See Table 530-19(a) for feeder demand factors for stage set lighting.

530-20 Pendant and portable lamps and special portable equipment need not be grounded if the operating supply voltage is not over 150 V DC to ground.

SOUND-RECORDING AND SIMILAR EQUIPMENT – 640

A. 640-2 An audio amplifier or pre-amplifier is a device that increases or decreases the current, voltage, or both potential of an audio signal.

B. 640-6 Equipment and cabling shall be installed in a neat and workmanlike manner. Cables shall be supported by the building structure in a manner which ensures that the cable will not be damaged by normal building use.

C. 640-7(b) Separately derived systems with 60 V to ground shall be grounded per 530-72.

D. 640-9(a)(1) Wiring and equipment from the source of power to and between the premises wiring system shall comply with the requirements of Chapters 1 – 4, unless modified by 640.

E. 640-9(c) Output wiring for amplifiers shall be permitted to be Class 1, Class 2, or Class 3 wiring as stated in 725. Overcurrent protection must be provided and may be inherent to the amplifier.

F. 640-24 Wiring within wireways, gutters, and auxiliary gutters shall comply with the requirements for the applicable article. For example, nonmetallic wireway fill is per 362-19.

FILM STORAGE VAULTS – 530, PART E

A. 530-51 Lamps in cellulose nitrate film storage vaults shall be rigid, glass-enclosed, and gasketed. A switch with pilot light shall be located outside the vault.

B. 530-52 No electrical equipment other than permanent lighting fixtures shall be located in the vault.

PIPE ORGANS – 650

A. 650-3 Source shall not be over 30 V.

640-4 Rectifiers used as source of supply for organs shall be grounded per 250-112(b).

B. 650-5(a) Conductors for electromagnetic valve supply shall be a minimum of No. 26 AWG. The common return shall not be smaller than No. 14.

650-5(b) Conductors shall have thermoplastic or thermosetting insulation.

650-5(c) Wires shall be cabled except for the common return conductor.

650-5(d) Cable shall have flame-retardant braid or close-wound fireproof tape unless in metal raceways.

C. 650-7 Overcurrent protection shall not be over 6 A except for the main supply and common return conductors.

AUXILIARY GUTTERS – RECESSED

A. 374-9(b) Gutters shall be constructed in a substantial manner and shall be installed as a complete enclosure for the contained conductors.

MOTION PICTURE PROJECTORS – 540

A. 540-11(a) Motor generators and similar equipment shall be located in a separate room where possible.

B. 540-11(b) In general, switches or overcurrent devices foreign to projector operation are not permitted in projection rooms.

540-11(b), Ex. 1 Rooms approved for safety film only require signs inside and outside.

540-11(b), Ex. 2 Remote control for auditorium lights or curtain motors.

C. 540-12 Clear working space of at least 30″ on each side and at rear of projector, floodlight, spotlight, or similar equipment.

540-13 Circuit conductors for arc projectors shall not be smaller than No. 8. Conductors for incandescent-type projectors shall conform to general provisions of the NEC®. See 210-34.

540-14 Wiring on lamps to have 200°C insulation where the ambient temperature is over 50°C.

D. 540-20 Projectors and associated equipment shall be listed.

EMERGENCY SYSTEMS – 700

EXIT LIGHTS

Ⓑ STATE-OF-CHARGE INDICATOR LIGHTS

Ⓐ

ILLUMINATED EXIT SIGN Ⓓ

FRONT

BRANCH-CIRCUIT FOR OVERCURRENT DEVICE

Ⓔ EMERGENCY CIRCUIT

JUNCTION BOX FOR CONDUIT CONNECTION Ⓒ

BACK

A. 700-1 Emergency systems are normally installed in buildings where large numbers of persons assemble or occupy such buildings, such as theaters, assembly halls, hotels, sports arenas, hospitals, and similar institutions. Emergency systems also provide power for hospital equipment that is essential to maintain life and for illumination and power for operating rooms, fire alarm systems, fire pumps, critical public address systems, industrial processes where interruption would cause hazards, and similar functions. See *Life Safety Code*, NFPA 101, 1997 (ANSI) for critical emergency illumination and exit lights and lighting. 700-1 applies when emergency systems or circuits are legally required by municipal, state, federal, or other codes or by any governmental agency having jurisdiction.

B. 700-7 Audible and visual derangement signals recommended to ensure constant readiness of emergency system. See 700-7(a – d).

C. 700-9 Wiring to be entirely separate from all other wiring and boxes and raceways to be identified. See 700-9(a – d).

700-12(e) Unit requirement for emergency lighting to consist of: 1. rechargeable battery; 2. means to charge battery; 3. one or more lamps on unit (terminals are

permitted for remote lamp location); 4. a relay automatically energized by failure of general lighting system.

Battery shall be able to maintain at least 87½% of nominal voltage at full lamp load for at least 1½ hours or maintain at least 60% of rated illumination for at least 1½ hours. Unit shall be approved for emergency service. Unit not to be portable and to be wired per Ch 3. Flexible cord and plug permitted if not over 3′ long. Unit wired to same branch circuit as other normal lighting in area and connected ahead of any local switches. Remote location unit equipment lamps shall be wired per Ch 3. Branch circuit feeding unit equipment shall be clearly identified at distribution panel.

700-15 Only emergency light outlet may be served by emergency lighting system.

D. 700-16 Emergency lighting includes exit lights and illuminated exit signs.

E. 700-20 Only authorized persons have access to emergency switch. Series, 3-, or 4-way switches not permitted. See Ex. 1, 2.

700-22 Exterior emergency light may be switched by light-actuated device.

STORAGE BATTERY

STORAGE BATTERY

700-12(a) Storage batteries are permitted for emergency service if their capacity is adequate to supply and maintain 87½% of the system voltage when supplying a full emergency load for 1½ hours. See 480 for installation of storage batteries.

GENERATOR SET

700-12(b)(1) A generator with automatic transfer and automatically-started prime mover with 15-minute time delay before restoring normal power is permitted for emergency power.

700-12(b)(2) Where an internal combustion engine is the prime mover, a 2-hour fuel supply is required.

700-12(b)(3) Where the prime mover fuel is natural gas from the utility company, a second fuel supply is required with automatic transfer to second fuel supply. See 700-12(b)(3), Ex.

700-12(b)(4) Where a battery is used to start the prime mover, it shall be suitable for the purpose and have automatic charging means.

SEPARATE SERVICE

SERVICE NO. 1

SERVICE NO. 2

700-12(d) Two distinct services with separate service drops or laterals, widely separated physically and electrically, is permitted. Each service to be treated as a separate service and installed per 230 shall be approved by the AHJ.

TRANSFER SWITCHES

EMERGENCY SYSTEM

NORMAL SERVICE SUPPLY

AUTOMATIC TRANSFER EQUIPMENT

EMERGENCY SERVICE SUPPLY

700-6 Transfer equipment with automatic transfer switches shall be automatic and identified for use or approved by the AHJ. Equipment shall be designed and installed so connection of both normal and emergency power is not possible. See 250-5(d) for grounding of separately derived systems.

700-18 Branch circuits for emergency use shall be automatically transferred upon failure of the normal power supply.

CONNECTION AHEAD OF DISCONNECTING MEANS

MAIN SERVICE

EMERGENCY SERVICE

700-12(a – d) Emergency power sources can be: (a) storage batteries; (b) generator set; (c) UPS, (d) separate service; or (e) unit equipment. Connections ahead of or on the line side of a service disconnecting means are not permitted to serve as an emergency power source.

CIRCUITS AND EQUIPMENT LESS THAN 50 VOLTS

720-1 Wiring shall be under 50 V, AC or DC, except per 411, 551, 650, 669, 690, 725, and 740.

720-4 Conductors shall not be smaller than No. 12 AWG Cu and not smaller than No. 10 AWG Cu if supplying more than one appliance receptacle.

720-5 Lampholders shall be rated at least 660 V.

720-6 Rating of receptacles shall not be less than 15 A.

720-7 20 A receptacles shall be used for kitchen, laundry, and other locations where portable appliances could be used.

720-10 Grounding shall be per 250.

CLASS 1 CIRCUITS

WIRING PER CH 3
C

OVERCURRENT DEVICE
B

CONDUCTORS PROTECTED WITHOUT DERATING FACTORS
A

725-6 Grounding of Class 1 circuits and equipment shall be per 250.

725-21(a)(b) Class 1 Circuits – Power Limitations: Source rated output not over 30 V and 1000 VA and having overcurrent protection not over 167% of VA rating of source divided by source-rated volts. Overcurrent devices to be noninterchangeable and approved for purpose. May be integral part of power supply. (1) Article 450 applies where transformers are used to supply Class 1 circuits. (2) If source is other than transformer, it shall have maximum output of 2500 VA and maximum amps times maximum volts shall not exceed 10,000 VA with overcurrent protection bypassed. (b) Class 1 remote-control and signaling circuits not over 600 V but power output need not be limited.

A. **725-23** Conductors (No. 14 or larger) to be protected per Tables 310-16 through 310-19 without derating factors. Maximum overcurrent protection for No. 18 is 7 A and 10 A for No. 16. See Ex. and FPN.

B. **725-24** Overcurrent device at point where conductor receives supply.

725-24, Ex. 1 Where overcurrent device protecting larger conductor also protects the smaller conductor.

725-24, Ex. 2 Transformer secondary conductors.

C. **725-25** Wiring installed per Ch 3 except where permitted by 725-26 through 725-28 and other articles of the NEC®.

725-26 Two or more Class 1 circuits may be in the same box cable or raceway if all wires are insulated for the maximum voltage. Also, Class 1 and power supply conductors shall be intermingled only when supplying equipment to which Class 1 conductors are functionally associated. See Ex. 1, 2.

725-27(a) No. 18 and No. 16 conductors permitted if in raceways or listed cables and if leads do not exceed ampacity per 402-5.

725-27(b) Conductors rated 600 V. Conductors No. 16 and larger per 310. No. 16 and No. 18 shall be TF or similar. See Types listed in 725-27(b).

725-28(a)(b) Ampacity of Class 1 wires shall be reduced for over 3 conductors if continuous load. Ampacity of all reduced when power supply and Class 1 in raceway and more than three Class 1 carry continuous loads or ampacity of power supply conductors only when more than three.

CLASS 2 AND CLASS 3 CIRCUITS

TRANSFORMER PRIMARY

A 20 A OVERCURRENT DEVICE

INPUT LEADS **B**

TRANSFORMER

TRANSFORMER SECONDARY

OVERCURRENT DEVICE **C**

725-6 Class 2 and 3 circuits and equipment shall be grounded per 250.

725-26(b) Class 1 circuits and power supply circuits can only occupy the same cable, enclosure, or raceway where the equipment powered is functionally associated. See Ex. 1 for factory or field-assembled control centers and Ex. 2 for conductors in manholes.

725-41(b) Power supplies shall not be interconnected or paralleled unless listed for such interconnection.

A. **725-51** Transformer is protected (maximum) by a 20 A overcurrent device.

B. **725-51, Ex.** Input leads shall not be smaller than No. 14 except may be No. 18 if not over 12″ long and insulation per 725-27(b).

725-54(b)(4) Class 2 and Class 3 circuits may be run in same cable as communication circuits, but shall be classed as communication circuits and conform to 800.

C. **Ch 9, Table 11(a), Notes** Overcurrent devices shall not be interchangeable with devices of higher ratings and they may be an integral part of power supply. Overcurrent devices shall be located at a point where conductor to be protected receives its supply.

COMBINING CLASS 2 AND CLASS 3 CIRCUITS

A CLASS 2 CIRCUIT

C CLASS 2 CIRCUIT

B CLASS 3 CIRCUIT

A. **725-54(b)(1)** Two or more Class 2 circuits are permitted in the same raceway, cable, or enclosure.

B. **725-54(b)(2)** Two or more Class 3 circuits are permitted in the same raceway, cable, or enclosure.

C. **725-54(b)(3)** One or more Class 2 circuits are permitted in same raceway, cable, or enclosure with Class 3 circuits if Class 2 conductor insulation is at least equal to that required for Class 3 circuits.

725-54(c) Where Class 2 or Class 3 circuits extend beyond building and are subject to possible contact with over 300 V to ground, light, or power circuits, they shall conform to 800-10, 800-12, 800-13, 800-30, 800-31, 800-32, 800-33, and 800-40.

725-71(a – h) Class 2 circuit conductor insulation and size shall be suitable for application. Class 3 circuit conductors per Table 725-71.

SIGNS – GENERAL RULES – 600 V OR LESS

A. 600-3 All signs shall be listed and installed per listing unless otherwise approved by special permission.

600-5(a) Commercial buildings shall have an outside sign outlet on a 20 A circuit. Interior halls or corridors are not considered accessible to pedestrians.

B. 600-5(b)(1) Rating of branch circuits that supply signs and other outline lighting systems with either fluorescent, incandescent, or high-intensity discharge illumination shall not exceed 20 A.

600-6, Ex. 1 A disconnect is not required for an exit directional sign.

C. 600-6(a) The disconnect shall be within sight of the sign it controls.

600-6(b) Devices such as switches or flashers shall have twice the ampere rating of the transformer, or shall be approved for the purpose, except AC general-use snap switches are permitted on AC inductive loads not exceeding the switch rating.

600-7 All metal equipment of outline lighting systems and signs shall be grounded. Listed flexible metal conduit is permitted to serve as equipment grounding means in lengths not exceeding 100' if it supplies secondary wiring of a transformer or power supply for electrical discharge tubing.

600-8 Live parts other than lamps and neon tubing shall be enclosed. See Exception.

600-21(a) Ballasts, transformers, and power supplies shall be accessible.

600-21(c) Ballasts, transformers, and power supplies shall be weatherproof or protected from the weather.

600-21(d) A 3' × 3' × 3' working space shall be provided.

NEON TUBING – 600-41

A. 600-41(a) Neon tubing shall be designed so the length of the tubing shall not cause a continuous overcurrent of transformer or electronic power supply.

B. 600-41(b) Neon tubing shall be supported by listed tube supports.

600-41(c) A minimum of ¼" spacing shall be provided between neon tubing and any surface other than its support.

600-42(a) Terminal connections of neon tubing electrode shall be accessible only to qualified personnel.

C. 600-42(b) Neon tubing terminal connections shall be made by a connection device, twisting conductors together, or an electrode receptacle. Connections shall be made in an enclosure listed for purpose.

D. 600-42(c) Neon tubing shall be supported no more than 6" from the electrode termination.

600-42(e) Bushings listed for the purpose shall be used where electrodes penetrate an enclosure, unless receptacles are used.

SKELETON TUBING – PART B

A. 600-2: **Skeleton Tubing** – Neon tubing that is the sign of outline lighting. It does not contain, nor is attached to a sign body or enclosure.

B. 600-31(a) Conductors for skeleton tubing shall be installed per Ch 3.

LOCATION – 600-9

100: Outline Lighting – An arrangement of either incandescent or electric-discharge lighting components designed to highlight certain elements of a building such as a window.

100 Electric signs can be either fixed, stationary, or portable, self-contained utilization equipment that are electrically illuminated. Signs call attention to or provide information about something.

600-9(a – d) Signs shall be installed so they are suitable for weather conditions they may encounter. Unless protected from physical damage, signs and outline lighting equipment shall be installed a minimum of 14' above ground where accessible to vehicles.

PORTABLE SIGNS – 600-10

A. 600-10(c)(1)(2) Portable signs installed in wet or damp locations shall be equipped with cords rated for junior-hard service or hard service per 400-4. All portable or mobile signs shall have factory-installed GFCI protection.

NEON SIGNS – OVER 1000 V – 600-32

A. 600-32(a) Conductors over 1000 V shall be installed in FMC, LTFMC, LTFNMC, metal enclosures, EMT, RNMC, RMC, IMC, or as concealed conductors on insulators, or as Type MC cable.

B. 600-32(b) Conductors shall be a minimum of No. 18 AWG with a temperature rating of at least 105°C. They shall be insulated and listed for purpose.

C. 600-32(c) Conductors shall be installed so they are not subject to physical damage.

D. 600-32(d) Sharp bends in conductors are not permitted.

600-32(h) If listed for purpose and provided with terminals at midpoint, conductors shall be permitted to run from neon tubing ends to grounded midpoint of a transformer.

600-32(i) In dwellings, open circuit voltage for neon conductors shall not exceed 1000 V.

OUTSIDE BRANCH CIRCUITS AND FEEDERS – 225

CONDUCTORS 1′ MINIMUM APART

CIRCUITS 4″ MINIMUM APART

8 FLOODLIGHTS – EACH ON SEPARATE CIRCUIT

LIGHT BELOW OTHER ELECTRICAL EQUIPMENT

NO. 12 FESTOON CONDUCTORS

6′ MINIMUM FIRE LADDER ZONE

225-4 Festoons or open conductors within 10′ of building shall be insulated or covered.

225-4, Ex. EGC and grounded conductor where elsewhere permitted.

225-6(a)(1) Overhead spans for 600 V or less and not over 50′, No. 10 Cu or No. 8 Al; over 50′, No. 8 Cu or No. 6 Al.

225-6(a)(2) Overhead spans for over 600 V, No. 6 Cu or No. 4 Al for separate conductors and No. 8 Cu or No. 6 Al for cable.

A. **225-6(b)** Festoon conductors shall be not less than No. 12 unless messenger supported. Messenger wire is required where the span exceeds 40′.

225-7(b) A common neutral may be used for circuits on a pole installation. The neutral size shall not be less than the total load connected to any one phase of the supply circuit.

B. **225-14(c)** Circuits shall be at least 4″ apart.

C. **225-14(d)** Conductors on poles shall be at least 1′ apart where not on racks or brackets. For horizontal climbing space, see 225-14(d).

D. **225-19(e)** Fire ladder zone of 6′ adjacent to building or beginning not more than 8′ therefrom if building is over 3 stories or 50′ high.

225-24 Stranded conductors are required where pin-type outdoor lampholders are used.

E. **225-25** Lights shall be below other electrical equipment unless safeguards or clearances are provided for relamping or equipment is controlled by lockable disconnect.

800-10(a)(1)(2) Telephone conductors shall be below power lines and not on the same crossarm with light or power.

800-10(b) Telephone conductors shall be 8′ minimum over roofs.

COMMUNICATION CIRCUITS – 800

AERIAL CABLE

NONMETALLIC CONDUCTORS

LIGHTNING CONDUCTOR

COMMUNICATION CONDUCTORS

LISTED PRIMARY

FUSE

ARRESTER

WATER PIPE

PROTECTOR GROUND

OPEN CONDUCTORS

A. **800-11(a)** Underground communication conductors shall be in separate compartment from light, power conductors, Class 1, or nonpower-limited fire alarm conductors.

B. **800-12(a)** Nonmetallic cables shall be listed for the purpose and have current capacity per 800-30(a)(1) b. or 800-30(a)(1) c.

800-12(c) Where protector is installed in a building, wires shall enter (in general) through an insulating bushing or metal raceway.

C. **800-13** Six foot separation from lightning conductors where practical.

D. **800-30(a)(1)** Fuseless type protectors may be used under special conditions outlined in paragraphs a., b., c., d., and e.

800-30(a)(2) Protector has fuse in each line wire and arrester between each wire and ground.

800-30(b) Protectors shall be located where the conductors enter the building.

800-30(c) Protectors shall not be located in any hazardous area.

E. **800-33** Metal sheath of aerial or underground cables entering buildings shall be grounded or interrupted at the building by an insulating joint or equivalent device.

F. **800-40(a)(b)** The protector ground shall be at least No. 14 Cu, 30 mil rubber insulation, run in straight line to the nearest accessible grounding electrode system per 250-50. Also see 800-40(c)(d).

G. **800-52(a)(2)** Open conductors shall be separated from Class 1 or light or power wiring not in raceways or in metal-clad, metal-sheathed, nonmetallic-sheathed, or UF cables a distance of 2″. See Ex. 1, 2.

SHOW WINDOW SIGNS – OVER 600 V

SIGN TRANSFORMER

NEON TUBING

LOW-VOLTAGE LEADS

HIGH-VOLTAGE TERMINALS

A. **600-3(a)(b)** All electric signs shall be listed and installed per listing unless excepted by special permission.

B. **600-32(c)** Conductors shall be installed as they are not subject to physical damage.

C. **600-32(d)** Sharp bends in installed conductors shall be avoided.

RECEIVING EQUIPMENT – ANTENNA SYSTEMS – 810, PART B

Ⓑ NOT OVER 250 V
BETWEEN CONDUCTORS

Ⓐ COPPER LEAD-IN
CONDUCTORS

OVER 2′
Ⓒ

LIGHT AND
POWER POLE

Ⓓ

EFFECTIVELY
GROUNDED

WATER PIPE

RECEIVING STATION

A. 810-11 Antenna and lead-in conductors shall be copper, bronze, or other corrosion-resistant material.

810-11, Ex. Soft- or medium-drawn copper shall be used for lead-in if the span is not over 35′.

B. 810-12 Antenna and associated conductors shall not be attached to a pole carrying light and power wires over 250 V between conductors or to electric service masts.

C. 810-13 Antenna and lead-in conductors shall not cross over light and power conductors or come within 2′ of them if 250 V or less.

D. 810-15 Masts shall be effectively grounded per 810-21.

810-16(a) Antenna conductors are sized per Table 810-16(a).

810-18(a) Lead-in conductors shall be supported so they cannot swing closer than 2′ to 250 V wires or 10′ to conductors over 250 V. Lead-ins shall not be closer than 6′ to the lightning rod system unless bonded per 250-60. Underground conductors shall be separated at least 12″ from light or power on Class I circuits.

810-18(b) Indoor antennas and lead-ins shall not be closer than 2″ to other wiring systems.

810-18(b), Ex. 1 Unless in metal raceway or cable armor.

810-18(b), Ex. 2 Effectively separated therefrom by continuous and firmly secured nonconductor such as porcelain tubes or flexible tubing.

810-20(a) Lead-in shall have a listed discharge unit.

810-20(a), Ex. Listing is not required where lead-in wires are enclosed in a continuous metal shield that is effectively grounded or protected by the antenna discharge unit.

810-21(b) Grounding conductor may be uninsulated.

810-21(e) Grounding conductor shall be run in a straight line from the mast or discharge unit as practicable.

810-21(f) The grounding conductor shall be connected to the grounding means per 810-21(f)(1)(2)(3).

810-21(g) Grounding conductor for receiving station may be inside or outside of the building.

810-21(h) The grounding conductor shall not be smaller than No. 10 Cu, No. 8 Al, or No. 17 Cu-clad steel or bronze.

810-21(i) A single grounding conductor may be used for both protective and operating purposes.

AMATEUR TRANSMITTING AND RECEIVING STATION – ANTENNA SYSTEMS – PART C

ANTENNA

Ⓐ

3″ CLEARANCE

AMATEUR TRANSMITTING
OR RECEIVING STATION

APPROVED
INSULATING
TUBE

Ⓑ

GROUND

Ⓒ

A. 810-54 Antenna and lead-in wires shall clear surface by 3″.

810-54, Ex. Lead-in may be attached to surface if encased in grounded shield.

B. 810-55 Except when shielded, lead-in conductors shall enter buildings through: (1) an approved insulating tube; (2) a hole giving 2″ of clearance all around; or (3) a drilled window pane.

810-57 An antenna discharge unit is required in transmitting stations or other suitable means to drain static charges.

810-57, Ex. 1 Where continuous metal shield is effectively grounded.

810-57, Ex. 2 Where antenna is effectively and permanently grounded.

C. 810-58(b) The protective ground for transmitting stations shall be as large as the lead-in but not less than No. 10 Cu, bronze, or Cu-clad steel.

810-58(c) The operating ground shall be not less than No. 14 Cu or equal.

INTERIOR INSTALLATION – TRANSMITTING STATION – PART D

OPEN
CONDUCTORS

4″ MINIMUM

SIGNAL WIRES

Ⓐ

METAL
ENCLOSURE

Ⓑ

TRANSMITTER

A. 810-70 Conductors inside the building shall be separated from other open conductors at least 4″ except: 1. as provided in 640; and 2. where conductors are separated by a fixed nonconductor or are in conduit.

B. 810-71(a) The transmitter shall be enclosed by metal or by installing a grounded barrier to separate the transmitter from operating space.

810-71(a)(b) All noncurrent metal parts including handles shall be grounded.

810-71(c) All doors giving access to voltages in excess of 350 V shall have interlocks.

Name _____

Date _____

Trade Competency Test 1

NEC®   **Answer**

_____ T   F   **1.** Seals shall be installed in conduit runs passing between two adjacent Class I, Division 2 rooms.

_____ T   F   **2.** Askarel-filled transformers may be used in Class II, Division 2 locations.

_____ _____   **3.** Locations where combustible dust is normally in the air are designated as _____.

    A. Class I, Division 2   C. Class II, Division 2
    B. Class II, Division 1   D. Class III, Division 1

_____ T   F   **4.** Locations where concentrations of combustible dust may be present are designated as Class II, Division 1.

_____ _____   **5.** Emergency lighting includes exit lights and illuminated exit _____.

_____ T   F   **6.** Open squirrel-cage induction motors that have no sliding contacts may be used in Class I, Division 1 locations.

_____ _____   **7.** Sealing compound is used with mineral-insulated cable in a Class I location for the purpose of _____.

    A. preventing passage of   C. limiting the extent of a
       gas or vapor   possible explosion
    B. excluding moisture   D. preventing the escape of
                              insulating powder

_____ T   F   **8.** Hazardous concentrations of gas or vapor are not normally present in Class I, Division 1 locations.

_____ _____   **9.** Mobile aircraft energizers shall carry a sign reading "WARNING — KEEP _____ FEET CLEAR OF AIRCRAFT ENGINES AND FUEL TANK AREAS."

_____ _____   **10.** In Class I, Division 1 locations, conduit seals shall be placed no farther than _____" from spark-producing devices.

    A. 12   C. 24
    B. 18   D. 30

_____ _____   **11.** In a conduit run leaving a Class I, Division 1 location, the seal may be placed on either side of the _____ within 10′ from the boundary.

_____ _____   **12.** Where gases and dust are present at the same time, both shall be considered when determining the operating _____ of equipment.

_____ _____   **13.** Equipment that has been approved for a Division 1 location shall be permitted in a Division 2 location of the same class and _____.

_____ _____   **14.** For limited flexibility of motor connections in a Class I, Division 2 location, flexible conduit with _____ fittings is permitted.

_____ _____   **15.** In a completed seal in a Class I location, the minimum thickness of the sealing compound for other than listed cable sealing fittings shall not be less than the trade size of the conduit and, in no case, less than _____".

16. Fused isolating switches for transformers not intended to interrupt current in a Class I, Division 2 location _____.

    A. shall be explosionproof   C. shall have interlocking devices

    B. may be installed in a general-   D. may not have doors or openings
       purpose type enclosure   in hazardous area

17. In Class I, Division 2 locations, concentrations of explosive gases are normally _____ in closed containers or systems.

18. Anesthetizing locations shall be Class I, Division 1 locations to a height of _____′ above the floor.

19. A flexible cord may be used in a Class I location between a portable appliance and the _____ portion of its supply circuit.

20. In Class I, Division 1 locations, pendant fixtures with rigid conduit stems over 12″ long shall be _____.

21. Locations where combustible fibers are stored are designated as _____.

    A. Class II, Division 2   C. Class III, Division 2

    B. Class III, Division 1   D. nonhazardous

22. In a Class II location, a horizontal run of conduit not less than _____′ long may be used as the sealing device.

23. The circuit to a 1ϕ motor on a gasoline-dispensing pump shall have a(n) _____ or disconnecting means.

24. A parking garage where only limited maintenance work is done is considered to be a(n) _____ area.

T   F   **25.** A location where cloth is woven is designated as Class III, Division 1.

26. Flexible cords suitable for the type of service and approved for _____ usage may be used as pendants in a commercial garage.

27. Underground conduit runs within 20′ of a gasoline dispensing location are considered to be in a Class I, Division _____ location.

28. Class III locations are those where _____ fibers are present.

29. Where electrodes penetrate an enclosure, _____ listed for the purpose shall be provided unless receptacles are installed.

30. The secondary voltage of an isolating transformer used in an anesthetizing location may not exceed _____ V.

31. All threaded joints in conduit used in Class I, Division 1 locations shall be made with at least _____ threads fully engaged.

32. Disappearing stage footlights shall have provisions for _____ disconnecting the current supply when the footlights are replaced in the recess designed for them.

33. Incandescent lamps less than 8′ from the floor of a theatrical dressing room shall be equipped with _____ guards.

34. In Class I, Division 1 locations, conduit seals are required adjacent to boxes containing splices if the conduit size is equal to or larger than _____″.

    A. ¾   C. 1½

    B. 1   D. 2

35. A switch controlling a plug receptacle in a theatrical dressing room shall be equipped with a(n) _____.

NEC®   **Answer**

_____ T   F   **1.** Dust-tight receptacles for attachment plugs are acceptable in Class II, Division 1 locations.

_____ T   F   **2.** In general, explosionproof portable electric lamps may be used inside a paint spray booth during operation.

_____ _____   **3.** Where magnesium, aluminum, or aluminum-bronze powders may be present, _____ are not permitted to be installed.

_____ T   F   **4.** Portable appliances used inside an aircraft hangar need be approved only for Class I, Division 2 locations.

_____ _____   **5.** Class 2 and 3 circuits can occupy the same raceway as communication circuits if they are _____ as such.

_____ _____   **6.** For general wiring in Class I, Division 1 locations, _____ may be used.
   A. RMC   C. FMC
   B. EMT   D. either A, B, or C

_____ T   F   **7.** In a commercial garage, a stockroom whose floor level is 18″ higher than that of the garage is normally considered a nonhazardous area.

_____ _____   **8.** Unit equipment to be used as emergency lighting can be cord-and-plug connected if the cord is not over _____′ long.

_____ _____   **9.** A permanently installed electric sign shall be controlled by a(n) _____ operable switch or circuit breaker.

_____ _____   **10.** Areas within 18″ horizontally of gas-dispensing pumps are considered Class I, Division 2 locations to the top of the _____.
   A. island   C. concrete
   B. curb   D. dispenser

_____ _____   **11.** All 120 V, 1ϕ, 15 A and 20 A circuits within _____′ of a lavatory or sink in a mobile home shall have GFCI protection.

_____ _____   **12.** Markings for signs and outline lighting systems shall be _____ during installation.

_____ _____   **13.** Locations with dried, flammable paints that have ventilating equipment interlocked with the electrical equipment may be designated as _____.
   A. Class I, Division 2   C. Class II, Division 2
   B. Class II, Division 1   D. unclassified

_____ _____   **14.** Portable lamps used in commercial garages shall be equipped with a hook and _____.

_____ _____   **15.** Neon tubing that is _____ to pedestrians shall be protected from physical damage.

_____ T   F   **16.** Open spaces within 20′ horizontally of a gas-dispensing pump are designated as Class I, Division 1 locations.

_____ T F **17.** Audible and visual signal devices shall be used, where practical, to give warning of derangement of emergency power.

_____ _____ **18.** If acceptable to the AHJ, a separate _____ is suitable for use as an emergency power source.

_____ _____ **19.** Internal combustion engines used as prime movers shall have at least a(n) _____ hour supply of fuel.

_____ T F **20.** Open outside areas within 20′ horizontally from the fill pipe of an underground gasoline tank shall be considered as a Class I, Division 2 location to a height of 18″.

_____ _____ **21.** Two or more Class 1 circuits can be in the same box, cable, or raceway where all wires are insulated for the _____ voltage.

_____ T F **22.** Spray booths may be illuminated through suitable glass panels.

_____ _____ **23.** Power supply for Class 2 and 3 circuits shall not be connected in _____ unless they are listed.

_____ _____ **24.** Storerooms and similar areas adjacent to aircraft hangars, but effectively isolated, shall be designated as _____.
A. Class I, Division 2   C. Class II, Division 2
B. Class II, Division 1   D. unclassified

_____ _____ **25.** The output to Class 1 remote-control and signaling circuits is unlimited for systems rated at _____ V or less.

_____ _____ **26.** Wiring to lamps in a motion picture projector shall have insulation with high ambient temperature if exposed to over _____°C.

_____ _____ **27.** Lampholders operating on systems rated less than 50 V are required to at least _____ W.

_____ _____ **28.** When a separate service is used as the emergency power source, it shall be electrically and physically _____ from the normal service.

_____ _____ **29.** In a Class II location where electrically conducting dust is present, flexible connections at motors may be made with _____.
A. FMC   C. hard-usage cord
B. Type AC armored cable   D. neither A, B, nor C

_____ _____ **30.** Branch circuits used for emergency lighting shall be _____ transferred upon failure of normal service.

_____ _____ **31.** OCPDs for feeders can be set at _____% of their conductor rating in buildings used for motion picture production.

_____ _____ **32.** Where a pool is installed for a dwelling, at least one receptacle is required between _____′ and 20′ from the pool.

_____ _____ **33.** No electrical equipment, other than permanent lighting fixtures, shall be located in a(n) _____ with motion picture film.

_____ _____ **34.** In commercial garages, the floor area at a height of 18″ above grade is designated as _____.
A. Class I, Division 1   C. Class II, Division 1
B. Class I, Division 2   D. Class II, Division 2

_____ _____ **35.** For other than simple apparatus, all intrinsically safe apparatus and associated apparatus shall be _____.

Name _____

Date _____

Trade Competency Test 3

NEC®   **Answer**

_____ T   F   **1.** Resistance and reactor dimmers for stage lighting may be placed in the grounded conductor of a circuit.

_____ _____   **2.** If the AHJ can determine that flammable liquids with a flash point below 100°F, such as gasoline, will not be handled, the location is not required to be _____.

_____ T   F   **3.** Branch circuits that supply only neon tubing installations shall be rated at 30 A or less.

_____ _____   **4.** A well-ventilated room where flammable anesthetics are stored in approved containers shall be considered a _____ area.
          A. Class I, Division 1   C. Class II, Division 2
          B. Class I, Division 2   D. nonhazardous

_____ T   F   **5.** Storage batteries, motor-generator sets, or isolating transformers may be used to provide ungrounded circuits in anesthetizing locations.

_____ _____   **6.** Cranes operating over combustible fibers shall be supplied by a(n) _____ power source.

_____ _____   **7.** For calculating the load for the required branch circuit for signs or outline lighting in commercial occupancies, _____ VA is used.

_____ _____   **8.** Conduits installed below the surface of a gasoline dispenser are considered to be in a(n) _____ location.

_____ T   F   **9.** A low-voltage isolating transformer supplying circuits within an anesthetizing location shall be grounded.

_____ _____   **10.** Pendant lampholders are not permitted in _____ rooms of theaters.

_____ _____   **11.** Each commercial building and occupancy with ground floor footage shall have at least one outside sign outlet rated at _____ A.
          A. 15   C. 25
          B. 20   D. 30

_____ _____   **12.** Boxes and fittings used in Class III, Division 1 locations are required to be _____.

_____ _____   **13.** A feeder supplying power to stage lighting units with a load of 70,000 VA can have a demand factor of _____% applied.

_____ T   F   **14.** Lamps in cellulose nitrate film vaults may be installed in general-use fixtures provided they are equipped with suitable guards.

_____ _____   **15.** The circuit supplying an autotransformer dimmer used in theaters shall not exceed _____ V between conductors.

_____ _____   **16.** Dimmers used for theater lighting shall have protection set at _____% of their rating.

_____ _____   **17.** In motion picture studios, the ampacity of feeder conductors to the stage may be protected at a maximum value of _____%.
     A. 200   C. 400
     B. 250   D. 500

_____ _____   **18.** Cable tray runs shall be _____ before conductors are installed.

_____ T   F   **19.** Surgical fixtures more than 5′ above the floor of an operating room may be installed on a grounded circuit.

_____ _____   **20.** All circuits routed through a(n) _____ pump shall be switched by a double-pole switch.

_____ _____   **21.** Lamps used in Class III, Division 2 locations shall be protected against the _____ of lint and fibers.

_____ _____   **22.** In aircraft hangars, lighting fixtures less than _____′ above wings and engine enclosures shall be totally enclosed.
     A. 10   C. 20
     B. 15   D. neither A, B, nor C

_____ _____   **23.** All metal parts around a dispensing pump shall be _____.

_____ _____   **24.** Switches and controllers used in Class III locations shall be installed in _____ enclosures.

_____ _____   **25.** In general, live parts of electric signs and outline lighting other than lamps and neon tubing shall be _____.

_____ _____   **26.** The _____ is the first fitting after a conduit emerges from the earth or concrete in a dispenser pump.

_____ _____   **27.** A storage battery supplying emergency lighting and power shall maintain not less than $87\frac{1}{2}$% of full voltage at total load for a period of at least _____ hour(s).
     A. $\frac{1}{4}$ C. $1\frac{1}{2}$
     B. 1   D. 2

_____ _____   **28.** In a Class II location, a 50 kVA transformer containing askarel shall be located at least _____″ from combustible material.

_____ _____   **29.** Storeroom office and control rooms are considered _____ if proper ventilation is provided.

_____ _____   **30.** Portable lamps used in Class III locations shall be of the _____ type.

_____ _____   **31.** Motor-driven utilization equipment shall be dust-_____ if installed in a Class II location.

_____ _____   **32.** Dry-type transformers with open terminals that are installed in Class II locations are required to be installed in _____.

_____ _____   **33.** The primary voltage of an isolating transformer used in an anesthetizing location shall not exceed _____ V.
     A. 150   C. 300
     B. 250   D. 600

_____ _____   **34.** A 10′ or longer length of horizontal conduit can serve as a(n) _____ in Class II locations.

_____ _____   **35.** The entire floor area in a commercial garage can be classified as nonhazardous with ventilation providing _____ air changes per hour.

Name _____

Date _____

Trade Competency Test 4

NEC®   **Answer**

_____ T F **1.** A plug receptacle exclusively for cleaning purposes may be tapped from the emergency circuit wires.

_____ _____ **2.** Conductors on poles, where not placed on racks or brackets, shall be separated by no less than _____".
   A. 6   C. 18
   B. 12   D. 24

_____ T F **3.** The area containing paint spray is considered a Class I, Division 1 location.

_____ _____ **4.** The color _____ can be used to identify intrinsically safe circuits.

_____ T F **5.** Metal raceways shall not be required where AC cable is used as a wiring method when the metal jacket is identified as a grounding means of a listed cable assembly.

_____ _____ **6.** If the normal power source is lost, the alternate power source shall be automatically connected within _____ seconds.

_____ _____ **7.** An anesthetizing location is considered a Class I, Division 1 location to a height of _____' above the floor.

_____ T F **8.** Testing each level is required when equipment ground fault protection is first installed.

_____ _____ **9.** The rating of a lampholder on a circuit operating at less than 50 V shall be at least _____ VA.
   A. 220   C. 550
   B. 330   D. 660

_____ T F **10.** A connection on the supply side of the service disconnect may be permitted as an emergency service.

_____ _____ **11.** On circuits of 600 V or less, overhead feeder spans up to 50' in length shall have copper conductors no smaller than No. _____.
   A. 6   C. 10
   B. 8   D. 12

_____ _____ **12.** Lighting fixtures installed over _____' above finished grade in commercial garages can be of the general purpose type.

_____ T F **13.** Business offices in health care facilities are not considered as patient care areas.

_____ _____ **14.** Electrical equipment that is approved to be installed in hazardous locations shall show _____, group, and operating temperature.

_____ _____ **15.** A pit in a commercial garage with at least _____ changes of air per hour is a Class I, Division 2 location where a dispenser is present.

_____ _____ **16.** In operating rooms using flammable anesthetics, the entire area is a Class I, Division 1 location up to a level of _____′ above the floor.
   A. 4   C. 6
   B. 5   D. 7

_____ _____ **17.** Surge arrestors installed in Class I, Division 1 locations are to be in _____ enclosures.

_____ _____ **18.** The alternate source of power shall have capacity to supply connected loads for a minimum of _____ hours.
   A. 1.5   C. 2.5
   B. 2   D. 3

_____ _____ **19.** Class I locations shall not extend beyond a(n) _____ partition that has no communicating opening.

_____ T   F **20.** In patient care areas, the grounding terminals of all receptacles shall be connected to an insulated copper equipment grounding conductor.

_____ _____ **21.** Plug receptacles installed in Class I, Division I locations shall be _____ for the location.

_____ _____ **22.** Signaling apparatus used in Class II locations shall be in _____ enclosures.

_____ _____ **23.** Open-type generators shall not be used in Class I, Division 1 locations if they have _____ devices exposed.

_____ _____ **24.** Boxes and fittings used in Class II locations shall have close-fitting covers to contain _____.

_____ _____ **25.** In unattended self-service stations, emergency controls shall be located at least _____′ from dispensers.

_____ _____ **26.** Rubber cord used for flexible connections in a Class II location shall be of the _____ type.

_____ _____ **27.** The isolated circuit conductors for a 3ϕ system supplying an anesthetizing location shall have the third conductor identified as _____.
   A. yellow   C. red
   B. black   D. blue

_____ _____ **28.** A minimum of _____″ spacing shall be provided between neon tubing and any surface other than its support.

_____ _____ **29.** In general, the space adjacent to enclosed coating and dipping operations shall be considered _____.

_____ _____ **30.** Portable lamps shall be approved for Class II locations and marked for _____ lamp wattage.

_____ _____ **31.** No transformer or capacitor is permitted in a Class II location where _____ is present.

_____ _____ **32.** A Class I, Division 1 location is an area where flammable gases and vapors are _____ present.

_____ T   F **33.** Telephone wires are permitted on the same crossarm with light or power wires if separated by at least 2′.

_____ _____ **34.** Explosionproof equipment contains an explosion without rupturing and _____ escaping gases.

_____ _____ **35.** Where a Class II, Group E motor operates without being subject to overload, its operating temperature is _____°C.

Name _____

Date _____

Final Test 1

NEC®	Answer	

_____ T F **1.** The smallest permissible hard-drawn copper service-drop conductor is No. 10.

_____ _____ **2.** A No. _____ Al EGC is required for an automatic OCPD set at 200 A.

A. 2   C. 6
B. 4   D. 8

_____ T F **3.** The GEC for a small, one-family dwelling shall be permitted to be spliced with listed means or an exothermic weld.

_____ _____ **4.** The neutral conductor for a high-impedance grounded system shall be at least No. _____ Cu.

_____ _____ **5.** Gaps in plaster, drywall, or plasterboard shall be repaired so they do not exceed _____″ at the edge of the box.

_____ T F **6.** Any available wall space in a dwelling shall be within 7′ of a plug receptacle.

_____ T F **7.** Under certain conditions, switch enclosures may be used as junction boxes.

_____ T F **8.** GFCI protection is required for all receptacles serving countertop surfaces and within 6′ of the wet bar sink.

_____ _____ **9.** The service disconnecting means rating shall not be less than _____ A for a one-family dwelling.

A. 50   C. 70
B. 60   D. 100

_____ _____ **10.** Conductors of dissimilar metals shall not be intermixed in a terminal or splicing connector where physical contact occurs, unless the device is _____ for the purpose and conditions of use.

A. marked   C. intended
B. stamped   D. identified

_____ T F **11.** The service disconnecting means may consist of not more than six switches or CBs.

_____ _____ **12.** The service or feeder capacity necessary for four 10 kVA electric ranges is _____ kVA.

A. 15   C. 19
B. 17   D. 21

_____ _____ **13.** When initially installed, flexible cords shall be used only in _____ lengths.

_____ _____ **14.** The clearance of 120/240 V service-drop conductors over residential driveways not subject to truck traffic shall be no less than _____′.

_____ _____ **15.** The first 20,000 VA of computed load in a hotel shall be assessed at _____%.

16. Unless otherwise specified, conductors used to carry current shall be of _____.

    A. copper   C. copper-clad aluminum
    B. aluminum   D. neither A, B, nor C

17. The Cu GEC for a 350 kcmil, 120/240 V service installation, if run to a waterpipe electrode, shall be no smaller than No. _____.

18. The smallest permissible driven ground rod electrode made of nonferrous materials is _____″ in diameter.

    A. ³⁄₈   C. ⅝
    B. ½   D. ¾

T   F   **19.** It is permissible to install ten No. 12 conductors in a 4″ by 1½″ square box that also contains one switch.

20. The minimum clearance from ground of service drop conductors operating at 600 V or less, which pass over public streets, alleys, roads, and driveways, on other-than-residential property, shall be _____′.

    A. 10   C. 15
    B. 12   D. 18

21. Under the optional calculation for a one-family dwelling, the central space heating load is assessed at _____%.

22. In general, Type AC cable shall be supported at intervals not exceeding _____′.

23. Heating cables installed on drywall ceilings shall be secured at intervals not exceeding _____″.

    A. 10   C. 14
    B. 12   D. 16

24. The first 3000 VA of computed load in a multifamily dwelling shall be assessed at _____%.

25. Receptacles for a 20 A circuit shall be rated no less than _____ A.

26. The leads of a 120 V heating cable are marked with the color identification _____ to indicate the circuit voltage on which it is to be used.

27. Except in dwellings, each _____′ or fraction of multioutlet assembly strip is usually considered as a load of 180 VA.

T   F   **28.** The cross-sectional area of all conductors in any part of a metal auxiliary gutter shall not exceed 15% of the cross-sectional area of the gutter.

29. A 5 HP, 230 V, 3ϕ squirrel-cage motor with no code letter is marked 40°C and has a nameplate rating of 15 A. The setting of an adjustable overload device shall not exceed _____ A.

    A. 19   C. 21
    B. 20   D. 25

30. The ampacity of No. 2 THWN-THHN, dual-rated copper conductors in rigid conduit and installed in a wet location is _____ A.

T   F   **31.** A header used with precast cellular concrete floor raceways shall not be run at right angles to the cells.

T   F   **32.** An AC general-use snap switch controlling an inductive load shall have an ampere rating equal to 150% of the load.

33. LTFNMC is limited in length to _____′.

34. Metal surface raceways used for signaling and lighting circuits shall have two separate compartments, each identified by contrasting colors of the _____ finish.

35. Splices and taps in underfloor raceway installations shall be made only in _____.

36. Nonmetallic underground conduit with conductors is permitted to be used _____.

    A. for direct burial installations   C. encased or embedded in concrete
    B. in cinder fill   D. either A, B, or C

37. Sheet metal auxiliary gutters installed in wet locations shall be _____ for the location.

38. The motor _____ shall be in sight of the motor and the driven machinery location or shall be individually capable of being locked in the open position.

T   F   39. The maximum rating of a 250 V cartridge fuse is 400 A.

T   F   40. A No. 18 Cu conductor may be used to ground a portable device connected to a 30 A circuit.

T   F   41. Conductors in underfloor raceway systems shall not be tapped at individual outlets.

42. The minimum size Cu GEC required for a 120/240 V service consisting of one 250 kcmil conductor per phase is No. _____. The grounding electrode is a ground rod and it is a sole connection to the electrode.

43. The allowable ampacity of a conductor is 55 A. It is considered protected by a(n) _____ A fuse or CB.

T   F   44. Secondary circuits of wound-rotor induction motors require separate overload protection.

45. Two or more services are permitted when the rating of the service is over _____ A.

    A. 1500   C. 2500
    B. 2000   D. 3000

T   F   46. Transformers are never required to have overcurrent protection devices in both primary and secondary circuits.

47. A wood building, without water, is built on the earth without footings. If a 3000 A, 460 V service is installed and grounded by ground rods, the minimum size copper GEC is _____.

    A. No. 3/0   C. No. 6
    B. No. 4   D. 400 kcmil

48. The conductor fill of an underfloor raceway shall not exceed _____% of the interior area of the raceway.

    A. 20   C. 75
    B. 40   D. 100

T   F   49. A motor branch-circuit, short-circuit device may sometimes be rated in excess of 300% of the full load motor current.

T   F   50. The ampacity of the ungrounded conductors from the generator to the first overcurrent device shall not be less than 115% of the nameplate current of the generator.

_____ _____ **51.** Branch-circuit conductors supplying a 50 HP, 460 V, 3ϕ induction motor shall be rated for at least _____ A.

    A. 65   C. 91

    B. 82   D. 130

_____   T   F **52.** A ½ HP pump motor controlled by a float switch requires running overcurrent protection.

_____   T   F **53.** The primary windings of potential transformers installed indoors shall always be fused.

_____ _____ **54.** The largest size conductor permitted in an underfloor raceway is _____.

    A. 250 kcmil   C. 500 kcmil

    B. No. 0   D. determined by design

_____   T   F **55.** An 8 A, 115 V, 1ϕ sealed (hermetic) motor compressor shall show the LRC on its nameplate.

_____   T   F **56.** The minimum size of Type THW Cu conductors supplying a 70 A, 480 V transformer arc welder used at 30% duty application is No. 8.

_____ _____ **57.** A lighting and power service has two No. 3/0 ungrounded copper conductors and a No. 1/0 neutral. A Cu GEC attached to a waterpipe electrode shall not be smaller than No. _____.

    A. 2   C. 6

    B. 4   D. 8

_____   T   F **58.** All receptacles on the property shall be located a minimum of 10′ from the inside wall of a permanently installed swimming pool.

_____ _____ **59.** The maximum initial rating of branch circuit NTDFs for a 5 HP, 230 V, 3ϕ squirrel-cage motor with no code letter is _____ A.

    A. 40   C. 50

    B. 45   D. 60

_____   T   F **60.** The color coding for intrinsically safe conductors is light green.

_____ _____ **61.** A CB disconnecting means for a 50 HP, 460 V, 3ϕ motor shall have an ampere rating of at least _____ A.

    A. 63   C. 91

    B. 75   D. 126

_____ _____ **62.** The maximum size of NTDFs for branch circuit overcurrent protection for a 50 HP, 460 V, 3ϕ motor is _____ A.

    A. 90   C. 250

    B. 200   D. 300

_____ _____ **63.** Duct heaters must include a fan circuit interlock which ensures that the fan circuit is _____ when any heater circuit is energized.

    A. locked out   C. energized

    B. locked in   D. neither A, B, nor C

_____ _____ **64.** A 5 HP, 230 V, 3ϕ squirrel-cage motor with no code letter is started at full voltage. The minimum ampacity of the branch circuit conductor is _____ A.

    A. 15.2   C. 19

    B. 18   D. 20

Name _____

Date _____

Final Test **2**

NEC®   **Answer**

_____   T   F   **1.** When the power for a fire pump is provided by a tap ahead of the service disconnecting means, the tap is permitted to be made within the service disconnecting means compartment.

_____   _____   **2.** A demand factor of _____% may be applied to the neutral feeder load for electric ranges.

A. 40   C. 70
B. 60   D. 80

_____   T   F   **3.** FMC used to provide flexibility at equipment shall be equipped with an EGC.

_____   T   F   **4.** A tap for an individual lampholder or fixture more than 12″ long may not be smaller than the branch circuit conductors.

_____   T   F   **5.** Under certain conditions, the neutral feeder conductor is permitted to be smaller than the ungrounded conductors.

_____   _____   **6.** In general, wiring for emergency systems shall be kept entirely _____ of all other wiring and equipment and shall not occupy the same raceway, box, cable or cabinet with other wiring.

_____   T   F   **7.** SE conductors, in general, shall not be spliced.

_____   _____   **8.** The ampacity of a No. 3/0 THW Cu conductor at a room temperature of 50°C is _____ A.

A. 125   C. 175
B. 150   D. 200

_____   _____   **9.** The rating of conductors supplying power to a capacitor shall be at least one-third of the motor circuit conductors and _____% of the FLC rating of the capacitor.

_____   _____   **10.** Conduit bodies shall not contain splices, taps, or devices unless they are marked by the manufacturer to indicate their _____ capacity.

_____   _____   **11.** For electrical equipment rated over _____ A or more than 6′ wide, that contains overcurrent devices, there shall be at least one entrance 24″ wide and 6½′ high at each end of the work space.

_____   T   F   **12.** Embedded electric space heating cables shall not be spliced.

_____   T   F   **13.** Nonflammable, fluid-insulated transformers installed indoors and rated at 15,000 V shall be installed in vaults.

_____   _____   **14.** A transformer with a primary FLC rating of less than 2 A shall be protected at _____% of its primary current.

_____   _____   **15.** The ampacity of conductors supplying a 20 A oscillator induction heating unit shall not be less than _____ A.

_____ _____   **16.** The primary OCPD, when properly selected, may protect the secondary conductors only if the secondary of the transformer is 1φ and feeds a(n) _____-wire circuit.

_____ _____   **17.** The feeder load for three electric ranges (13 kVA, 18 kVA, and 20 kVA) is _____ kVA.

         A. 15   C. 19

         B. 17.5   D. 21

_____ _____   **18.** Where thermal cutouts are used for motor overload protection for a 3φ motor, a total of _____ thermal cutouts is required.

_____ _____   **19.** Where explosion proof enclosures cannot be used in Class I, Division 1 locations due to grouping, _____ and pressurized enclosures can be used.

_____ _____   **20.** Construction site receptacles mounted on a _____ kW or less generator do not require GFCI protection.

         A. 5   C. 15

         B. 10   D. 20

_____ _____   **21.** To supply a feeder for night lights in an apartment, the required size of THW Cu conductors for a continuous 100 A load shall not be smaller than No. _____.

         A. 0   C. 2

         B. 1   D. 3

_____   T   F   **22.** The screw shell of a fuseholder shall not be connected to the line side of the circuit.

_____ _____   **23.** Inside locations of spray booths are classified as a Class I, _____ location.

_____ _____   **24.** The general lighting load for an office building is calculated at _____ VA per sq ft.

         A. 2   C. 3.5

         B. 2.5   D. 5

_____ _____   **25.** Transformers shall not be installed in locations where _____ dust is present.

_____ _____   **26.** In general, raceways shall not be used as a means of support for other _____, cables, or nonelectric equipment.

_____ _____   **27.** The maximum rating of a cord- and plug-connected appliance used on a 20 A branch circuit shall be _____ A.

         A. 10   C. 16

         B. 12   D. 20

_____ _____   **28.** The area within _____′ horizontally from aircraft fuel tanks is classified as a Class I, Division 2 location.

_____ _____   **29.** The electrical systems in aircraft shall be _____ when they are stored in hangars.

_____ _____   **30.** PVC can be used to supply power to a dispenser pump if buried at least _____′ in the earth.

_____ _____   **31.** Under certain conditions, an unbroken run of conduit can pass through a Class I, Division 1 location without _____ being provided.

_____ _____   **32.** Approved sealing _____ shall be used to seal fittings in a conduit run.

_____ _____   **33.** Receptacles in marinas and boatyards that provide shore power to boats shall be rated not less than 20 A and shall be of the locking and _____ types.

_____ _____ **34.** Nonmedical X-ray, high-voltage _____ cables shall be grounded.

_____ _____ **35.** Spas and hot tubs with associated electrical components shall be _____-protected.

_____ _____ **36.** The screw shell of a plug-type fuseholder shall be connected to the _____ side of the circuit.

_____ _____ **37.** The full load current of a 5 HP, 208 V, 3φ induction motor is _____ A.

_____ _____ **38.** The radius of bends in mineral-insulated cable with a ¾″ external diameter shall be not less than _____ times the cable diameter.

_____ _____ **39.** All exposed noncurrent-carrying metal parts of information technology equipment shall be _____.

_____ _____ **40.** Horsepower rated switches are not required as a disconnecting means for 300 V stationary motors of _____ HP or less.

_____ _____ **41.** In general, the controller disconnect shall be in sight of the _____.

_____ T F **42.** Open wiring on insulators is permitted to be used for branch circuit temporary wiring on construction sites.

_____ _____ **43.** Open transformers installed indoors and rated less than 112½ kVA shall have a clearance of at least _____″ from combustible material.

_____ _____ **44.** A psychiatric hospital is used exclusively for such care on a 24-hour basis for _____ or more inpatients.

_____ _____ **45.** The isolated circuit conductors for a 2-wire system supplying an anesthetizing location shall be identified by using _____ conductors.
A. orange and brown   C. purple and pink
B. black and red   D. neither A, B, nor C

_____ _____ **46.** Flexible cords and cables shall not be used as a substitute for the _____ wiring of a structure.

_____ _____ **47.** Intrinsic safety wiring methods shall be marked and identified every _____′.

_____ _____ **48.** Unshielded lead-in conductors of amateur transmitting stations shall clear the building surface that is wired over by a distance of at least _____″.
A. 1   C. 3
B. 2   D. 4

_____ _____ **49.** General purpose enclosures can be used in Class I, Division 2 locations if the contacts are immersed in _____.

_____ _____ **50.** A basic parking garage is not considered a(n) _____ location.

_____ _____ **51.** Flexible cords feeding 120 V equipment within a hazardous anesthetizing location shall be of the _____ type.

_____ _____ **52.** A 10′ or longer length of horizontal conduit can serve as a(n) _____ in Class II locations.

_____ _____ **53.** The entire floor area in a commercial garage can be classified as nonhazardous with ventilation providing _____ air changes per hour.

_____ _____ **54.** Ten 10 kVA ovens located in an apartment complex can have a demand factor of _____ kVA.

_____ _____ **55.** In general, junction boxes for swimming pools shall be located at least 4′ from the inside wall of the pool and they shall be located not less than 4″ from the pool deck or 8″ from the maximum level of the pool water, whichever provides the _____ elevation.

_____ _____ **56.** With no overcurrent protection in the secondary, the primary overcurrent device for a 600 V dry-type transformer with a rated primary current of 100 A shall not exceed _____ A.

    A. 125   C. 250
    B. 150   D. 600

_____ _____ **57.** The use of a continuous white or natural gray covering on a conductor or termination is reserved for the _____ conductor.

_____ _____ **58.** A bathroom is an area with a(n) _____ and one or more of the following: toilet, tub, or shower.

_____ _____ **59.** Receptacles installed over kitchen countertops 12″ or wider shall be not more than _____″ apart.

_____ _____ **60.** A service with phase conductors rated at 300 kcmil per phase shall have a minimum grounded conductor of No. _____ Cu.

_____ T   F **61.** Two to six disconnecting means can serve to disconnect the service equipment of a multifamily dwelling.

_____ _____ **62.** Fixtures over _____ lb shall be supported independently of the box unless listed for the weight to be supported.

_____ T   F **63.** The lowest point of attachment for a service drop to a building is 12′.

_____ _____ **64.** THHN Cu conductors in a circuit routed through an ambient temperature of 121°F shall be derated _____%.

_____ _____ **65.** A No. 6 Cu conductor rated at 60°C terminated to an OCPD has an ampacity of _____ A.

_____ _____ **66.** The smallest size EGC permitted in cable trays is No. _____ or larger.

_____ T   F **67.** The ungrounded conductors of a 3-pole OCPD shall be disconnected simultaneously.

_____ _____ **68.** Each length of RNMC shall be marked within every _____′.

_____ _____ **69.** A 6′ fence with 1′ of barbed wire on top is equivalent to a(n) _____′ fence for guarding equipment of over 600V.

_____ _____ **70.** A(n) _____ VA rating is required for each linear foot of show window.

_____ _____ **71.** A control transformer with a primary FLC of 1 A can be protected with a(n) _____ A fuse.

_____ _____ **72.** A disconnecting means located within _____′ and within sight can serve as the disconnecting switch for a motor.

_____ _____ **73.** A 4¹¹⁄₁₆″ by 1¼″ square box has a(n) _____ cu in. rating.

Appendix

THREE-PHASE VOLTAGE VALUES

For 208 V × 1.732, use 360
For 230 V × 1.732, use 398
For 240 V × 1.732, use 416
For 440 V × 1.732, use 762
For 460 V × 1.732, use 797
For 480 V × 1.732, use 831

POWER FORMULA ABBREVIATIONS AND SYMBOLS

P = Watts	V = Volts
I = Amps	VA = Volt Amps
A = Amps	φ = Phase
R = Ohms	√ = Square Root
E = Volts	

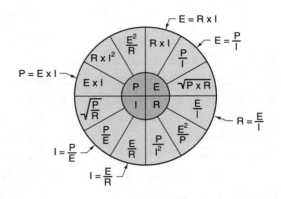

VALUES IN INNER CIRCLE
ARE EQUAL TO VALUES
IN CORRESPONDING
OUTER CIRCLE

OHM'S LAW AND POWER FORMULA

POWER FORMULAS – 1φ, 3φ

Phase	To Find	Use Formula	Example		
			Given	Find	Solution
1φ	I	$I = \dfrac{VA}{V}$	32,000 VA, 240 V	I	$I = \dfrac{VA}{V}$ $I = \dfrac{32{,}000 \text{ VA}}{240 \text{ V}}$ **I = 133 A**
1φ	VA	$VA = I \times V$	100 A, 240 V	VA	$VA = I \times V$ $VA = 100 \text{ A} \times 240 \text{ V}$ **VA = 24,000 VA**
1φ	V	$V = \dfrac{VA}{I}$	42,000 VA, 350 A	V	$V = \dfrac{VA}{I}$ $V = \dfrac{42{,}000 \text{ VA}}{350 \text{ A}}$ **V = 120 V**
3φ	I	$I = \dfrac{VA}{V \times \sqrt{3}}$	72,000 VA, 208 V	I	$I = \dfrac{VA}{V \times \sqrt{3}}$ $I = \dfrac{72{,}000 \text{ VA}}{360}$ **I = 200 A**
3φ	VA	$VA = I \times V \times \sqrt{3}$	2 A, 240 V	VA	$VA = I \times V \times \sqrt{3}$ $VA = 2 \times 416$ **VA = 832 VA**

AC/DC FORMULAS

To Find	DC	AC		
		1φ, 115 or 220 V	1φ, 208, 230, or 240 V	3φ – All Voltages
I, HP known	$\dfrac{HP \times 746}{E \times E_{ff}}$	$\dfrac{HP \times 746}{E \times E_{ff} \times PF}$	$\dfrac{HP \times 746}{E \times E_{ff} \times PF}$	$\dfrac{HP \times 746}{1.73 \times E \times E_{ff} \times PF}$
I, kW known	$\dfrac{kW \times 1000}{E}$	$\dfrac{kW \times 1000}{E \times PF}$	$\dfrac{kW \times 1000}{E \times PF}$	$\dfrac{kW \times 1000}{1.73 \times E \times PF}$
I, kVA known		$\dfrac{kVA \times 1000}{E}$	$\dfrac{kVA \times 1000}{E}$	$\dfrac{kVA \times 1000}{1.763 \times E}$
kW	$\dfrac{I \times E}{1000}$	$\dfrac{I \times E \times PF}{1000}$	$\dfrac{I \times E \times PF}{1000}$	$\dfrac{I \times E \times 1.73 \times PF}{1000}$
kVA		$\dfrac{I \times E}{1000}$	$\dfrac{I \times E}{1000}$	$\dfrac{I \times E \times 1.73}{1000}$
HP (output)	$\dfrac{I \times E \times E_{ff}}{746}$	$\dfrac{I \times E \times E_{ff} \times PF}{746}$	$\dfrac{I \times E \times E_{ff} \times PF}{746}$	$\dfrac{I \times E \times 1.73 \times E_{ff} \times PF}{746}$

HORSEPOWER FORMULAS

To Find	Use Formula	Example		
		Given	Find	Solution
HP	$HP = \dfrac{I \times E \times E_{ff}}{746}$	240 V, 20 A, 85% E_{ff}	HP	$HP = \dfrac{I \times E \times E_{ff}}{746}$ $HP = \dfrac{20\,A \times 240\,V \times 85\%}{746}$ $HP = \mathbf{5.5}$
I	$I = \dfrac{HP \times 746}{E \times E_{ff} \times PF}$	10 HP, 240 V, 90% E_{ff}, 88% PF	I	$I = \dfrac{HP \times 746}{E \times E_{ff} \times PF}$ $I = \dfrac{10\,HP \times 746}{240\,V \times 90\% \times 88\%}$ $I = \mathbf{39\ A}$

VOLTAGE DROP FORMULAS – 1φ, 3φ

Phase	To Find	Use Formula	Example		
			Given	Find	Solution
1φ	VD	$VD = \dfrac{2 \times R \times L \times I}{1000}$	240 V, 40 A, 60′ L, .764 R	VD	$VD = \dfrac{2 \times R \times L \times I}{1000}$ $VD = \dfrac{2 \times .764 \times 60 \times 40}{1000}$ $VD = \mathbf{3.67\ V}$
3φ	VD	$VD = \dfrac{2 \times R \times L \times I}{1000} \times .866$	208 V, 110 A, 75′ L, .194 R, .866 multiplier	VD	$VD = \dfrac{2 \times R \times L \times I}{1000} \times .866$ $VD = \dfrac{2 \times .194 \times 75 \times 110}{1000} \times .866$ $VD = \mathbf{2.77\ V}$

VOLTAGE DROP VARIABLES

V = Voltage (in V)	VD = Voltage drop (in V)	$\%VD$ = Percent voltage drop (in V)	V_l = Voltage loss (in V)
V_s = Supply voltage (in V)	K = Resistivity of conductor (in Ω)	R = Resistance of conductor (in Ω/kft)	CM = Circular mils (in area)
I = Current (In A)	L = Length of conductor (in ft)	1000 = 1000′ or less of conductor	$.866 = \dfrac{\sqrt{3}}{2}$

POWERS OF 10

1×10^4	=	10,000	= $10 \times 10 \times 10 \times 10$	Read ten to the fourth power
1×10^3	=	1000	= $10 \times 10 \times 10$	Read ten to the third power or ten cubed
1×10^2	=	100	= 10×10	Read ten to the second power or ten squared
1×10^1	=	10	= 10	Read ten to the first power
1×10^0	=	1	= 1	Read ten to the zero power
1×10^{-1}	=	.1	= 1/10	Read ten to the minus first power
1×10^{-2}	=	.01	= $1/(10 \times 10)$ or 1/100	Read ten to the minus second power
1×10^{-3}	=	.001	= $1/(10 \times 10 \times 10)$ or 1/1000	Read ten to the minus third power
1×10^{-4}	=	.0001	= $1/(10 \times 10 \times 10 \times 10)$ or 1/10,000	Read ten to the minus fourth power

UNITS OF ENERGY

Energy	Btu	ft-lb	J	kcal	kWh
British thermal unit	1	777.9	1.056	0.252	2.930×10^{-4}
Foot-pound	1.285×10^{-3}	1	1.356	3.240×10^{-4}	3.766×10^{-7}
Joule	9.481×10^{-4}	0.7376	1	2.390×10^{-4}	2.778×10^{-7}
Kilocalorie	3.968	3.086	4.184	1	1.163×10^{-3}
Kilowatt-hour	3.413	2.655×10^6	3.6×10^6	860.2	1

UNITS OF POWER

Power	W	ft-lb/s	HP	kW
Watt	1	0.7376	1.341×10^{-3}	0.001
Foot-pound/sec	1.356	1	1.818×10^{-3}	1.356×10^{-3}
Horsepower	745.7	550	1	0.7457
Kilowatt	1000	736.6	1.341	1

STANDARD SIZES OF FUSES AND CBs

NEC® 240-6(a) lists standard ampere ratings of fuses and fixed-trip CBs as follows:
15, 20, 25, 30, 35, 40, 45,
50, 60, 70, 80, 90, 100, 110,
125, 150, 175, 200, 225,
250, 300, 350, 400, 450,
500, 600, 700, 800,
1000, 1200, 1600,
2000, 2500, 3000, 4000, 5000, 6000

VOLTAGE CONVERSIONS

To Convert	To	Multiply By
rms	Average	.9
rms	Peak	1.414
Average	rms	1.111
Average	Peak	1.567
Peak	rms	.707
Peak	Average	.637
Peak	Peak-to-peak	2

WIRING MATERIALS

AC	Armored Cable	NM	Nonmetallic-Sheathed Cable (dry)
BX	Trade Name for AC	NMC	Nonmetallic-Sheathed Cable (dry or damp)
EMT	Electrical Metallic Tubing	NMS	Nonmetallic-Sheathed Cable (per 780)
ENT	Electrical Nonmetallic Tubing	PVC	Polyvinyl Chloride
FMC	Flexible Metal Conduit	RMC	Rigid Metallic Conduit
FMT	Flexible Metallic Tubing	RNMC	Rigid Nonmetallic Conduit
IGS	Integrated Gas Spacer Cable	Romex	Trade Name for NM
IMC	Intermediate Metal Conduit	SE	Service-Entrance Cable
LTFMC	Liquidtight Flexible Metal Conduit	SNM	Shielded Nonmetallic-Sheathed Cable
LTFNMC	Liquidtight Flexible Nonmetallic Conduit	TC	Tray Cable
MC	Metal-Clad Cable	UF	Underground Feeder
MI	Mineral Insulated Metal-Sheathed Cable	USE	Underground Service-Entrance Cable

DECIMAL EQUIVALENTS OF AN INCH

Fraction	Decimal	Fraction	Decimal	Fraction	Decimal	Fraction	Decimal
1/64	0.015625	17/64	0.265625	33/64	0.515625	47/64	0.765625
1/32	0.03125	9/32	0.28125	17/32	0.53125	25/32	0.78125
3/64	0.046875	19/64	0.296875	35/64	0.546875	51/64	0.796875
1/16	0.0625	5/16	0.3125	9/16	0.5625	13/16	0.8125
5/64	0.078125	21/64	0.328125	37/64	0.578125	53/64	0.828125
3/32	0.09375	11/32	0.34375	19/32	0.59375	27/32	0.84375
7/64	0.109375	23/64	0.359375	39/64	0.609375	55/64	0.859375
1/8	0.125	3/8	0.375	5/8	0.625	7/8	0.875
9/64	0.140625	25/64	0.390625	41/64	0.640625	57/64	0.890625
5/32	0.15625	13/32	0.40625	21/32	0.65625	29/32	0.90625
11/64	0.171875	27/64	0.421875	43/64	0.671875	59/64	0.921875
3/16	0.1875	7/16	0.4375	11/16	0.6875	15/16	0.9375
13/64	0.203125	29/64	0.453125	45/64	0.703125	61/64	0.953125
7/32	0.21875	15/32	0.46875	23/32	0.71875	31/32	0.96875
15/64	0.234375	31/64	0.484375	47/64	0.734375	63/64	0.984375
1/4	0.250	1/2	0.500	3/4	0.750	1	1.000

ENGLISH SYSTEM

	Unit	Abbr	Equivalents
LENGTH	mile	mi	5280′, 320 rd, 1760 yd
	rod	rd	5.50 yd, 16.5′
	yard	yd	3′, 36″
	foot	ft or ′	12″, .333 yd
	inch	in. or ″	.083′, .028 yd
AREA A = l x w	square mile	sq mi or mi²	640 A, 102,400 sq rd
	acre	A	4840 sq yd, 43,560 sq ft
	square rod	sq rd or rd²	30.25 sq yd, .00625 A
	square yard	sq yd or yd²	1296 sq in., 9 sq ft
	square foot	sq ft or ft²	144 sq in., .111 sq yd
	square inch	sq in. or in²	.0069 sq ft, .00077 sq yd
VOLUME V = l x w x t	cubic yard	cu yd or yd³	27 cu ft, 46,656 cu in.
	cubic foot	cu ft or ft³	1728 cu in., .0370 cu yd
	cubic inch	cu in. or in³	.00058 cu ft, .000021 cu yd

CAPACITY

U.S. liquid measure — WATER, FUEL, ETC.

Unit	Abbr	Equivalents
gallon	gal.	4 qt (231 cu in.)
quart	qt	2 pt (57.75 cu in.)
pint	pt	4 gi (28.875 cu in.)
gill	gi	4 fl oz (7.219 cu in.)
fluidounce	fl oz	8 fl dr (1.805 cu in.)
fluidram	fl dr	60 min (.226 cu in.)
minim	min	1/6 fl dr (.003760 cu in.)

U.S. dry measure — VEGETABLES, GRAIN, ETC.

Unit	Abbr	Equivalents
bushel	bu	4 pk (2150.42 cu in.)
peck	pk	8 qt (537.605 cu in.)
quart	qt	2 pt (67.201 cu in.)
pint	pt	½ qt (33.600 cu in.)

British imperial liquid and dry measure — DRUGS

Unit	Abbr	Equivalents
bushel	bu	4 pk (2219.36 cu in.)
peck	pk	2 gal. (554.84 cu in.)
gallon	gal.	4 qt (277.420 cu in.)
quart	qt	2 pt (69.355 cu in.)
pint	pt	4 gi (34.678 cu in.)
gill	gi	5 fl oz (8.669 cu in.)
fluidounce	fl oz	8 fl dr (1.7339 cu in.)
fluidram	fl dr	60 min (.216734 cu in.)
minim	min	1/60 fl dr (.003612 cu in.)

MASS AND WEIGHT

avoirdupois — COAL, GRAIN, ETC.

Unit	Abbr	Equivalents
ton	t	2000 lb
short ton		2000 lb
long ton		2240 lb
pound	lb or #	16 oz, 7000 gr
ounce	oz	16 dr, 437.5 gr
dram	dr	27.344 gr, .0625 oz
grain	gr	.037 dr, .002286 oz

troy — GOLD, SILVER, ETC.

Unit	Abbr	Equivalents
pound	lb	12 oz, 240 dwt, 5760 gr
ounce	oz	20 dwt, 480 gr
pennyweight	dwt or pwt	24 gr, .05 oz
grain	gr	.042 dwt, .002083 oz

apothecaries' — DRUGS

Unit	Abbr	Equivalents
pound	lb ap	12 oz, 5760 gr
ounce	oz ap	8 dr ap, 480 gr
dram	dr ap	3 s ap, 60 gr
scruple	s ap	20 gr, .333 dr ap
grain	gr	.05 s, .002083 oz, .0166 dr ap

METRIC SYSTEM

	Unit	Abbreviation	Number of Base Units
LENGTH	kilometer	km	1000
	hectometer	hm	100
	dekameter	dam	10
	meter*	m	1
	decimeter	dm	.1
	centimeter	cm	.01
	millimeter	mm	.001
AREA $A = l \times w$	square kilometer	sq km *or* km^2	1,000,000
	hectare	ha	10,000
	are	a	100
	square centimeter	sq cm *or* cm^2	.0001
VOLUME $V = l \times w \times t$	cubic centimeter	cu cm, cm^3, *or* cc	.000001
	cubic decimeter	dm^3	.001
	cubic meter*	m^3	1
CAPACITY WATER, FUEL, ETC. VEGETABLES, GRAIN, ETC. DRUGS	kiloliter	kl	1000
	hectoliter	hl	100
	dekaliter	dal	10
	liter*	l	1
	cubic decimeter	dm^3	1
	deciliter	dl	.10
	centiliter	cl	.01
	milliliter	ml	.001
MASS AND WEIGHT COAL, GRAIN, ETC. GOLD, SILVER, ETC. DRUGS	metric ton	t	1,000,000
	kilogram	kg	1000
	hectogram	hg	100
	dekagram	dag	10
	gram*	g	1
	decigram	dg	.10
	centigram	cg	.01
	milligram	mg	.001

* base units

METRIC PREFIXES

Multiples and Submultiples	Prefixes	Symbols	Meaning
$1,000,000,000,000 = 10^{12}$	tera	T	trillion
$1,000,000,000 = 10^{9}$	giga	G	billion
$1,000,000 = 10^{6}$	mega	M	million
$1000 = 10^{3}$	kilo	k	thousand
$100 = 10^{2}$	hecto	h	hundred
$10 = 10^{1}$	deka	d	ten
Unit $1 = 10^{0}$			
$1 = 10^{-1}$	deci	d	tenth
$.01 = 10^{-2}$	centi	c	hundredth
$.001 = 10^{-3}$	milli	m	thousandth
$.000001 = 10^{-6}$	micro	μ	millionth
$.000000001 = 10^{-9}$	nano	n	billionth
$.000000000001 = 10^{-12}$	pico	p	trillionth

METRIC CONVERSIONS

Initial Units	Final Units											
	giga	mega	kilo	hecto	deka	base	deci	centi	milli	micro	nano	pico
giga		3R	6R	7R	8R	9R	10R	11R	12R	15R	18R	21R
mega	3L		3R	4R	5R	6R	7R	8R	9R	12R	15R	18R
kilo	6L	3L		1R	2R	3R	4R	5R	6R	9R	12R	15R
hecto	7L	4L	1L		1R	2R	3R	4R	5R	8R	11R	14R
deka	8L	5L	2L	1L		1R	2R	3R	4R	7R	10R	13R
base	9L	6L	3L	2L	1L		1R	2R	3R	6R	9R	12R
deci	10L	7L	4L	3L	2L	1L		1R	2R	5R	8R	11R
centi	11L	8L	5L	4L	3L	2L	1L		1R	4R	7R	10R
milli	12L	9L	6L	5L	4L	3L	2L	1L		3R	6R	9R
micro	15L	12L	9L	8L	7L	6L	5L	4L	3L		3R	6R
nano	18L	15L	12L	11L	10L	9L	8L	7L	6L	3L		3R
pico	21L	18L	15L	14L	13L	12L	11L	10L	6L	6L	3L	

R = move the decimal point to the right
L = move the decimal point to the left

COMMON PREFIXES

Symbol	Prefix	Equivalent
G	giga	1,000,000,000
M	mega	1,000,000
k	kilo	1000
base unit	—	1
m	milli	0.001
μ	micro	0.000001
n	nano	0.000000001
p	pico	0.000000000001
Z	impedance	ohms — Ω

METRIC TO ENGLISH EQUIVALENTS

LENGTH

Unit	English Equivalent
kilometer	.62 mi
hectometer	109.36 yd
dekameter	32.81′
meter	39.37″
decimeter	3.94″
centimeter	.39″
millimeter	.039″

AREA

A = l x w

Unit	English Equivalent
square kilometer	.3861 sq mi
hectacre	2.47 A
are	119.60 sq yd
square centimeter	.155 sq in.

VOLUME

V = l x w x t

Unit	English Equivalent
cubic centimeter	.061 cu in.
cubic decimeter	61.023 cu in.
cubic meter	1.307 cu yd

CAPACITY

WATER, FUEL, ETC.

VEGETABLES, GRAIN, ETC.

DRUGS

Unit	cubic	dry	liquid
kiloliter	1.31 cu yd		
hectoliter	3.53 cu ft		2.84 bu
dekaliter	.35 cu ft	1.14 pk	2.64 gal.
liter	61.02 cu in.	.908 qt	1.057 qt
cubic decimeter	61.02 cu in.	.908 qt	1.057 qt
deciliter	6.1 cu in.	.18 pt	.21 pt
centiliter	.61 cu in.		338 fl oz
milliliter	.061 cu in.		.27 fl dr

MASS AND WEIGHT

COAL, GRAIN, ETC.

GOLD, SILVER, ETC.

DRUGS

Unit	English Equivalent
metric ton	1.102 t
kilogram	2.2046 lb
hectogram	3.527 oz
dekagram	.353 oz
gram	.035 oz
decigram	1.543 gr
centigram	.154 gr
milligram	.015 gr

ENGLISH TO METRIC EQUIVALENTS

		Unit	Metric Equivalent
LENGTH		mile	1.609 km
		rod	5.029 m
		yard	.9144 m
		foot	30.48 cm
		inch	2.54 cm
AREA		square mile	2.590 k^2
		acre	.405 hectacre, 4047 m^2
A = l x w		square rod	25.293 m^2
		square yard	.836 m^2
		square foot	.093 m^2
		square inch	6.452 cm^2
VOLUME		cubic yard	.765 m^3
V = l x w x t		cubic foot	.028 m^3
		cubic inch	16.387 cm^3
CAPACITY	*U.S. liquid measure*	gallon	3.785 l
		quart	.946 l
		pint	.473 l
WATER, FUEL, ETC.		gill	118.294 ml
		fluidounce	29.573 ml
		fluidram	3.697 ml
		minim	.061610 ml
	U.S. dry measure	bushel	35.239 l
VEGETABLES, GRAIN, ETC.		peck	8.810
		quart	1.101 l
		pint	.551 l
	British imperial liquid and dry measure	bushel	.036 m^3
		peck	.0091 m^3
		gallon	4.546 l
DRUGS		quart	1.136 l
		pint	568.26 cm^3
		gill	142.066 cm^3
		fluidounce	28.412 cm^3
		fluidram	3.5516 cm^3
		minim	.059194 cm^3
MASS AND WEIGHT	*avoirdupois*	short ton	.907 t
		long ton	1.016 t
		pound	.454 kg
COAL, GRAIN, ETC.		ounce	28.350 g
		dram	1.772 g
		grain	.0648 g
	troy	pound	.373 kg
		ounce	31.103 g
GOLD, SILVER, ETC.		pennyweight	1.555 g
		grain	.0648 g
	apothecaries'	pound	.373 kg
		ounce	31.103 g
DRUGS		dram	3.888 g
		scruple	1.296 g
		grain	.0648 g

ABBREVIATIONS . . .

A

abbreviation	ABBR
acoustic	ACST
acoustical tile	ACT. or AT.
adhesive	ADD. or ADH
adjustable	ADJ
adjustable-trip circuit breaker	ATCB
aggregate	AGGR
aileron	AIL
air conditioner	A/C
air handler	A/H
air tight	AT
alarm	ALM
alloy	ALY
alternating current	AC
aluminum	Al
ambient	AMB
Ambulatory Health Care Center	AHCC
American National Standards Institute	ANSI
American Wire Gauge	AWG
ammeter	A or AM
ampere	A or AMP
ampere interrupting rating	AIR.
amps	A
anchor bolt	AB
anode	A
antenna	ANT.
apartment	APT
appliance	APPL
approved	APPD or APVD
approximate	APPROX
approximately	APPROX
architectural	ARCH.
architecture	ARCH.
area	A
area drain	AD
armature	A or ARM.
asphalt	ASPH
asphalt tile	AT.
as required	AR
Assured Equipment Grounding Program	AEGP
astragal	A
Authority Having Jurisdiction	AHJ
automatic	AU or AUTO
automatic sprinkler	AS
auxiliary	AUX
avenue	AVE
azimuth	AZ

B

basement	BSMT
bathroom	B
bathtub	BT
battery (electric)	BAT.
beam	BM
bearing	BRG
bearing plate	BPL or BRG PL
bedroom	BR
benchmark	BM
black	BK
block	BLK
blocking	BLKG
blue	BL
board	BD
board foot	BF
bonding jumper	BJ
boulevard	BLVD
brake relay	BR
brass	BRS
brick	BRK

bridge	BRDG
bronze	BRZ
brown	BR
building	BL or BLDG
building line	BL
built-up roofing	BUR
bypass	BYP

C

cabinet	CAB.
cable	CA
Cable Antenna Television	CATV
cantilever	CANTIL
capacitor	CAP.
cased opening	CO
casement	CSMT
casing	CSG
cast iron	CI
cast-iron pipe	CIP
cast steel	CS
cast stone	CS or CST
catch basin	CB
cathode	K
ceiling	CLG
cellar	CEL
Celsius	°C
cement	CEM
cement floor	CF
center	CTR
centerline	CL
center-to-center	C to C
centigrade	C
central processing unit	CPU
ceramic	CER
ceramic tile	CT
ceramic-tile floor	CTF
channel	CHAN
chapter	CH
chimney	CHM
circuit	CIR or CKT
circuit breaker	CB
circuit interrupter	CI
circular mils	CM
cleanout	CO
clockwise	CW
closet	CLO
coarse	CRS
coated	CTD
coaxial	COAX
Code-Making Panel	CMP
cold air	CA
cold water	CW
column	COL
compacted	COMP
concrete	CONC
concrete block	CCB or CONC BLK
concrete floor	CCF
concrete pipe	CP
condenser	COND
conductor resistivity	K
conduit	C or CND
construction joint	CJ
continuous	CONT
contour	CTR
control	CONT
control joint	CJ or CLJ
control relay	CR
control relay master	CRM
copper	Cu
corner	COR
cornice	COR

corrugated	CORR
counterclockwise	CCW
counter electromotive force	CEMF
county	CO
cubic	CU
cubic foot	CU FT
cubic foot per minute	CFM
cubic foot per second	CFS
cubic inch	CU IN.
cubic yard	CU YD
current	I
current transformer	CT
cut out	CO
cycles per second	CPS

D

damper	DMPR
dampproofing	DP
dead load	DL
decibel	DB
deck	DK
demolition	DML
depth	DP
detail	DET or DTL
diagonal	DIAG
diagram	DIAG
diameter	D or DIA
dimension	DIM.
dimmer	DMR
dining room	DR
diode	DIO
direct current	DC
disconnect switch	DS
dishwasher	DW
distribution	DISTR
distribution panel	DPNL
division	DIV
door	DR
dormer	DRM
double-acting	DBL ACT
double hung window	DHW
double-pole	DP
double-pole double-throw	DPDT
double-pole double-throw switch	DPDT SW
double-pole single-throw	DPST
double-pole single-throw switch	DPST SW
double-pole switch	DP SW
double strength glass	DSG
double-throw	DT
down	D or DN
downspout	DS
drain	DR
drain tile	DT
drawing	DWG
drinking fountain	DF
drum switch	DS
dryer	D
drywall	DW
duplex	DX
dust tight	DT
dutch door	DD
duty cycle	DTY CY
dynamic braking contactor or relay	DB

E

each	EA
east	E
efficiency	Eff
ejector pump	EP
electric	ELEC

. . . ABBREVIATIONS . . .

electrical	ELEC	frequency	FREQ	inch	IN.
electrical metallic tubing	EMT	front view	FV	inch per second	IPS
electric panel	EP	full-load amps	FLA	inch-pound	IN. LB
electromechanical	ELMCH	full-load current	FLC	infrared	IR
electromotive force	EMF	full-load torque	FLT	inside diameter	ID
electronic	ELEK	furnace	FUR.	instantaneous overload	IOL
elevation	EL	furring	FUR.	instantaneous-trip circuit breaker	ITB
elevator	ELEV	fuse	FU	insulation	INSUL
enamel	ENAM	fuse block	FB	integrated circuit	IC
entrance	ENTR	fuse box	FUBX	interior	INT or INTR
equipment	EQPT	fuse holder	FUHLR	interlock	INTLK
equipment bonding jumper	EBJ	fusible	FSBL	intermediate	INT
equipment grounding conductor	EGC	future	FUT	intermediate metal conduit	IMC
equivalent	EQUIV			International Electrotechnical	
estimate	EST			Commission	IEC
excavate	EXC	**G**		interrupt	INT
exception	Ex.	gallon per hour	GPH	inverse time breaker	ITB
exhaust	EXH	gallon per minute	GPM	inverse-time circuit breaker	ITCB
existing	EXIST. or EXST	galvanized	GALV	iron	I
expanded metal	EM	garage	GAR	iron pipe	IP
expansion joint	EXP JT	gas	G	isolated ground	IG
explosionproof	EP	gate	G		
exterior	EXT	gauge	GA	**J**	
exterior grade	EXT GR	generator	GEN	jamb	JB or JMB
		glass	GL or GLS	joint	JT
		glass block	GLB	joist	J or JST
F		glaze	GLZ	junction	JCT
face brick	FB	gold	Au	junction box	JB
Fahrenheit	°F	grade	GR		
fast	F	grade line	GL	**K**	
field	F	gravel	GVL	key way	KWY
figure	FIG.	gray	GY	kick plate	KPL
fine print note	FPN	green	G or GR	kiln-dried	KD
finish	FNSH	gross vehicle weight	GVW	kilo (1000)	k
finish all over	FAO	ground	G, GND, GRD, or GRND	1000 circular mils	kcmil
finished floor	FNSH FL	grounded (outlet)	G	1000'	kFT
finished grade	FG or FIN GR	ground-fault circuit interrupter	GFCI	kilovolt amps	kVA
finish one side	F1S	ground fault protection of equipment	GFPE	kilowatt	kW
finish two sides	F2S	grounding electrode conductor	GEC	kilowatt-hour	kWh
firebrick	FBCK	grounding electrode system	GES	kitchen	K or KIT.
fire door	FDR	gypsum	GYP	knife switch	KN SW
fire extinguisher	FEXT	gypsum sheathing board	GSB	knockout	KO
fire hydrant	FHY				
fireplace	FP	**H**		**L**	
fireproof	FPRF	hand-off-auto	HOA		
fire wall	FW	handrail	HNDRL	lamp	LT
fixed window	FX WDW	hardboard	HBD	lath	LTH
fixture	FXTR	hardware	HDW	laundry	LAU
flammable	FLMB	hardwood	HDWD	laundry tray	LT
flashing	FL	hazardous	HAZ	lavatory	LAV
flat	FL	header	HDR	left	L
flexible metallic conduit	FMC	heating	HTG	left hand	LH
float switch	FS	heating, air-conditioning, refrigeration	HACR	less-flammable, liquid-insulated	LFLI
floor	FL	heating, ventilating, and air		library	LBRY or LIB
floor drain	FD	conditioning	HVAC	light	LT
flooring	FLG or FLR	heavy-duty	HD	lighting	LTG
floor line	FL	hertz	Hz	lighting panel	LP
flow switch	FLS	highway	HWY	lights	LTS
fluorescent	FLUOR or FLUR	hollow core	HC	limit switch	LS
flush	FL	hollow metal door	HMD	line	L
flush mount	FLMT	horizontal	HOR	linoleum	LINO or LINOL
footing	FTG	horsepower	HP	lintel	LNTL
foot per minute	FPM	hose bibb	HB	live load	LL
foot per second	FPS	hot water	HW	living room	LR
foot switch	FTS	hot water heater	HWH	load	L
forward	F or FWD	hours	HRS	location	LOC
foundation	FDN	hydraulic	HYDR	locked-rotor ampacity	LRA
four-pole	4P			locked-rotor current	LRC
four-pole double-throw switch	4PDT SW	**I**		louver	LVR or LV
four-pole single-throw switch	4PST SW			lumber	LBR
four-pole switch	4PSW	immersion detection circuit interrupter	IDCI		
frame	FR				

. . . ABBREVIATIONS . . .

M

magnetic brake	MB
main	MN
main bonding jumper	MBJ
main control center	MCC
manhole	MH
manual	MAN., MN, or MNL
manufacturer	MFR
marble	MRB
masonry	MSNRY
masonry opening	MO
material	MATL or MTL
maximum	MAX
maximum working pressure	MWP
mechanical	MECH
medicine cabinet	MC
medium	MED
memory	MEM
metal	MET or MTL
metal anchor	MA
metal door	METD
metal flashing	METF
metal jalousie	METJ
metal lath and plaster	MLP
metal threshold	MT
mezzanine	MEZZ
miles per gallon	MPG
miles per hour	MPH
minimum	MIN
mirror	MIR
miscellaneous	MISC
molding	MLDG
monolithic	ML
mortar	MOR
motor	M, MOT, or MTR
motor branch-circuit, short-circuit, ground-fault	MBCSCGF
motor circuit switch	MCS
motor control center	MCC
motor starter	M
motor switch	MS
mounted	MTD

N

nameplate	NPL
National Electrical Code®	NEC®
National Electrical Manufacturers Association	NEMA
National Electrical Safety Code	NESC
National Fire Protection Association	NFPA
negative	NEG
net weight	NTWT
neutral	N or NEUT
nominal	NOM
nonadjustable-trip circuit breaker	NATCB
non-time delay fuse	NTDF
normally closed	NC
normally open	NO
north	N
nosing	NOS
not to scale	NTS
number	NO.

O

Occupational Safety and Health Administration	OSHA
ohmmeter	OHM.
on center	OC
opaque	OPA

opening	OPNG
open web joist	OJ or OWJ
orange	O
ounces per inch	OZ/IN.
outlet	OUT.
outside diameter	OD
overall	OA
overcurrent	OC
overcurrent protection device	OCPD
overhang	OVHG
overhead	OH.
overload	OL
overload relay	OL

P

paint	PNT
painted	PTD
pair	PR
panel	PNL
pantry	PAN.
parallel	PRL
peak-to-peak	P-P
perpendicular	PERP
personal computer	PC
phase	PH
piece	PC
piling	PLG
pillar	PLR
pilot light	PL
piping	PP
pitch	P
plank	PLK
plaster	PLAS
plastic	PLSTC
plate	PL
plate glass	PLGL
plugging switch	PLS
plumbing	PLMB or PLBG
plywood	PLYWD
pneumatic	PNEU
point of beginning	POB
pole	P
polyvinyl chloride	PVC
porcelain	PORC
porch	P
positive	POS
pound(s)	LB
pounds per feet	LB-FT
pounds per inch	LB-IN.
pounds per square foot	PSF
pounds per square inch	PSI
power	P or PWR
power consumed	P
power factor	PF
precast	PRCST
prefabricated	PFB or PREFAB
prefinished	PFN
pressure switch	PS
primary switch	PRI.
property line	PL
pull box	PB
pull switch	PS
pull-up torque	PUT
pushbutton	PB

Q

quadrant	QDRNT
quantity	QTY
quarry tile	QT

quarry-tile roof	QTR
quart	QT
quick-acting	QA

R

radius	R
raintight	RT
random	RDM or RNDM
range	R or RNG
receipt of comments	ROC
receipt of proposals	ROP
receptacle	RECEPT or RCPT
recess	REC
recessed	REC
rectifier	REC
red	R
reference	REF
refrigerator	REF
register	REG or RGTR
reinforce	RE
reinforced concrete	RC
reinforcing steel	RST
reinforcing steel bar	REBAR
required	REQD
resistance	R
resistor	R or RES
return	RTN
reverse	R or REV
reverse-acting	RACT
revision	REV
revolutions per minute	RPM
revolutions per second	RPS
rheostat	RH
ribbed	RIB
right	R
right hand	RH
rigid	RGD
riser	R
road	RD
roll roofing	RR
roof	RF
roof drain	RD
roofing	RFG
room	R or RM
root mean square	RMS
rotor	RTR
rough	RGH
rough opening	RO
rough sawn	RS

S

safety switch	SSW
sanitary	SAN
scale	SC
schedule	SCH or SCHED
screen	SCR
scuttle	S
secondary	SEC
section	SECT
selector switch	SS
series	S
service	SERV
service entrance	SE
service factor	SF
sewer	SEW.
shake	SHK
sheathing	SHTHG
sheet	SH or SHT
sheet metal	SM
shelf and rod	SH & RD

. . . ABBREVIATIONS

shelving	SHELV
shingle	SHGL
shower	SH
shutter	SHTR
siding	SDG
silicon controlled rectifier	SCR
sill cock	SC
silver	Ag
single-phase	1PH
single-pole	SP
single-pole circuit breaker	SPCB
single-pole double-throw	SPDT
single-pole double-throw switch	SPDT SW
single-pole single-throw	SPST
single-pole single-throw switch	SPST SW
single-pole switch	SP SW
single strength glass	SSG
sink	S or SK
skylight	SLT
slate	S, SL, or SLT
sliding door	SLD
slope	SLP
slow	S
smoke detector	SD
socket	SOC
soffit	SF
soil pipe	SP
solenoid	SOL
solid core	SC
south	S
spare	SP
specification	SPEC
splash block	SB
square	SQ
square foot	SQ FT
square inch	SQ IN.
square yard	SQ YD
stack	STK
stainless steel	SST
stairs	ST
standard	STD
standpipe	SP
starter	START or STR
steel	STL
stone	STN
storage	STOR
street	ST
structural glass	SG
sump pump	SP
supply	SPLY
surface four sides	S4S
surface one side	S1S
switch	S or SW
switched disconnect	SWD

T

telephone	TEL
television	TV
temperature	TEMP
tempered	TEMP
terazzo	TER
terminal	T or TERM.
terminal board	TB
terra cotta	TC
thermal	THRM
thermally protected	TP
thermostat	THERMO
thermostat switch	THS
three-phase	3PH
three-pole	3P
three-pole double-throw	3PDT
three-pole single-throw	3PST
three-way	3WAY
three-wire	3W
threshold	TH
time	T
time delay	TD
time-delay fuse	TDF
time delay relay	TR
toilet	T
tongue-and-groove	T & G
torque	T
transformer	T, TRANS, or XFMR
transformer, primary side	H
transformer, secondary side	X
tread	TR
triple-pole double-throw	3PDT
triple-pole double-throw switch	3PDT SW
triple-pole single-throw	3PST
triple-pole single-throw switch	3PST SW
triple-pole switch	3P SW
truss	TR
two-phase	2PH
two-pole	DP
two-pole double-throw	DPDT
two-pole single-throw	DPST
typical	TYP

U

unclamp	UCL
underground	UGND
underground feeder	UF
undervoltage	UV
Underwriters Laboratories Inc.	UL
unexcavated	UNEXC
unfinished	UNFIN
up	U
utility room	U RM

V

valley	VAL
valve	V
vapor seal	VS
vaportight	VT
vent	V
ventilation	VENT.
vent pipe	VP
vent stack	VS
vertical	V or VERT
vinyl tile	VTILE or VT
violet	V
volt	V
voltage	E or V
voltage drop	VD
volt amps	VA
volts	V
volts alternating current	VAC
volts direct current	VDC
volume	VOL

W

wainscot	WAIN
walk-in closet	WIC
warm air	WA
washing machine	WM
waste pipe	WP
waste stack	WS
water	WTR
water closet	WC
water heater	WH
water meter	WM
waterproof	WP
watt(s)	W
weatherproof	WP
welded	WLD
welded wire fabric	WWF
west	W
white	W
wide flange	WF
wire gauge	WG
wire mesh	WM
with	W/
without	W/O
wood	WD
wrought iron	WI

Y

yellow	Y

ALPHABET OF LINES

NAME AND USE	CONVENTIONAL REPRESENTATION	EXAMPLE	
OBJECT LINE Define shape. Outline and detail objects.	THICK	OBJECT LINE	
HIDDEN LINE Show hidden features.	$\frac{1}{8}''$ (3 mm) THIN — $\frac{1}{32}''$ (0.75 mm)	HIDDEN LINE	
CENTER LINE Locate centerpoints of arcs and circles.	$\frac{1}{16}''$ (1.5 mm) THIN — $\frac{1}{8}''$ (3 mm) — $\frac{3}{4}''$ (18 mm) TO $1\frac{1}{2}''$ (36 mm)	CENTER LINE CENTERPOINT	
DIMENSION LINE Show size or location. **EXTENSION LINE** Define size or location.	DIMENSION LINE — DIMENSION 2'-6" THIN — EXTENSION LINE	DIMENSION LINE $1\frac{3}{4}$ EXTENSION LINE	
LEADER Call out specific features.	OPEN ARROWHEAD THIN X 3X CLOSED ARROWHEAD	$1\frac{1}{2}$ DRILL LEADER	
CUTTING PLANE Show internal features.	$\frac{1}{8}''$ (3 mm) $\frac{1}{16}''$ (1.5 mm) A A $\frac{3}{4}''$ (18 mm) TO $1\frac{1}{2}''$ (36 mm)	A A' LETTER IDENTIFIES SECTION VIEW CUTTING PLANE LINE	
SECTION LINE Identify internal features.	$\frac{1}{16}''$ (1.5 mm) THIN	SECTION LINES	
BREAK LINE Show long breaks. **BREAK LINE** Show short breaks.	$\frac{3}{4}''$ (18 mm) TO $1\frac{1}{2}''$ (36 mm) THIN FREEHAND THICK	LONG BREAK LINE SHORT BREAK LINE	

ARCHITECTURAL SYMBOLS . . .

Material	Elevation	Plan	Section
EARTH			
BRICK	WITH NOTE INDICATING TYPE OF BRICK (COMMON, FACE, ETC.)	COMMON OR FACE / FIREBRICK	SAME AS PLAN VIEWS
CONCRETE		LIGHTWEIGHT / STRUCTURAL	SAME AS PLAN VIEWS
CONCRETE BLOCK		OR	OR
STONE	CUT STONE / RUBBLE	CUT STONE / RUBBLE / CAST STONE (CONCRETE)	CUT STONE / CAST STONE (CONCRETE) / RUBBLE OR CUT STONE
WOOD	SIDING / PANEL	WOOD STUD / REMODELING / DISPLAY	ROUGH MEMBERS / FINISHED MEMBERS / PLYWOOD
PLASTER		WOOD STUD, LATH, AND PLASTER / METAL LATH AND PLASTER / SOLID PLASTER	LATH AND PLASTER
ROOFING	SHINGLES	SAME AS ELEVATION VIEW	
GLASS	OR / GLASS BLOCK	GLASS / GLASS BLOCK	SMALL SCALE / LARGE SCALE

. . . ARCHITECTURAL SYMBOLS

Material	Elevation	Plan	Section
FACING TILE	CERAMIC TILE	FLOOR TILE	CERAMIC TILE LARGE SCALE / CERAMIC TILE SMALL SCALE
STRUCTURAL CLAY TILE			SAME AS PLAN VIEW
INSULATION		LOOSE FILL OR BATTS / RIGID / SPRAY FOAM	SAME AS PLAN VIEWS
SHEET METAL FLASHING		OCCASIONALLY INDICATED BY NOTE	
METALS OTHER THAN FLASHING	INDICATED BY NOTE OR DRAWN TO SCALE	SAME AS ELEVATION	SMALL SCALE / STEEL / CAST IRON / ALUMINUM / BRONZE OR BRASS
STRUCTURAL STEEL	INDICATED BY NOTE OR DRAWN TO SCALE	OR	REBARS / LI SMALL SCALE / LI LARGE SCALE / L-ANGLES, S-BEAMS, ETC.

PLOT PLAN SYMBOLS

N	NORTH	●	FIRE HYDRANT		WALK	E OR	ELECTRIC SERVICE
	POINT OF BEGINNING (POB)	⊠	MAILBOX		IMPROVED ROAD	G OR	NATURAL GAS LINE
▲	UTILITY METER OR VALVE	○	MANHOLE		UNIMPROVED ROAD	W OR	WATER LINE
●	POWER POLE AND GUY	+	TREE	BL	BUILDING LINE	T OR	TELEPHONE LINE
	LIGHT STANDARD		BUSH	PL	PROPERTY LINE		NATURAL GRADE
	TRAFFIC SIGNAL		HEDGE ROW		PROPERTY LINE		FINISH GRADE
	STREET SIGN	—•—•—	FENCE		TOWNSHIP LINE	+ XX.00′	EXISTING ELEVATION

ELECTRICAL SYMBOLS . . .

LIGHTING OUTLETS

OUTLET BOX AND INCANDESCENT LIGHTING FIXTURE	CEILING WALL
INCANDESCENT TRACK LIGHTING	
BLANKED OUTLET	B B
DROP CORD	D
EXIT LIGHT AND OUTLET BOX. SHADED AREAS DENOTE FACES.	
OUTDOOR POLE-MOUNTED FIXTURES	
JUNCTION BOX	J J
LAMPHOLDER WITH PULL SWITCH	L_{PS} L_{PS}
MULTIPLE FLOODLIGHT ASSEMBLY	
EMERGENCY BATTERY PACK WITH CHARGER	B
INDIVIDUAL FLUORESCENT FIXTURE	
OUTLET BOX AND FLUORESCENT LIGHTING TRACK FIXTURE	
CONTINUOUS FLUORESCENT FIXTURE	
SURFACE-MOUNTED FLUORESCENT FIXTURE	

PANELBOARDS

FLUSH-MOUNTED PANELBOARD AND CABINET	
SURFACE-MOUNTED PANELBOARD AND CABINET	

CONVENIENCE OUTLETS

SINGLE RECEPTACLE OUTLET	
DUPLEX RECEPTACLE OUTLET	
TRIPLEX RECEPTACLE OUTLET	
SPLIT-WIRED DUPLEX RECEPTACLE OUTLET	
SPLIT-WIRED TRIPLEX RECEPTACLE OUTLET	
SINGLE SPECIAL-PURPOSE RECEPTACLE OUTLET	
DUPLEX SPECIAL-PURPOSE RECEPTACLE OUTLET	
RANGE OUTLET	R
SPECIAL-PURPOSE CONNECTION	DW
CLOSED-CIRCUIT TELEVISION CAMERA	
CLOCK HANGER RECEPTACLE	C
FAN HANGER RECEPTACLE	F
FLOOR SINGLE RECEPTACLE OUTLET	
FLOOR DUPLEX RECEPTACLE OUTLET	
FLOOR SPECIAL-PURPOSE OUTLET	
UNDERFLOOR DUCT AND JUNCTION BOX FOR TRIPLE, DOUBLE, OR SINGLE DUCT SYSTEM AS INDICATED BY NUMBER OF PARALLEL LINES	

BUSDUCTS AND WIREWAYS

SERVICE, FEEDER, OR PLUG-IN BUSWAY	B B B
CABLE THROUGH LADDER OR CHANNEL	C C C
WIREWAY	W W W

SWITCH OUTLETS

SINGLE-POLE SWITCH	S
DOUBLE-POLE SWITCH	S_2
THREE-WAY SWITCH	S_3
FOUR-WAY SWITCH	S_4
AUTOMATIC DOOR SWITCH	S_D
KEY-OPERATED SWITCH	S_K
CIRCUIT BREAKER	S_{CB}
WEATHERPROOF CIRCUIT BREAKER	S_{WCB}
DIMMER	S_{DM}
REMOTE CONTROL SWITCH	S_{RC}
WEATHERPROOF SWITCH	S_{WP}
FUSED SWITCH	S_F
WEATHERPROOF FUSED SWITCH	S_{WF}
TIME SWITCH	S_T
CEILING PULL SWITCH	S
SWITCH AND SINGLE RECEPTACLE	S
SWITCH AND DOUBLE RECEPTACLE	S
A STANDARD SYMBOL WITH AN ADDED LOWERCASE SUBSCRIPT LETTER IS USED TO DESIGNATE A VARIATION IN STANDARD EQUIPMENT	$_{a.b}$ $_{a.b}$ $S_{a.b}$

. . . ELECTRICAL SYMBOLS

COMMERCIAL AND INDUSTRIAL SYSTEMS

PAGING SYSTEM DEVICE

FIRE ALARM SYSTEM DEVICE

COMPUTER DATA SYSTEM DEVICE

PRIVATE TELEPHONE SYSTEM DEVICE

SOUND SYSTEM

FIRE ALARM CONTROL PANEL — FACP

SIGNALING SYSTEM OUTLETS FOR RESIDENTIAL SYSTEMS

PUSHBUTTON

BUZZER

BELL

BELL AND BUZZER COMBINATION

COMPUTER DATA OUTLET

BELL RINGING TRANSFORMER — BT

ELECTRIC DOOR OPENER — D

CHIME — CH

TELEVISION OUTLET — TV

THERMOSTAT — T

UNDERGROUND ELECTRICAL DISTRIBUTION OR ELECTRICAL LIGHTING SYSTEMS

MANHOLE — M

HANDHOLE — H

TRANSFORMER- MANHOLE OR VAULT — TM

TRANSFORMER PAD — TP

UNDERGROUND DIRECT BURIAL CABLE

UNDERGROUND DUCT LINE

STREET LIGHT STANDARD FED FROM UNDERGROUND CIRCUIT

ABOVE-GROUND ELECTRICAL DISTRIBUTION OR LIGHTING SYSTEMS

POLE

STREET LIGHT AND BRACKET

PRIMARY CIRCUIT

SECONDARY CIRCUIT

DOWN GUY

HEAD GUY

SIDEWALK GUY

SERVICE WEATHERHEAD

PANEL CIRCUITS AND MISCELLANEOUS

LIGHTING PANEL

POWER PANEL

WIRING – CONCEALED IN CEILING OR WALL

WIRING – CONCEALED IN FLOOR

WIRING EXPOSED

HOME RUN TO PANEL BOARD
Indicate number of circuits by number of arrows. Any circuit without such designation indicates a two-wire circuit. For a greater number of wires indicate as follows: —///— (3 wires)
—////— (4 wires), etc.

FEEDERS
Use heavy lines and designate by number corresponding to listing in feeder schedule

WIRING TURNED UP

WIRING TURNED DOWN

GENERATOR — G

MOTOR — M

INSTRUMENT (SPECIFY) — I

TRANSFORMER — T

CONTROLLER

EXTERNALLY-OPERATED DISCONNECT SWITCH

PULL BOX

PLUMBING SYMBOLS . . .

FIXTURES...

STANDARD BATHTUB	
OVAL BATHTUB	
WHIRLPOOL BATH	
SHOWER STALL	
SHOWER HEAD	
TANK-TYPE WATER CLOSET	
WALL-MOUNTED WATER CLOSET	
FLOOR-MOUNTED WATER CLOSET	
LOW-PROFILE WATER CLOSET	
BIDET	
WALL-MOUNTED URINAL	
FLOOR-MOUNTED URINAL	
TROUGH-TYPE URINAL	
WALL-MOUNTED LAVATORY	
PEDESTAL LAVATORY	
BUILT-IN LAVATORY	
WHEELCHAIR LAVATORY	
CORNER LAVATORY	
FLOOR DRAIN	
FLOOR SINK	

...FIXTURES

LAUNDRY TRAY	
BUILT-IN SINK	
DOUBLE OR TRIPLE BUILT-IN SINK	
COMMERCIAL KITCHEN SINK	
SERVICE SINK	SS
CLINIC SERVICE SINK	
FLOOR-MOUNTED SERVICE SINK	
DRINKING FOUNTAIN	DF
WATER COOLER	
HOT WATER TANK	HWT
WATER HEATER	WH
METER	M
HOSE BIBB	HB
GAS OUTLET	G
GREASE SEPARATOR	G
GARAGE DRAIN	
FLOOR DRAIN WITH BACKWATER VALVE	

PIPING...

SOIL, WASTE, OR LEADER – ABOVE GRADE	
SOIL, WASTE, OR LEADER – BELOW GRADE	
VENT	
COMBINATION WASTE AND VENT	SV
STORM DRAIN	SD
COLD WATER	

...PIPING

CHILLED DRINKING WATER SUPPLY	DWS
CHILLED DRINKING WATER RETURN	DWR
HOT WATER	
HOT WATER RETURN	
SANITIZING HOT WATER SUPPLY (180° F)	
SANITIZING HOT WATER RETURN (180° F)	
DRY STANDPIPE	DSP
COMBINATION STANDPIPE	CSP
MAIN SUPPLIES SPRINKLER	S
BRANCH AND HEAD SPRINKLER	
GAS – LOW PRESSURE	G G
GAS – MEDIUM PRESSURE	MG
GAS – HIGH PRESSURE	HG
COMPRESSED AIR	A
OXYGEN	O
NITROGEN	N
HYDROGEN	H
HELIUM	HE
ARGON	AR
LIQUID PETROLEUM GAS	LPG
INDUSTRIAL WASTE	INW
CAST IRON	CI
CULVERT PIPE	CP
CLAY TILE	CT
DUCTILE IRON	DI
REINFORCED CONCRETE	RCP
DRAIN – OPEN TILE OR AGRICULTURAL TILE	

... PLUMBING SYMBOLS
PIPE FITTING AND VALVE SYMBOLS

	FLANGED	SCREWED	BELL & SPIGOT		FLANGED	SCREWED	BELL & SPIGOT		FLANGED	SCREWED	BELL & SPIGOT
BUSHING				REDUCING FLANGE				AUTOMATIC BY-PASS VALVE			
CAP				BULL PLUG				AUTOMATIC REDUCING VALVE			
REDUCING CROSS				PIPE PLUG				STRAIGHT CHECK VALVE			
STRAIGHT-SIZE CROSS				CONCENTRIC REDUCER				COCK			
CROSSOVER				ECCENTRIC REDUCER				DIAPHRAGM VALVE			
45° ELBOW				SLEEVE				FLOAT VALVE			
90° ELBOW				STRAIGHT-SIZE TEE				GATE VALVE			
ELBOW – TURNED DOWN				TEE – OUTLET UP				MOTOR-OPERATED GATE VALVE			
ELBOW – TURNED UP				TEE – OUTLET DOWN				GLOBE VALVE			
BASE ELBOW				DOUBLE-SWEEP TEE				MOTOR-OPERATED GLOBE VALVE			
DOUBLE-BRANCH ELBOW				REDUCING TEE				ANGLE HOSE VALVE			
LONG-RADIUS ELBOW				SINGLE-SWEEP TEE							
REDUCING ELBOW				SIDE OUTLET TEE – OUTLET DOWN				GATE VALVE			
SIDE OUTLET ELBOW – OUTLET DOWN				SIDE OUTLET TEE – OUTLET UP				GLOBE VALVE			
SIDE OUTLET ELBOW – OUTLET UP				UNION				LOCKSHIELD VALVE			
STREET ELBOW				ANGLE CHECK VALVE				QUICK-OPENING VALVE			
CONNECTING PIPE JOINT				ANGLE GATE VALVE – ELEVATION				SAFETY VALVE			
EXPANSION JOINT				ANGLE GATE VALVE – PLAN				GOVERNOR-OPERATED AUTOMATIC VALVE			
LATERAL				ANGLE GLOBE VALVE – ELEVATION							
ORIFICE FLANGE				ANGLE GLOBE VALVE – PLAN							

HVAC SYMBOLS

EQUIPMENT SYMBOLS	DUCTWORK	HEATING PIPING
EXPOSED RADIATOR	DUCT (1ST FIGURE, WIDTH; 2ND FIGURE, DEPTH) 12 X 20	HIGH-PRESSURE STEAM ——HPS——
RECESSED RADIATOR	DIRECTION OF FLOW	MEDIUM-PRESSURE STEAM ——MPS——
FLUSH ENCLOSED RADIATOR	FLEXIBLE CONNECTION	LOW-PRESSURE STEAM —LPS—
PROJECTING ENCLOSED RADIATOR	DUCTWORK WITH ACOUSTICAL LINING	HIGH-PRESSURE RETURN ——HPR—
UNIT HEATER (PROPELLER) – PLAN	FIRE DAMPER WITH ACCESS DOOR FD AD	MEDIUM-PRESSURE RETURN ——MPR—
UNIT HEATER (CENTRIFUGAL) – PLAN	MANUAL VOLUME DAMPER — VD	LOW-PRESSURE RETURN —LPR—
UNIT VENTILATOR – PLAN	AUTOMATIC VOLUME DAMPER	BOILER BLOW OFF —— BD —
STEAM	EXHAUST, RETURN OR OUTSIDE AIR DUCT – SECTION 20 X 12	CONDENSATE OR VACUUM PUMP DISCHARGE ——VPD—
DUPLEX STRAINER	SUPPLY DUCT – SECTION 20 X 12	FEEDWATER PUMP DISCHARGE ——PPD—
PRESSURE-REDUCING VALVE	CEILING DIFFUSER SUPPLY OUTLET 20" DIA CD 1000 CFM	MAKEUP WATER —— MU —
AIR LINE VALVE	CEILING DIFFUSER SUPPLY OUTLET 20 X 12 CD 700 CFM	AIR RELIEF LINE —— V —
STRAINER	LINEAR DIFFUSER 96 X 6-LD 400 CFM	FUEL OIL SUCTION ——FOS—
THERMOMETER	FLOOR REGISTER 20 X 12 FR 700 CFM	FUEL OIL RETURN ——FOR—
PRESSURE GAUGE AND COCK	TURNING VANES	FUEL OIL VENT ——FOV—
RELIEF VALVE	FAN AND MOTOR WITH BELT GUARD	COMPRESSED AIR —— A —
AUTOMATIC 3-WAY VALVE		HOT WATER HEATING SUPPLY —— HW —
AUTOMATIC 2-WAY VALVE		HOT WATER HEATING RETURN ——HWR—
SOLENOID VALVE S	LOUVER OPENING 20 X 12-L 700 CFM	

AIR CONDITIONING PIPING

REFRIGERANT LIQUID	— RL —
REFRIGERANT DISCHARGE	— RD —
REFRIGERANT SUCTION	—— RS ——
CONDENSER WATER SUPPLY	—CWS—
CONDENSER WATER RETURN	—CWR—
CHILLED WATER SUPPLY	—— CHWS——
CHILLED WATER RETURN	——CHWR——
MAKEUP WATER	—— MU —
HUMIDIFICATION LINE	—— H —
DRAIN	—— D —

REFRIGERATION SYMBOLS

GAUGE	PRESSURE SWITCH	DRYER
SIGHT GLASS	HAND EXPANSION VALVE	FILTER AND STRAINER
HIGH SIDE FLOAT VALVE	AUTOMATIC EXPANSION VALVE	COMBINATION STRAINER AND DRYER
LOW SIDE FLOAT VALVE	THERMOSTATIC EXPANSION VALVE	EVAPORATIVE CONDENSOR
IMMERSION COOLING UNIT	CONSTANT PRESSURE VALVE, SUCTION	HEAT EXCHANGER
COOLING TOWER	THERMAL BULB	AIR-COOLED CONDENSING UNIT
NATURAL CONVECTION, FINNED TYPE EVAPORATOR	SCALE TRAP	WATER-COOLED CONDENSING UNIT
FORCED CONVECTION EVAPORATOR	SELF-CONTAINED THERMOSTAT	

HAZARDOUS LOCATIONS

Class	Division	Group	Material
I	1 or 2	A	Acetylene
	1 or 2	B	Hydrogen, butadiene, ethylene oxide, propylene oxide
	1 or 2	C	Carbon monoxide, ether, ethylene, hydrogen sulfide, morpholine, cyclopropane
	1 or 2	D	Gasoline, benzene, butane, propane, alcohol, acetone, ammonia, vinyl chloride
II	1 or 2	E	Metal dusts
	1 or 2	F	Carbon black, coke dust, coal
	1 or 2	G	Grain dust, flour, starch, sugar, plastics
III	1 or 2	No groups	Wood chips, cotton, flax, nylon

ENCLOSURES . . .

An enclosure is a case or housing of apparatus to prevent personnel from accidentally contacting energized parts, or to protect the equipment from physical damage. See 100. Enclosures for hazardous (classified) locations are marked with a ✓.

Type 1 Enclosure

Type 1 enclosures are intended for indoor use primarily to provide a degree of protection against limited amounts of flying dirt. NEMA Standard 7-15-1991.

General Electric Company

General duty safety switches from General Electric are available in Type 1 and 3R enclosures that are designed for residential or high commercial applications where duty is not severe.

Type 2 Enclosure

Type 2 enclosures are intended for indoor use primarily to provide a degree of protection against limited amounts of falling water and dirt. NEMA Standard 1-10-1979.

Type 3 Enclosure

Type 3 enclosures are intended for outdoor use primarily to provide a degree of protection against windblown dust, rain, sleet, and external ice formation. NEMA Standard 7-15-1991.

Type 3R Enclosure

Type 3R enclosures are intended for outdoor use primarily to provide a degree of protection against falling rain, sleet, and external ice formation. NEMA Standard 7-15-1991.

Type 3S Enclosure

Type 3S enclosures are intended for outdoor use primarily to provide a degree of protection against windblown dust, rain, sleet, and provide for operation of external mechanisms when ice laden. NEMA 7-15-1991.

Square D Company

Heavy-duty safety switches from Square D Company have visible blades and are available in NEMA Type 3R rainproof enclosures.

Type 4 Enclosure

Type 4 enclosures are intended for indoor or outdoor use primarily to provide a degree of protection against windblown dust and rain, splashing water, hose-directed water, and damage from external ice formation. NEMA Standard 1-10-1979.

Type 4X Enclosure

Type 4X enclosures are intended for indoor or outdoor use primarily to provide a degree of protection against corrosion, windblown dust and rain, splashing water, hose-directed water, and damage from external ice formation. NEMA Standard 1-10-1979.

Crouse-Hinds Division, Cooper Industries, Inc.

Crouse-Hinds manufactures enclosures made from high-impact strength fiberglass-reinforced polyester material that is corrosion-resistant, dust tight, water-tight, and weatherproof, and rated as NEMA Type 3, 4X, and 12.

Type 5 Enclosure

Type 5 enclosures are intended for indoor use primarily to provide a degree of protection against settling airborne dust, falling dirt, and dripping noncorrosive liquids. NEMA Standard 5-25-1988.

Type 6 Enclosure

Type 6 enclosures are intended for indoor or outdoor use primarily to provide a degree of protection against hose-directed water, the entry of water during occasional temporary submersion at a limited depth, and damage from external ice formation. NEMA Standard 7-15-1991.

. . . ENCLOSURES

Type 6P Enclosure

Type 6P enclosures are intended for indoor or outdoor use primarily to provide a degree of protection against hose-directed water, the entry of water during prolonged submersion at a limited depth, and damage from external ice formation. NEMA Standard 7-15-1991.

Type 7 Enclosure ✓

Type 7 enclosures are intended for indoor use in locations classified as Class I, Groups A, B, C, or D, as defined in the National Electrical Code. NEMA Standard 7-15-1991.

Rockwell Automation, Allen-Bradley Company, Inc.

Bulletin 609U manual starting switches from Allen-Bradley are available with undervoltage protection in Type 3R, 7, and 9 enclosures.

Type 8 Enclosure ✓

Type 8 enclosures are intended for indoor or outdoor use in locations classified as Class I, Groups A, B, C, and D, as defined in the National Electrical Code. NEMA Standard 7-15-1991.

Type 9 Enclosure ✓

Type 9 enclosures are for use in indoor locations classified as Class II, Groups E, F, or G, as defined in the National Electrical Code. NEMA Standard 7-15-1991.

Type 10 Enclosure ✓

Type 10 enclosures are constructed to meet the applicable requirements of the Mine Safety and Health Administration. NEMA Standard 1-10-1979.

Type 12 Enclosure

Type 12 enclosures are intended for indoor use primarily to provide a degree of protection against circulating dust, falling dirt, and dripping noncorrosive liquids. NEMA Standard 7-15-1991.

Type 8 Enclosure ✓

Type 8 enclosures are intended for indoor or outdoor use in locations classified as Class I, Groups A, B, C, and D, as defined in the National Electrical Code. NEMA Standard 7-15-1991.

Type 9 Enclosure ✓

Type 9 enclosures are for use in indoor locations classified as Class II, Groups E, F, or G, as defined in the National Electrical Code. NEMA Standard 7-15-1991.

Type 10 Enclosure ✓

Type 10 enclosures are constructed to meet the applicable requirements of the Mine Safety and Health Administration. NEMA Standard 1-10-1979.

Type 12 Enclosure

Type 12 enclosures are intended for indoor use primarily to provide a degree of protection against circulating dust, falling dirt, and dripping noncorrosive liquids. NEMA Standard 7-15-1991.

Rockwell Automation, Allen-Bradley Company, Inc.

Bulletin 100 molded enclosed starters from Allen-Bradley are available in Type 1, 4/4X, and 12 construction.

Type 12K Enclosure

Type 12K enclosures with knockouts are intended for indoor use primarily to provide a degree of protection against circulating dust, falling dirt, and dripping noncorrosive liquids. See NEMA Standard 7-15-1991.

Type 13 Enclosure

Type 13 enclosures are intended for indoor use primarily to provide a degree of protection against dust, spraying of water, oil, and noncorrosive coolant. NEMA Standard 1-10-1979.

ENCLOSURE TESTING

Enclosures for electrical equipment up to 1000 V maximum are designed and tested per NEMA 250-1991, *Enclosures for Electrical Equipment (1000 V Maximum)*. Section 4 of NEMA 250-1991 includes descriptions, applications, features, and test criteria for nonhazardous location enclosures (Type 1, 2, 3, 3R, 3S, 4, 4X, 5, 6, 6P, 12, 12K, and 13). Section 5 of NEMA 250-1991 includes descriptions, applications, features, and test criteria for hazardous (classified) location enclosures (Type 7, 8, 9, and 10).

Nonhazardous location enclosures are designed according to 11 design tests. See Enclosures. The design tests are given in Section 6 of NEMA 250-1991. For example, the external icing test is used to test Type 3, 3R, 3S, 4, 4X, 6, and 6P enclosures. The enclosures must be manually operated by one person while ice-laden and must be undamaged after the ice is melted. See External Icing Test. For additional design tests, see NEMA 250-1991, Section 6.

ENCLOSURES													
Design Test	**1**	**2**	**3**	**3R**	**3S**	**4**	**4X**	**5**	**6**	**6P**	**12**	**12K**	**13**
Rod Entry	•	•		•									
Drip		•						•			•	•	
Rain			•	•	•								
Dust			•		•			•			•	•	
External Icing			•	•	•	•	•		•	•			
Hosedown						•	•		•	•			
Rust Resistance	•	•						•			•	•	•
Corrosion Protection			•	•	•	•	•		•	•			
Submersion									•				
Air Pressure										•			
Oil Exclusion													•

EXTERNAL ICING

ROOM TEMP
BEGIN: 32°F – 37°F
AFTER 1 HR: 20°F – 27°F
AFTER $\frac{3}{4}''$ ICE BUILDUP: 20°F – 27°F FOR 3 HR

WATER TEMP 32°F – 37°F

1 – 2 GAL./SQ FT PER HR

ICE BUILDUP RATE APPROXIMATELY $\frac{1}{4}''$ PER HR CONTINUED UNTIL $\frac{3}{4}''$

1″ DIA TEST BAR 2′ LONG TO RECEIVE SAME WATER SPRAY AS ENCLOSURE

EVALUATION

6.6.2.1
- Enclosure manually operated by one person while ice laden without damage
- Access can be gained to enclosure without damage to enclosure

6.6.2.2
- Enclosure undamaged after ice melted

LOCKING WIRING DEVICES

2-POLE, 3-WIRE

WIRING DIAGRAM	NEMA ANSI	RECEPTACLE CONFIGURATION	RATING
	ML2 C73.44		15 A 125 V
	L5-15 C73.42		15 A 125 V
	L5-20 C73.72		20 A 125 V
	L6-15 C73.74		15 A 250 V
	L6-20 C73.75		20 A 250 V
	L6-30 C73.76		30 A 250 V
	L7-15 C73.43		15 A 277 V
	L7-20 C73.77		20 A 277 V
	L8-20 C73.79		20 A 480 V
	L9-20 C73.81		20 A 600 V

3-POLE, 4-WIRE

WIRING DIAGRAM	NEMA ANSI	RECEPTACLE CONFIGURATION	RATING
	L14-20 C73.83		20 A 125/250 V
	L14-30 C73.84		30 A 125/250 V
	L15-20 C73.85		20 A 3φ 250 V
	L15-30 C73.86		30 A 3φ 250 V
	L16-20 C73.87		20 A 3φ 480 V
	L16-30 C73.88		30 A 3φ 480 V
	L17-30 C73.89		30 A 3φ 600 V

3-POLE, 3-WIRE

WIRING DIAGRAM	NEMA ANSI	RECEPTACLE CONFIGURATION	RATING
	ML3 C73.30		15 A 125/250 V
	L10-20 C73.96		20 A 125/250 V
	L10-30 C73.97		30 A 125/250 V
	L11-15 C73.98		15 A 3φ 250 V
	L11-20 C73.99		20 A 3φ 250 V
	L12-20 C73.101		20 A 3φ 480 V
	L12-30 C73.102		30 A 3φ 480 V
	L13-30 C73.103		30 A 3φ 600 V

4-POLE, 4-WIRE

WIRING DIAGRAM	NEMA ANSI	RECEPTACLE CONFIGURATION	RATING
	L18-20 C73.104		20 A 3φ Y 120/208 V
	L18-30 C73.105		30 A 3φ Y 120/208 V
	L19-20 C73.106		20 A 3φ Y 277/480 V
	L20-20 C73.108		20 A 3φ Y 347/600 V

4-POLE, 5-WIRE

WIRING DIAGRAM	NEMA ANSI	RECEPTACLE CONFIGURATION	RATING
	L21-20 C73.90		20 A 3φ Y 120/208 V
	L22-20 C73.92		20 A 3φ Y 277/480 V
	L23-20 C73.94		20 A 3φ Y 347/600 V

NON-LOCKING WIRING DEVICES

2-POLE, 3-WIRE

WIRING DIAGRAM	NEMA ANSI	RECEPTACLE CONFIGURATION	RATING
	5-15 C73.11		15 A 125 V
	5-20 C73.12		20 A 125 V
	5-30 C73.45		30 A 125 V
	5-50 C73.46		50 A 125 V
	6-15 C73.20		15 A 250 V
	6-20 C73.51		20 A 250 V
	6-30 C73.52		30 A 250 V
	6-50 C73.53		50 A 250 V
	7-15 C73.28		15 A 277 V
	7-20 C73.63		20 A 277 V
	7-30 C73.64		30 A 277 V
	7-50 C73.65		50 A 277 V

3-POLE, 3-WIRE

WIRING DIAGRAM	NEMA ANSI	RECEPTACLE CONFIGURATION	RATING
	10-20 C73.23		20 A 125/250 V
	10-30 C73.24		30 A 125/250 V
	10-50 C73.25		50 A 125/250 V
	11-15 C73.54		15 A 3φ 250 V
	11-20 C73.55		20 A 3φ 250 V
	11-30 C73.56		30 A 3φ 250 V
	11-50 C73.57		50 A 3φ 250 V

3-POLE, 4-WIRE

WIRING DIAGRAM	NEMA ANSI	RECEPTACLE CONFIGURATION	RATING
	14-15 C73.49		15 A 125/250 V
	14-20 C73.50		20 A 125/250 V
	14-30 C73.16		30 A 125/250 V
	14-50 C73.17		50 A 125/250 V
	14-60 C73.18		60 A 125/250 V
	15-15 C73.58		15 A 3φ 250 V
	15-20 C73.59		20 A 3φ 250 V
	15-30 C73.60		30 A 3φ 250 V
	15-50 C73.61		50 A 3φ 250 V
	15-60 C73.62		60 A 3φ 250 V

4-POLE, 4-WIRE

WIRING DIAGRAM	NEMA ANSI	RECEPTACLE CONFIGURATION	RATING
	18-15 C73.15		15 A 3φ Y 120/208 V
	18-20 C73.26		20 A 3φ Y 120/208 V
	18-30 C73.47		30 A 3φ Y 120/208 V
	18-50 C73.48		50 A 3φ Y 120/208 V
	18-60 C73.27		60 A 3φ Y 120/208 V

STANDARD CALCULATION: ONE-FAMILY DWELLING

1. GENERAL LIGHTING: *Table 220-3(a)*

_____ sq ft × 3 VA =   _____ VA

Small appliances: *220-16(a)*

_____ VA × _____ circuits =   _____ VA

Laundry: *220-16(b)*

_____ VA × 1 =   _____ VA

                                                            _____ VA

Applying Demand Factors: *Table 220-11*

First 3000 VA × 100% =   3000 VA

Next _____ VA × 35% =   _____ VA   **PHASES**   **NEUTRAL**

Remaining _____ VA × 25% =   _____ VA
Total   _____ VA   _____ VA   _____ VA

2. FIXED APPLIANCES: *220-17*

Dishwasher =   _____ VA

Disposer =   _____ VA

Compactor =   _____ VA

Water heater =   _____ VA

_____ = _____ VA

_____ = _____ VA

_____ = _____ VA

Total   _____ VA × 75% = _____ VA   _____ VA   _____ VA

(120 V Loads × 75%)

3. DRYER: *220-18; Table 220-18*

_____ VA × _____ % =   _____ VA   _____ VA × 70% = _____ VA

4. COOKING EQUIPMENT: *Table 220-19; Notes*

Col A _____ VA × _____ % =   _____ VA

Col B _____ VA × _____ % =   _____ VA

Col C _____ VA × _____ % =   _____ VA

Total   _____ VA   _____ VA × 70% = _____ VA

5. HEATING or A/C: *220-21*

Heating unit = _____ VA × 100% =   _____ VA

A/C unit = _____ VA × 100% =   _____ VA

Heat pump = _____ VA × 100% =   _____ VA

Largest Load   _____ VA   _____ VA   _____ VA

6. LARGEST MOTOR: *220-14*

φ _____ VA × 25% =   _____ VA   _____ VA

N _____ VA × 25% =   _____ VA   _____ VA

1φ service: PHASES $I = \dfrac{\text{VA}}{\text{V}} =$ _____ A

NEUTRAL $I = \dfrac{\text{VA}}{\text{V}} =$ _____ A   [ ] VA   [ ] VA

220-22; First 200 A × 100% =   200 A

Remaining _____ A × 70% =   _____ A
Total   _____ A

OPTIONAL CALCULATION: ONE-FAMILY DWELLING

1. HEATING and A/C: *220-30(c)(1 – 6)*

Heating units (3 or less) = _____ VA × 65% = _____ VA

Heating units (4 or more) = _____ VA × 40% = _____ VA

A/C unit = _____ VA × 100% = _____ VA

Heat pump = _____ VA × 100% = _____ VA   **PHASES**

Largest Load _____ VA

Total _____ VA   _____ VA

2. GENERAL LOADS: *220-30(b)(1 – 4)*

General lighting: *220-30(b)(2)*

_____ sq ft × 3 VA   _____ VA

Small appliance and laundry loads: *220-30(b)(1)*

_____ VA × _____ circuits =   _____ VA

Special loads: *220-30(b)(3)(4)*

Dishwasher = _____ VA

Disposer = _____ VA

Compactor = _____ VA

Water heater = _____ VA

_____ = _____ VA

_____ = _____ VA

_____ = _____ VA

_____ = _____ VA

_____ = _____ VA

_____ VA   _____ VA

Total _____ VA

Applying Demand Factors: *220-30(b)*

First 10,000 VA × 100% =   10,000 VA

Remaining _____ VA × 40% =   _____ VA

Total _____ VA   _____ VA

NEUTRAL (Loads from Standard Calculation)

1. General lighting = _____ VA

2. Fixed appliances = _____ VA

3. Dryer = _____ VA

4. Cooking equipment = _____ VA

5. Heating or A/C = _____ VA

6. Largest motor = _____ VA

Total _____ VA

1φ service: PHASES $I = \dfrac{\text{VA}}{\text{V}} =$ _____ A

NEUTRAL $I = \dfrac{\text{VA}}{\text{V}} =$ _____ A   _____ VA

CSI MASTERFORMAT™ – LEVEL TWO NUMBERS AND TITLES

INTRODUCTORY INFORMATION
00001 PROJECT TITLE PAGE
00005 CERTIFICATIONS PAGE
00007 SEALS PAGE
00010 TABLE OF CONTENTS
00015 LIST OF DRAWINGS
00020 LIST OF SCHEDULES

BIDDING REQUIREMENTS
00100 BID SOLICITATION
00200 INSTRUCTIONS TO BIDDERS
00300 INFORMATION AVAILABLE TO BIDDERS
00400 BID FORMS AND SUPPLEMENTS
00490 BIDDING ADDENDA

CONTRACTING REQUIREMENTS
00500 AGREEMENT
00600 BONDS AND CERTIFICATES
00700 GENERAL CONDITIONS
00800 SUPPLEMENTARY CONDITIONS
00900 ADDENDA AND MODIFICATIONS

FACILITIES AND SPACES
FACILITIES AND SPACES

SYSTEMS AND ASSEMBLIES
SYSTEMS AND ASSEMBLIES

CONSTRUCTION PRODUCTS AND ACTIVITIES

DIVISION 1 GENERAL REQUIREMENTS
01100 SUMMARY
01200 PRICE AND PAYMENT PROCEDURES
01300 ADMINISTRATIVE REQUIREMENTS
01400 QUALITY REQUIREMENTS
01500 TEMPORARY FACILITIES AND CONTROLS
01600 PRODUCT REQUIREMENTS
01700 EXECUTION REQUIREMENTS
01800 FACILITY OPERATION
01900 FACILITY DECOMMISSIONING

DIVISION 2 SITE CONSTRUCTION
02050 BASIC SITE MATERIALS AND METHODS
02100 SITE REMEDIATION
02200 SITE PREPARATION
02300 EARTHWORK
02400 TUNNELING, BORING, AND JACKING
02450 FOUNDATION AND LOAD-BEARING ELEMENTS
02500 UTILITY SERVICES
02600 DRAINAGE AND CONTAINMENT
02700 BASES, BALLASTS, PAVEMENTS, AND APPURTENANCES
02800 SITE IMPROVEMENTS AND AMENITIES
02900 PLANTING
02950 SITE RESTORATION AND REHABILITATION

DIVISION 3 CONCRETE
03050 BASIC CONCRETE MATERIALS AND METHODS
03100 CONCRETE FORMS AND ACCESSORIES
03200 CONCRETE REINFORCEMENT
03300 CAST-IN-PLACE CONCRETE
03400 PRECAST CONCRETE
03500 CEMENTITIOUS DECKS AND UNDERLAYMENT
03600 GROUTS
03700 MASS CONCRETE
03900 CONCRETE RESTORATION AND CLEANING

DIVISION 4 MASONRY
04050 BASIC MASONRY MATERIALS AND METHODS
04200 MASONRY UNITS
04400 STONE
04500 REFRACTORIES
04600 CORROSION-RESISTANT MASONRY
04700 SIMULATED MASONRY
04800 MASONRY ASSEMBLIES
04900 MASONRY RESTORATION AND CLEANING

DIVISION 5 METALS
05050 BASIC METAL MATERIALS AND METHODS
05100 STRUCTURAL METAL FRAMING
05200 METAL JOISTS
05300 METAL DECK
05400 COLD-FORMED METAL FRAMING
05500 METAL FABRICATIONS
05600 HYDRAULIC FABRICATIONS
05650 RAILROAD TRACK AND ACCESSORIES
05700 ORNAMENTAL METAL
05800 EXPANSION CONTROL
05900 METAL RESTORATION AND CLEANING

DIVISION 6 WOOD AND PLASTICS
06050 BASIC WOOD AND PLASTIC MATERIALS AND METHODS
06100 ROUGH CARPENTRY
06200 FINISH CARPENTRY

06400 ARCHITECTURAL WOODWORK
06500 STRUCTURAL PLASTICS
06600 PLASTIC FABRICATIONS
06900 WOOD AND PLASTIC RESTORATION AND CLEANING

DIVISION 7 THERMAL AND MOISTURE PROTECTION
07050 BASIC THERMAL AND MOISTURE PROTECTION MATERIALS AND METHODS
07100 DAMPPROOFING AND WATERPROOFING
07200 THERMAL PROTECTION
07300 SHINGLES, ROOF TILES, AND ROOF COVERINGS
07400 ROOFING AND SIDING PANELS
07500 MEMBRANE ROOFING
07600 FLASHING AND SHEET METAL
07700 ROOF SPECIALTIES AND ACCESSORIES
07800 FIRE AND SMOKE PROTECTION
07900 JOINT SEALERS

DIVISION 8 DOORS AND WINDOWS
08050 BASIC DOOR AND WINDOW MATERIALS AND METHODS
08100 METAL DOORS AND FRAMES
08200 WOOD AND PLASTIC DOORS
08300 SPECIALTY DOORS
08400 ENTRANCES AND STOREFRONTS
08500 WINDOWS
08600 SKYLIGHTS
08700 HARDWARE
08800 GLAZING
08900 GLAZED CURTAIN WALL

DIVISION 9 FINISHES
09050 BASIC FINISH MATERIALS AND METHODS
09100 METAL SUPPORT ASSEMBLIES
09200 PLASTER AND GYPSUM BOARD
09300 TILE
09400 TERRAZZO
09500 CEILINGS
09600 FLOORING
09700 WALL FINISHES
09800 ACOUSTICAL TREATMENT
09900 PAINTS AND COATINGS

DIVISION 10 SPECIALTIES
10100 VISUAL DISPLAY BOARDS
10150 COMPARTMENTS AND CUBICLES
10200 LOUVERS AND VENTS
10240 GRILLES AND SCREENS
10250 SERVICE WALLS
10260 WALL AND CORNER GUARDS
10270 ACCESS FLOORING
10290 PEST CONTROL
10300 FIREPLACES AND STOVES
10340 MANUFACTURED EXTERIOR SPECIALTIES
10350 FLAGPOLES
10400 IDENTIFICATION DEVICES
10450 PEDESTRIAN CONTROL DEVICES
10500 LOCKERS
10520 FIRE PROTECTION SPECIALTIES
10530 PROTECTIVE COVERS
10550 POSTAL SPECIALTIES
10600 PARTITIONS
10670 STORAGE SHELVING
10700 EXTERIOR PROTECTION
10750 TELEPHONE SPECIALTIES
10800 TOILET, BATH, AND LAUNDRY ACCESSORIES
10880 SCALES
10900 WARDROBE AND CLOSET SPECIALTIES

DIVISION 11 EQUIPMENT
11010 MAINTENANCE EQUIPMENT
11020 SECURITY AND VAULT EQUIPMENT
11030 TELLER AND SERVICE EQUIPMENT
11040 ECCLESIASTICAL EQUIPMENT
11050 LIBRARY EQUIPMENT
11060 THEATER AND STAGE EQUIPMENT
11070 INSTRUMENTAL EQUIPMENT
11080 REGISTRATION EQUIPMENT
11090 CHECKROOM EQUIPMENT
11100 MERCANTILE EQUIPMENT
11110 COMMERCIAL LAUNDRY AND DRY CLEANING EQUIPMENT
11120 VENDING EQUIPMENT
11130 AUDIO-VISUAL EQUIPMENT
11140 VEHICLE SERVICE EQUIPMENT
11150 PARKING CONTROL EQUIPMENT
11160 LOADING DOCK EQUIPMENT
11170 SOLID WASTE HANDLING EQUIPMENT
11190 DETENTION EQUIPMENT
11200 WATER SUPPLY AND TREATMENT EQUIPMENT
11280 HYDRAULIC GATES AND VALVES
11300 FLUID WASTE TREATMENT AND DISPOSAL EQUIPMENT

11400 FOOD SERVICE EQUIPMENT
11450 RESIDENTIAL EQUIPMENT
11460 UNIT KITCHENS
11470 DARKROOM EQUIPMENT
11480 ATHLETIC, RECREATIONAL, AND THERAPEUTIC EQUIPMENT
11500 INDUSTRIAL AND PROCESS EQUIPMENT
11600 LABORATORY EQUIPMENT
11650 PLANETARIUM EQUIPMENT
11660 OBSERVATORY EQUIPMENT
11680 OFFICE EQUIPMENT
11700 MEDICAL EQUIPMENT
11780 MORTUARY EQUIPMENT
11850 NAVIGATION EQUIPMENT
11870 AGRICULTURAL EQUIPMENT
11900 EXHIBIT EQUIPMENT

DIVISION 12 FURNISHINGS
12050 FABRICS
12100 ART
12300 MANUFACTURED CASEWORK
12400 FURNISHINGS AND ACCESSORIES
12500 FURNITURE
12600 MULTIPLE SEATING
12700 SYSTEMS FURNITURE
12800 INTERIOR PLANTS AND PLANTERS
12900 FURNISHINGS RESTORATION AND REPAIR

DIVISION 13 SPECIAL CONSTRUCTION
13010 AIR-SUPPORTED STRUCTURES
13020 BUILDING MODULES
13030 SPECIAL PURPOSE ROOMS
13080 SOUND, VIBRATION, AND SEISMIC CONTROL
13090 RADIATION PROTECTION
13100 LIGHTNING PROTECTION
13110 CATHODIC PROTECTION
13120 PRE-ENGINEERED STRUCTURES
13150 SWIMMING POOLS
13160 AQUARIUMS
13165 AQUATIC PARK FACILITIES
13170 TUBS AND POOLS
13175 ICE RINKS
13185 KENNELS AND ANIMAL SHELTERS
13190 SITE-CONSTRUCTED INCINERATORS
13200 STORAGE TANKS
13220 FILTER UNDERDRAINS AND MEDIA
13230 DIGESTER COVERS AND APPURTENANCES
13240 OXYGENATION SYSTEMS
13260 SLUDGE CONDITIONING SYSTEMS
13280 HAZARDOUS MATERIAL REMEDIATION
13400 MEASUREMENT AND CONTROL INSTRUMENTATION
13500 RECORDING INSTRUMENTATION
13550 TRANSPORTATION CONTROL INSTRUMENTATION
13600 SOLAR AND WIND ENERGY EQUIPMENT
13700 SECURITY ACCESS AND SURVEILLANCE
13800 BUILDING AUTOMATION AND CONTROL
13850 DETECTION AND ALARM
13900 FIRE SUPPRESSION

DIVISION 14 CONVEYING SYSTEMS
14100 DUMBWAITERS
14200 ELEVATORS
14300 ESCALATORS AND MOVING WALKS
14400 LIFTS
14500 MATERIAL HANDLING
14600 HOISTS AND CRANES
14700 TURNTABLES
14800 SCAFFOLDING
14900 TRANSPORTATION

DIVISION 15 MECHANICAL
15050 BASIC MECHANICAL MATERIALS AND METHODS
15100 BUILDING SERVICES PIPING
15200 PROCESS PIPING
15300 FIRE PROTECTION PIPING
15400 PLUMBING FIXTURES AND EQUIPMENT
15500 HEAT-GENERATION EQUIPMENT
15600 REFRIGERATION EQUIPMENT
15700 HEATING, VENTILATING, AND AIR CONDITIONING EQUIPMENT
15800 AIR DISTRIBUTION
15900 HVAC INSTRUMENTATION AND CONTROLS
15950 TESTING, ADJUSTING, AND BALANCING

DIVISION 16 ELECTRICAL
16050 BASIC ELECTRICAL MATERIALS AND METHODS
16100 WIRING METHODS
16200 ELECTRICAL POWER
16300 TRANSMISSION AND DISTRIBUTION
16400 LOW-VOLTAGE DISTRIBUTION
16500 LIGHTING
16700 COMMUNICATIONS
16800 SOUND AND VIDEO

Reprinted from the Construction Specifications Institute and Construction Specifications Canada MasterFormat™ 1995 Edition with permission from Construction Specifications Institute, 1997.

INDUSTRY AND STANDARDS ORGANIZATIONS

ACI	American Concrete Institute 22400 W. Seven Mile Rd Detroit, MI 48127	**IEEE**	Institute of Electrical and Electronics Engineers, Inc. 345 E 47th St New York, NY 10017
AIA	American Institute of Architects 1735 New York Ave NW Washington, DC 20006	**IBEW**	International Brotherhood of Electrical Workers 1125 15th St NW Washington, DC 20005
AISC	American Institute of Steel Construction, Inc. 400 N Michigan Ave Chicago, IL 60611	**ICBO**	International Conference of Building Officials 5360 S Workman Mill Rd Whittier, CA 90601
AISI	American Iron and Steel Institute 1000 16th NW Washington, DC 20036	**ISO**	International Organization for Standardization Case Postale 56 Ch 1211 Geneve 20 Switzerland
ANSI	American National Standards Institute 11 W 42nd St New York, NY 10036	**NAHB**	National Association of Home Builders of the U.S. 15th and M Streets NW Washington, DC 20005
APA	American Plywood Association PO Box 11700 Tacoma, WA 98411-0700	**NCMA**	National Concrete Masonry Association PO Box 781 Herndon, VA 22070
ASHRAE	American Society of Heating, Refrigerating, and Air-Conditioning Engineers, Inc. 1791 Tullie Circle NE Atlanta, GA 30329	**NECA**	National Electrical Contractors Association 3 Bethesda Metro Circle Suite 1100 Bethesda, MD 20814
ASME	American Society of Mechanical Engineers 345 E 47th St New York, NY 10017	**NEMA**	National Electrical Manufacturers Association 2101 L St NW Washington, DC 20037
ASTM	American Society for Testing and Materials 1916 Race St Philadelphia, PA 19103	**NFPA**	National Fire Protection Association, Inc. Batterymarch Park Quincy, MA 02269
AWS	American Welding Society 550 NW LeJuene Rd PO Box 351040 Miami, FL 33145	**PCA**	Portland Cement Association 5420 Old Orchard Rd Skokie, IL 60077
BIA	Brick Institute of America 11490 Commerce Park Dr Reston, VA 22091	**SBCC**	Southern Building Code Congress International, Inc. 900 Montclair Rd Birmingham, AL 35213
BOCA	Building Officials and Code Administrators International, Inc. 4051 W Flossmoor Rd Country Club Hills, IL 60478-5795	**SJI**	Steel Joist Institute 1703 Parham Rd Suite 204 Richmond, VA 23229
CRSI	Concrete Reinforcing Steel Institute 933 N Plum Grove Rd Schaumburg, IL 60195	**UL**	Underwriters' Laboratories, Inc. 333 Pfingsten Rd Northbrook, IL 60062
CSI	Construction Specifications Institute 601 Madison St Alexandria, VA 22314	**WRI**	Wire Reinforcement Institute 5024 Wisconsin Ave NE Washington, DC 20016

accessible: Equipment admits close approach and is not guarded by locked doors, elevation, etc.

adjustable-trip CBs (ATCBs): CBs whose trip setting can be changed by adjusting the ampere setpoint, trip time characteristics, or both, within a particular range.

ambient temperature: The temperature of air around a piece of equipment.

ampacity: The current that a conductor can carry continuously, under the conditions of use.

appliance: Any utilization equipment which performs one or more functions, such as clothes washing, air conditioning, cooking, etc.

appliance branch circuit: A branch circuit that supplies energy to one or more outlets to which appliances are to be connected.

approved: Acceptable to the AHJ.

armored cable (AC): A factory assembly that contains the conductors within a jacket made of a spiral wrap of steel.

array: An assembly of modules or panels mounted on a support structure to form a power-producing unit.

askarel: A group of nonflammable synthetic chlorinated hydrocarbons that were once used where nonflammable insulating oils were required.

autotransformers: Single-winding transformers that share a common winding between the primary and secondary circuits.

auxiliary gutter: A sheet-metal enclosure equipped with hinged or removable covers that is used to supplement wiring space.

bare conductor: A conductor with no insulation or covering of any type.

bathroom: An area with a basin and one or more of a toilet, tub, or shower.

bend: Any change in direction of a raceway.

bonding: Joining metal parts to form a continuous path to conduct safely any current that is commonly imposed.

box: A metallic or nonmetallic electrical enclosure used for equipment, devices, and pulling or terminating conductors.

branch circuit: The portion of the electrical circuit between the last overcurrent device (fuse or CB) and the outlets or utilization equipment.

branch-circuit conductor: The circuit conductor(s) between the final overcurrent device protecting the circuit and the outlet(s).

branch-circuit rating: The ampere rating or setting of the overcurrent device protecting the conductors.

building: A stand-alone structure or a structure which is separated from adjoining structures by fire walls.

bulk storage plant: A location where volatile flammable liquids are stored in tanks having a total capacity of one carload or more from which liquids are distributed by tank truck or other methods.

bushing: A fitting placed on the end of a conduit to protect the conductor's insulation from abrasion.

busway: A sheet metal enclosure that contains factory-assembled aluminum or copper busbars which are supported on insulators.

cable: A factory assembly with two or more conductors and an overall covering.

cable assembly: A flexible assembly containing multiconductors with a protective outer sheath.

cablebus: An assembly of insulated conductors and terminations in an enclosed, ventilated protective metal housing.

cable tray system (CTS): An assembly of sections and associated fittings which form a rigid structural system used to support cables and raceways.

cadwelding: A welding process used to make electrical connections of copper to copper or copper to steel in which no outside source of heat or power is required.

cartridge fuse: A fuse constructed of a metallic link(s) which is designed to open at predetermined current levels to protect circuit conductors and equipment.

cell: A single, enclosed, tubular space in a precast cellular concrete slab floor with the direction of the cell being parallel to the direction of the floor member.

cellular metal floor raceway: The hollow spaces of cellular metal floors, with fittings, approved as enclosures for electric conductors.

circuit breaker (CB): A device which opens and closes circuits by nonautomatic means and opens circuits automatically when a predetermined overcurrent exists.

circular mil: A measurement used to determine the cross-sectional area of a conductor.

Class I location: A hazardous location in which sufficient quantities of flammable gases and vapors are present in the air to cause an explosion or ignite the hazardous materials.

Class II location: A hazardous location in which sufficient quantities of combustible dust are present in the air to cause an explosion or ignite the hazardous materials.

Class III location: A hazardous location in which easily-ignitable fibers or flyings are present in the air but not in a sufficient quantity to cause an explosion or ignite the hazardous materials.

conductor: A slender rod or wire that is used to control the flow of electrons in an electrical circuit.

conduit body: A conduit fitting that provides access to the raceway system through a removable cover at a junction or termination point.

conduit seal: A fitting which is inserted into runs of conduit to isolate certain electrical apparatus from atmospheric hazards.

continuous load: A load in which the maximum current may continue for three hours or more.

controller: The device in a motor circuit which turns the motor ON or OFF.

cover: The shortest distance measured between a point on the top surface of any direct-buried conductor, cable, conduit, or other raceway and the top surface of finished grade, concrete, or similar cover.

covered conductor: A conductor not encased in a material recognized by the NEC®.

current-limiting fuses: Fuses that open a circuit in less than $\frac{1}{2}$ of a cycle to protect the circuit components from damaging short-circuit currents.

current transformer: A transformer that creates a constant ratio of primary to secondary current instead of attempting to maintain a constant ratio of primary to secondary voltage.

damp location: A partially protected area subject to some moisture.

dead front: A cover required for the operation of a plug or connector.

demand factor: The ratio of the maximum demand of a system, or part of a system, to the total connected load of a system or the part of the system under consideration.

device box: A box which houses an electrical device.

devices: Electrical components, such as receptacles and switches, that are designed to carry, but not utilize, electricity.

disconnecting means: A device, or group of devices, by which the circuit conductors are disconnected from their source of supply.

division: The classification assigned to each Class based upon the likelihood of the presence of the hazardous substance in the atmosphere.

Division 1 location: A hazardous location in which the hazardous substance is normally present in the air in sufficient quantities to cause an explosion or ignite the hazardous materials.

Division 2 location: A hazardous location in which the hazardous substance is not normally present in the air in sufficient quantities to cause an explosion or ignite the hazardous materials.

dry location: A location which is not normally damp or wet.

dry-niche lighting fixture: A lighting fixture installed in the pool wall in a niche sealed against pool water entry.

dry-type transformer: A transformer which provides air circulation based on the principle of heat transfer.

dust-ignitionproof: Enclosed in a manner which prevents the entrance of dusts and does not permit arcs, sparks, or excessive temperature to cause ignition of exterior accumulations of specified dust.

dustproof: Construction in which dust does not interfere with the successful operation of equipment.

dusttight: Construction that does not permit dust to enter the enclosing case under specified test conditions.

dwelling: A structure that contains eating, living, and sleeping space, and permanent provisions for cooking and sanitation.

dwelling unit: A dwelling with one or more rooms used by one or more people for housekeeping.

dynamic load: A load that produces a small but constant vibration.

Edison-base fuse: A plug fuse that incorporates a screw configuration which is interchangeable with fuses or other ampere ratings.

effectively grounded: Grounded with sufficient low impedance and current-carrying capacity to prevent hazardous voltage buildups.

electrical metallic tubing (EMT): A lightweight tubular steel raceway without threads on the ends.

electric-discharge lighting fixture: A lighting fixture that utilizes a ballast for the operation of the lamp.

enclosure: The case or housing of equipment or other apparatus which provides protection from live or energized parts.

equipment: Any device, fixture, apparatus, appliance, etc. used in conjunction with electrical installations.

equipment bonding jumper (EBJ): A conductor that connects two or more parts of the EGC.

equipment grounding conductor (EGC): An electrical conductor that provides a low-impedance path between electrical equipment and enclosures and the system grounded conductor and GEC.

equipotential plane: An area in which all conductive elements are bonded or otherwise connected together in a manner which prevents a difference of potential from developing within the plane.

explosionproof apparatus: Equipment which is enclosed in a case that is capable of withstanding any explosion that may occur within it, without permitting the ignition of flammable gases or vapors on the outside of the enclosure.

exposed: As applied to wiring methods, is on a surface or behind panels which allow access.

feeder: All circuit conductors between the service equipment or the source of a separately derived system and the final branch-circuit overcurrent device.

feeder neutral load: The maximum unbalance between any of the ungrounded conductors and the grounded conductor.

fitting: An electrical system accessory that performs a mechanical function.

flash point (fire point): The temperature at which liquids give off vapor sufficient to form an ignitable mixture with the air near the surface of the liquid.

flat conductor cable (FCC): A cable with three or more flat copper conductors edge-to-edge and separated and enclosed by an insulating material.

flexible cable: An assembly of one or more insulated conductors, with or without braids, contained within an overall outer covering and used for the connection of equipment to a power source.

flexible cord: An assembly of two or more insulated conductors, with or without braids, contained within an overall outer covering and used for the connection of equipment to a power source.

flexible metal conduit (FMC): A raceway of metal strips which are formed into a circular cross-sectional raceway.

forming shell: A structure mounted in a pool and designed to support a wet-niche lighting fixture assembly.

fuse: An overcurrent protection device with a fusible link that melts and opens the circuit when an overload condition or short circuit occurs.

general-purpose branch circuit: A branch circuit that supplies a number of outlets for lighting and appliances.

general-use snap switch: A form of general-use switch constructed so that it can be installed in device boxes, etc.

general-use switch: A switch for use in general distribution and branch circuits. The ampere-rated switch is capable of interrupting its rated current at its rated voltage.

generator: A device that is used to convert mechanical power to electrical power.

grade: The level or elevation of the earth on a job site.

ground: A conducting connection between electrical circuits or equipment and the earth.

grounded: Connected to the earth or a conducting body connected to the earth.

grounded conductor: A conductor that has been intentionally grounded.

ground fault: An unintentional connection between an ungrounded conductor and any grounded raceway, box, enclosure, fitting, etc.

ground-fault circuit interrupter (GFCI): A device whose function is to de-energize a circuit within established time when current to ground exceeds a predetermined value which is less than that required to operate the protective device of a supply circuit.

grounding electrode conductor (GEC): The conductor that connects the grounding electrode(s) to the grounded conductor and/or the EGC.

grounding receptacles: Receptacles which include a grounding terminal connected to a grounding slot in the receptacle configuration.

group: An atmosphere containing flammable gases or vapors or combustible dust.

hazardous location: A location where there is an increased risk of fire or explosion due to the presence of flammable gases, vapors, liquids, combustible dusts, or easily-ignitable fibers or flyings.

header: A transverse metal raceway for electric conductors with access to predetermined cells, permitting the installation of electric conductors from a distribution center to the floor cells.

health care facility: A location, either a building or a portion of a building, which contains occupancies such as, hospitals, nursing homes, limited or supervisory care facilities, clinics, medical and dental offices, and either movable or permanent ambulatory facilities.

heating panel: A complete assembly with a junction box or a length of flexible conduit for connection.

heating panel set: A rigid or nonrigid assembly with nonheating leads or a terminal junction suitable for connection.

heating system: A complete system of components such as heating elements, fasteners, nonheating circuit wiring, leads, temperature controllers, safety signs, junction boxes, raceways, and fittings.

hermetic refrigerant motor-compressor: A combination of a compressor and motor enclosed in the same housing, having no external shaft or shaft seals, with the motor operating in the refrigerant.

high-intensity discharge (HID) lighting fixture: A lighting fixture that generates light from an arc lamp contained within an outer tube.

hydromassage bathtub: A permanently installed bathtub with a recirculating pump designed to accept, circulate, and discharge water for each use.

identified: Recognized as suitable for the use, purpose, etc.

immersion detection circuit interrupter (IDCI): Circuit interrupter designed to provide protection against shock when appliances fall into a sink or bathtub.

impedance: The total opposition to the flow of current in a circuit.

impedance heating system: A system in which heat is generated by pipe(s) and rod(s) by an electrical source.

indicating device: A pilot light, buzzer, horn, or other type of alarm. Often, the wiring of a motor control circuit is very elaborate and requires ten times the amount of wiring as the motor power circuit.

individual branch circuit: A branch circuit that supplies only one piece of utilization equipment.

in sight from: Visible and not more than 50′ away.

instantaneous-trip CBs (ITBs): CBs with no delay between the fault or overload sensing element and the tripping action of the device.

instrument transformer: A transformer used to reduce higher voltage and current ratings to safer and more suitable levels for the purposes of control and measurement.

insulated conductor: A conductor covered with a material classified as electrical insulation.

interactive system: A system that operates in parallel with and is permitted to deliver power to a normal utility service connected to the same load.

intermittent load: A load in which the maximum current does not continue for three hours.

interrupting rating: The maximum amount of current that an OCPD can clear safely.

intrinsically safe system: A system with an assembly of intrinsically safe apparatus and associated apparatus which is interconnected and used in hazardous locations to supply equipment.

inverse-time CBs (ITCBs): CBs with an intentional delay between the time when the fault or overload is sensed and the time when the CB operates.

inverter: Equipment used to change voltage level or waveform, or both, of electrical energy.

isolated-ground receptacles: Receptacles in which the grounding terminal is isolated from the device yoke or strap.

isolating switch: A switch designed to isolate an electric circuit from the power source. It has no interrupting rating and is operated only after the circuit has been opened by other means.

isolation transformer: A transformer that utilizes a shield between the primary and secondary windings and a transformer ratio of 1 : 1 to ensure that the load is separated from the power source.

junction box: A box in which splices, taps, or terminations are made.

kick: A single bend in a raceway.

labeled: Equipment acceptable to the AHJ and to which a label has been attached.

lampholders: Devices designed to accommodate a lamp for the purpose of illumination.

less-flammable liquid: An insulating oil which is flammable but has reduced flammable characteristics and a higher fire point.

lighting and appliance branch-circuit panelboard: A panelboard with more than 10% of its branch-circuit fuses or CBs rated at 30 A or less (15 A, 20 A, 25 A, or 30 A).

lighting outlet: An outlet intended for the direct connection of a lampholder, a lighting fixture, or pendant cord terminating in a lampholder.

lighting outlets: Outlets that provide power for lighting fixtures.

lighting track: An assembly consisting of an energized track and lighting fixture heads which can be positioned in any location along the track.

line surge: A temporary increase in the circuit or system voltage or current that may occur as a result of fluctuations in the electrical distribution system.

liquid-filled transformer: A transformer that utilizes some form of insulating liquid to immerse the core and windings of the transformer to aid in the removal of heat generated by the transformer windings.

liquidtight flexible metal conduit (LTFMC): A raceway of circular cross section with an outer liquidtight, non-metallic, sunlight-resistant jacket over an inner helically-wound metal strip.

liquidtight flexible nonmetallic conduit (LTFNMC): A raceway of circular cross-section with an outer jacket which is resistant to oil, water, sunlight, corrosion, etc. The inner core varies based on intended use.

listed: Equipment or material approved by the AHJ in a list.

luminaire: A complete lighting fixture consisting of the lamp or lamps, reflector or other parts to distribute the light, lamp guards, and lamp power supply.

main bonding jumper (MBJ): The connection at the service equipment that ties together the EGC, the grounded conductor, and the GEC.

manufactured building: A building of closed construction, normally factory-made or assembled for installation or assembly and installation on building site. This does not include mobile homes or recreational vehicles.

metal-clad cable (MC): A factory assembly of one or more conductors with or without fiber-optic members.

metal wireway: A sheet metal trough with a hinged or removable cover that houses and protects wires and cables laid in place after the wireway has been installed.

mil: .001″.

milliampere (mA): $\frac{1}{1000}$ of an ampere (1000 mA = 1 A).

mineral oil: A chemically untreated insulating oil that is distilled from petroleum.

mobile home: A transportable factory-assembled structure or structures constructed on a permanent chassis for use as a dwelling. A mobile home is not constructed on a permanent foundation but is connected to the required utilities. The term "mobile home" includes manufactured homes.

module: The smallest complete, environmentally-protected assembly of solar cells, exclusive of tracking, designed to generate DC power under sunlight.

motor branch circuit: The point from the last fuse or CB in the motor circuit out to the motor.

motor control center (MCC): An assembly of one or more enclosed sections with a common power bus and primarily containing motor control units.

motor control circuit: The circuit of a control apparatus or system which carries electric signals directing the performance of the controller, but does not carry the main power current.

multifamily dwelling: A dwelling with three or more dwelling units.

multiple receptacle: A single device with two or more receptacles.

multioutlet assembly: A metal raceway with factory-installed conductors and attachment plug receptacles. A flush or surface raceway which contains conductors and receptacles.

multiwire branch circuit: A branch circuit with two or more ungrounded conductors having a potential difference between them, and is connected to the neutral or grounded conductor of the system.

nipple: A short piece of conduit or tubing that does not exceed 24″ in length.

nominal voltage: Any voltages within an acceptable range.

nonadjustable-trip CBs (NATCBs): Fixed CBs designed without provisions for adjusting either the ampere trip setpoint or the time-trip setpoint.

noncoincidental loads: Loads that are not on at the same time.

nonflammable liquid: A liquid that is noncombustible and does not burn when exposed to air.

nongrounding receptacles: Receptacles with two wiring slots for branch-circuit wiring systems that do not provide an equipment grounding conductor.

no-niche lighting fixture: A lighting fixture installed above or below the water without a niche.

nonlinear load: A load where the wave shape of the steady-state current does not follow the wave shape of the applied voltage.

nonmetallic extensions: Assemblies of two insulated conductors in a nonmetallic jacket or an extruded thermoplastic covering. Nonmetallic extensions may be either surface extensions or aerial cable.

nonmetallic-sheathed cable (NM): A factory assembly of two or more insulated conductors having an outer sheath of moisture-resistant, flame-retardant, nonmetallic material.

nonmetallic wireway: A flame-retardant nonmetallic trough with a removable cover that houses and protects wires and cables laid in place after the wireway has been installed.

non-selective system: A fault on an individual branch circuit not only opens the branch-circuit overcurrent protection device, but also opens the feeder overcurrent protection device.

non-time delay fuses (NTDFs): Fuses that may detect an overcurrent and open the circuit almost instantly.

offset: A double bend in a raceway, each containing the same number of degrees.

one-family dwelling: A dwelling with one dwelling unit.

outlet: Any point in the electrical system where current supplies utilization equipment.

outlet box: A box which houses a piece of utilization equipment.

overcurrent: Any current in excess of that for which the conductor or equipment is rated.

overlamping: Installing a lamp of a higher wattage than for which the fixture is designed.

overload: A small-magnitude overcurrent, that over a period of time, leads to an overcurrent which may operate the overcurrent protection device (fuse or CB).

oxidation: The process by which oxygen mixes with other elements and forms a rust-like material.

oxide: A thin, but highly resistive coating that forms on metal when exposed to the air.

panel: A collection of modules, secured together, wired, and designed to provide a field-installable unit.

panelboard: A single panel or group of assembled panels with buses and overcurrent devices, which may have switches to control light, heat, or power circuits.

panic hardware: Door hardware designed to open easily in an emergency situation.

parallel conductors: Two or more conductors that are electrically connected at both ends to form a single conductor.

pendants: Hanging light fixtures that use flexible cords to support the lampholder.

permanently-connected appliance: A hard-wired appliance that is not cord-and-plug connected.

permanently installed swimming pool: A pool constructed in ground or partially above ground and designed to hold over 42″ of water, and all indoor pools regardless of depth.

phase converter: An electrical device that converts 1ϕ power to 3ϕ power.

phase-to-ground voltage: The difference of potential between a phase conductor and ground.

phase-to-phase voltage: The maximum voltage between any two phases of an electrical distribution system.

photovoltaic output circuit: Conductors between photovoltaic source circuits and the inverter or DC utilization equipment.

photovoltaic power source: One or more arrays which generate DC power at system voltage and current.

photovoltaic source circuit: Conductors between modules and from modules to common junction point(s) of the DC system.

pilot device: A sensing device that controls the motor controller.

place of assembly: A building, structure, or portion of a building designed or intended for use by 100 or more persons.

plug fuse: A fuse that uses a metallic strip which melts when a predetermined amount of current flows through it.

potential transformer: A transformer which steps down higher voltages while allowing the voltage of the secondary to remain fairly constant from no-load to full-load conditions.

power panelboard: A panelboard with more than 10% of its branch-circuit fuses or CBs rated over 30 A or more.

premises wiring: Basically all interior and exterior wiring installed on the load side of the service point or the source of a separately derived system.

pull box: A box used as a point to pull or feed electrical conductors into the raceway system.

raceway: A metal or nonmetallic enclosed channel for conductors.

raceway system: An enclosed channel of metal or nonmetallic materials used to contain the wires or cables of an electrical system.

readily accessible: Capable of being reached quickly.

receptacle outlets: Outlets that provide power for cord-and-plug connected equipment.

receptacles: Contact devices installed at outlets for the connection of cord-connected electrical equipment.

recreational vehicle: A vehicular type unit which is self-propelled or is mounted on or pulled by another vehicle and is primarily designed as temporary living quarters for camping, travel, or recreational use.

redundant grounding: Grounding with two separate grounding paths.

resistance heating element: A specific element to generate heat that is embedded in or fastened to the surface.

rigid metal conduit (RMC): A conduit made of metal. It is the universal raceway.

rigid nonmetallic conduit (RNMC): A conduit made of materials other than metal.

selective system: A fault on an individual branch circuit opens only the branch-circuit overcurrent protective device and does not affect the feeder overcurrent protection device.

self-grounding receptacles: Grounding type receptacles which utilize a pressure clip around the 6 – 32 mounting screw to ensure good electrical contact between the receptacle yoke and the outlet box.

separately derived system: A system that supplies premises with electrical power derived or taken from storage batteries, solar photovoltaic systems, generators, transformers, or converter windings.

service: The electrical supply, in the form of conductors and equipment, that provides electrical power to the building or structure.

service conductors: The conductors from the service point or other source of power to the service disconnecting means.

service drop: The conductors that extend from the overhead utility supply system to the service-entrance conductors at the building or structure.

service-entrance cable (SE): A single or multiconductor assembly with or without an overall covering.

service-entrance conductors – overhead systems: Conductors that connect the service equipment for the building or structure with the electrical utility supply conductors.

service-entrance conductors – underground systems: Conductors that connect the service equipment with the service lateral.

service equipment: All of the necessary equipment to control the supply of electrical power to a building or a structure.

service lateral: The underground service conductors that connect the utility's electrical distribution system with the service-entrance conductors.

service mast: An assembly consisting of a service raceway, guy wires or braces, service head, and any fittings necessary for the support of service-drop conductors.

service point: The point of connection between the local electrical utility company and the premises wiring of the building or structure.

short circuit: The unintentional connection of two ungrounded conductors that have a potential difference between them. The condition that occurs when two ungrounded conductors (hot wires), or an ungrounded and a grounded conductor of a 1φ circuit, come in contact with each other.

single receptacle: A single contact device with no other contact device on the same yoke.

skeleton tubing: Neon tubing that is the sign of outline lighting. It does not contain, nor is attached to a sign body or enclosure.

solar cell: Basic photovoltaic device that generates electricity when exposed to light.

solar photovoltaic system: The system used to convert solar energy into electrical energy suitable for connection to a utilization load.

spa (hot tub): An indoor or outdoor hydromassage pool or tub that is not designed to have the water discharged after each use.

special permission: The written approval of the authority having jurisdiction.

step-down transformer: A transformer with more windings in the primary winding, which results in a load voltage which is less than the applied voltage.

step-up transformer: A transformer with more windings in the secondary winding, which results in a load voltage which is greater than the applied voltage.

storable swimming pool: A pool constructed on or above ground and designed to hold less than 42″ of water, or a pool with nonmetallic, molded polymeric walls or inflatable fabric walls regardless of size.

strut-type channel raceway: A surface raceway formed of moisture-resistant and corrosion-resistant metal.

supervised installation: An electrical installation in which the conditions of maintenance are such that only qualified persons monitor or service the electrical equipment.

surface raceway: An enclosed channel for conductors which is attached to a surface.

switch: A device, with a current and voltage rating, used to open or close an electrical circuit.

switchboard: A single panel or group of assembled panels with buses, overcurrent devices, and instruments.

temperature rise: The amount of heat that an electrical component produces above the ambient temperature.

thermally protected: Designed with an internal thermal protective device which senses excessive operating temperatures and opens the supply circuit to the fixtures.

time delay fuses (TDFs): Fuses that may detect and remove a short circuit almost instantly, but allow small overloads to exist for a short period of time.

torque: A turning or twisting force, typically measured in foot-pounds (ft-lb).

transformer: A device that converts or "transforms" electrical power at one voltage or current to another voltage or current.

transformers: Electrical devices that contain no moving parts and are used primarily to convert electrical power at one voltage and current rating to another voltage and current rating.

travel trailer: A vehicle that is mounted on wheels, has a trailer area less than 320 sq ft (excluding wardrobes, closets, cabinets, kitchen units, fixtures, etc.), is of such size and weight that a special highway use permit is not required, and is designed as temporary living quarters while camping or traveling.

two-family dwelling: A dwelling with two dwelling units.

Type I building: A building in which all structural members (walls, columns, beams, girders, trusses, arches, floors, and roofs) are constructed of approved noncombustible or limited-combustible materials.

Type II building: A building that does not qualify as Type I construction in which the structural members (walls, columns, beams, girders, trusses, arches, floors, roofs, etc.) are constructed of approved noncombustible or limited-combustible materials.

Type S fuse: A plug fuse that incorporates a screw and adapter configuration which is not interchangeable with fuses of another ampere rating.

unfinished basement: The portion of area of a basement which is not intended as a habitable room, but is limited to storage areas, work areas, etc.

unit switch: A switch designed to control a specific unit load.

USE: Underground service-entrance cable with a moisture-resistant covering that is not flame-resistant.

utilization equipment: Any electrical equipment which uses electrical energy for electronic, electromechanical, chemical, heating, lighting, etc. purposes.

voltage-to-ground: The difference of potential between a given conductor and ground.

wet location: Any location in which a conductor is subjected to excessive moisture or saturation from any type of liquids or water.

wet-niche lighting fixture: A lighting fixture installed in a forming shell and completely surrounded by water.

wireway: A metallic or nonmetallic trough with a hinged or removable cover designed to house and protect conductors and cables.

within sight: Visible and not more than 50′ away.

abbreviations, 12, 34
AC voltage measurement, 98
adapters, 55
addition to existing building, 94
adjustable trip circuit breaker, 53
aerial cable, 91
agricultural building, 143
air conditioning equipment, 97
aircraft hangars, 206
air plenum wiring methods, 134
aluminum conductor, 80, 118
aluminum wire and conduit, 117
amateur transmitting and receiving station, 217
ampacity of fixture wires, 80
amplifiers, 212
angle pulls, 138
antenna systems, 217
apartment
    details and elevations, 84 – 85
    floor plan, 81 – 82
    load calculations, 81 – 83
    service calculations, 83 – 84
appliance, 70
    branch circuit protection, 95
    panelboard, 55, 130
applying the NEC®, 4 – 5
area, 33
armored cable, 60
array, 144
askarel transformers, 179
authority having jurisdiction, 4 – 5
autotransformer, 179
    dimmer, 211
auxiliary gutter, 138
    recessed, 212
available fault current, 128

bathroom, 31
bonding jumpers, 133
bonding swimming pools, 189
box
    angle pull, 138
    device, 51, 52
    junction, 138
    metal, 51
    outlet, 51, 52
    pull, 138
box fill, 51
bracket fixture wiring, 211
braided-covered insulated conductors, 183
branch circuit, 100
    calculations, 34, 36
    conductor, 142

conductor identification, 100
load calculation, 142
loads, 100
protection, 95
tap rules, 133
taps, 142
building, 123
    addition to existing, 94
    opening service drop clearances, 41
bulk storage plant, 207
bushing, 60
busways, 166
    no plug-in, 166
    plug-in, 166
    trolley, 166
BX, 60

cable
    aerial, 91
    connectors, 211
    Type SE, 41
    Type UF, 67
cablebus, 167
cable fished into masonry walls, 67
cable limiter, 124
cable trays, 210
calculations, 31 – 39
capacitors
    motor circuits, 183
    over 600 V, 183
    600 V or less, 183
cartridge fuses, 56 – 57
    classes, 57
    dual-element, time-delay, 56
    dual-element, time-delay current-limiting, 56
    single-element, current-limiting, 56
cell, 136
cellular concrete floor raceway, 136
cellular metal floor raceway, 136
circuit breaker, 53, 128
circuits, 66
    Class 1, 214
    Class 2, 214
    Class 3, 214
    combining Class 2 and Class 3, 214
    communication, 216
    less than 50 volts, 213
Class I, Division 1 locations, 202
Class I, Division 2 locations, 203
Class II, Division 1 locations, 204
Class II, Division 2 locations, 204
Class III, Division 1 locations, 205
Class III, Division 2 locations, 205
clearances for service drops, 40
clearances for service drops from building
    openings, 41
clearances for service laterals, 40
coating processes, 208

Code-Making Panel, 2
    membership, 2
combustible, 200
commercial garages, 206
commercial locations, 109
    electrical plans, 110 – 118
    lighting system, 117
    service calculation, 115 – 116
    service conduit size, 116 – 117
common neutral for feeders, 142
communication circuits, 216
concrete floor raceways, 136
conductor, 66, 79, 109 – 110
    aluminum, 80
    application and insulations, 80
    copper, 80
    general rules, 66
    insulated, 62
    neutral, 45
    number in conduit, 139
    overcurrent protection, 128
    protection – pole, 126
    supports, 134
conduit
    body, 51, 52
    conductors in, 139
    rigid metal, 64
    rigid nonmetallic, 65
connection ahead of disconnecting means, 213
connection of fluorescent fixtures, 140
Construction Specifications Institute, 11
continuity – raceways and boxes, 134
control system, 175
copper conductor, 80, 118
cord-connected appliances, 95
cord-connected showcases, 140
cord overcurrent protection, 129
countertop ranges, 70
cove lighting, 140
crane motors, 156
cranes, 174
current limiting device, 124
current-limiting overcurrent protective device, 124

deicing equipment, 94
demand, 32
details, 9, 11, 16, 21
device box, 51, 52
dimmers, 211
dipping processes, 208
direct current motor, 171
disconnecting means, 43, 169, 182
    for appliances, 70
    for free-standing ranges, 70
    for motor-driven appliances, 95
distribution panelboards, 87
dressing rooms, 211

Electrical training materials . . .

JOURNEYMAN ELECTRICIAN'S EXAM WORKBOOK
Based on the 1999 NEC®
By Robert E. Chellew

This extremely practical workbook contains six chapters which discuss the Exam and provide examples of formulas, NEC®-related problems, and definitions. Contains 542 sample and test questions.

132 pages, illustrated, soft cover.

	ISBN
Text	1710-0
Instructor's Guide	1711-9

ELECTRICAL SYSTEMS
Based on the 1999 NEC®
By Michael I. Callanan, Bill Wusinich

This comprehensive, NEC®-based text covers dwellings, commercial, and industrial buildings. Includes wiring methods and materials, conductors and OCPDs, branch circuits and feeders, grounding, transformers, services, special locations, and calculations. Procedures cover all aspects of designing electrical systems.

424 pages, 351 illustrations, soft cover.

	ISBN
Text	1692-9
Instructor's Guide	1693-7

ELECTRICAL MOTOR CONTROLS
By Gary Rockis, Glen A. Mazur

This full-color text with state-of-the-art technology and applications covers the control devices used in modern industrial electrical systems. The devices, applications, and systems are chosen to represent the broad range of electrical uses found in different industries.

The *Workbook* contains worksheets for each major concept presented in the text. The 42 full-color transparencies illustrate the key elements of *Electrical Motor Controls*. The *Manual* may be used with the text or as a stand-alone instructional manual.

490 pages, 787 illustrations, hard cover.

	ISBN
Text	1671-6
Workbook	1672-4
Instructor's Guide	1673-2
Transparencies	1674-0
Manual	1668-6
Instructor's Guide	1670-8

Enhanced with full-color art.

TRANSFORMERS
By Kenneth R. Edwards

This text gives the apprentice or journeyman a strong background for understanding transformers. *Transformers* presents basic electrical and magnetic principles as applied to transformers as well as advanced principles of transformer operation.

The text covers operation, cooling methods, tap changers, transformer connections, autotransformers, and reactors.

316 pages, 188 illustrations, soft cover.

	ISBN
Text	1603-1

SOLID STATE FUNDAMENTALS FOR ELECTRICIANS
2nd Edition
By Gary Rockis

Comprehensive overview of solid state devices and systems, including fiber optics, integrated circuits, and light-activated components. The textbook is designed for electricians, students, and others who have some basic knowledge of electronics.

232 pages, 481 illustrations, hard cover.

	ISBN
Text	1631-7
Workbook	1632-5
Instructor's Guide	1633-3

Call TOLL-FREE 1-800-323-3471 or visit our web site at www.americantech.org.

132 pages,
90 illustrations,
soft cover.

DIGITAL MULTIMETER PRINCIPLES

By Glen A. Mazur

This comprehensive instructional program includes everything needed to train technicians in the safe operation of digital multimeters. Proper DMM usage while taking all basic measurements can help prevent injury to personnel and damage to equipment. *Digital Multimeter Principles*, authorized by Fluke Corporation, is a must for all technicians, from contractors to maintenance staffs. Successful completion of the tasks covered in the text is documented by a Certificate of Completion.

Contents Include:

- Safety
- DMM Abbreviations, Symbols, and Terminology
- DMM Displays
- DMM Advanced Features
- Measuring AC Voltage
- Measuring DC Voltage
- Measuring Resistance and Continuity Testing
- Measuring AC and DC Current

- Ohm's Law and Power Formula
- Measuring Frequency and Duty Cycle
- Testing Diodes
- Measuring Capacitance
- DMM Accessories
- DMM Selection
- Appendix
- Glossary
- Index

	ISBN
Text	1488-8

Effective training makes workers safer and more productive.

Digital Multimeter Principles Videotape

By Glen A. Mazur

This 29 minute videotape depicts step-by-step procedures for using digital multimeters. The easy-to-follow format simplifies instruction and aids student comprehension.

	ISBN
Videotape	1494-2

ABCs of DMM Safety Videotape

This 14 minute videotape covers DMM safety considerations. Clear descriptions show common industry practices used to reduce the potential of injury.

	ISBN
Videotape	9150-5

Components Kit

Kit includes all items needed to complete the *Resource Guide* activities and is also used with the *Digital Multimeter Principles Workbook*.

- DC Power Supply
- 4 Test leads
- Fuse and holder
- 2 Potentiometers
- SPDT switch
- 4 Resistors
- 2 Capacitors
- Diode
- Incandescent lamp
- Thermistor

	ISBN
Components	1495-0

Digital Multimeter Principles Set

This set includes 10 *Digital Multimeter Principles* texts and 10 *Components Kits*. Save money by buying this set.

	ISBN
DMP Set	1487-X

DIGITAL MULTIMETER PRINCIPLES RESOURCE GUIDE

By Glen A. Mazur

The *Digital Multimeter Principles Resource Guide* is a valuable instructional tool. Everything you need to teach the basics of digital multimeter usage is included. All Activities and Worksheets are reproducible in whatever quantities you require.

	ISBN
Resource Guide	1489-6

FLUKE®

Authorized by
Fluke Corporation

Resource Guide includes:
- *Digital Multimeter Principles* **text**
- *Instructor's Guide*
- *Instructional Outline*
- *Activity* and *Worksheet Masters*
- **42 full-color** *Transparencies*
- *Digital Multimeter Principles Videotape*
- *ABCs of DMM Safety Videotape*
- *Electrical Principles and Practices* **text**
- *Electrical/Electronic Systems* **text**
- *Components Kit*
- *Spare Components Kit* **(fuses, etc.)**

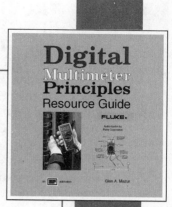

DIGITAL MULTIMETER PRINCIPLES WORKBOOK

By Glen A. Mazur

The *Digital Multimeter Principles Workbook* can be used by an individual for self study with the *Digital Multimeter Principles* text and the *Components Kit*. Instructions for using the *Workbook* are written for easy self study. All activities and applications are numbered to coincide with text chapters. Text references and components required are listed for each activity. Step-by-step instructions and training tips guide the user through completion of workbook tasks.

Step-by-step instructions and training tips.

	ISBN
Workbook	1484-5

Save with the Self-Study Package.

The package includes the Digital Multimeter Principles:
- **Text**
- **Components Kit**
- **Workbook**
- **Videotape**

	ISBN
Self-Study Package	1483-7

ELECTRICAL/ELECTRONIC FUNDAMENTALS VIDEO SERIES

By Glen A. Mazur

Electrical/Electronic Fundamentals Video Series consists of seven videos which present basic electrical theory, electrical safety, digital multimeter principles, and an introduction to fundamental troubleshooting procedures. The videos feature the latest electrical equipment used in various industrial applications.

		ISBN
Electrical/Electronic Fundamentals Video Series		1406-3
Electrical Principles 1		1407-1
Electrical Principles 2		1408-X
Symbols and Circuits		1409-8
Electrical Safety		1410-1
Digital Multimeters-Basic Measurements	set	1411-X
Digital Multimeters-Advanced Measurements		1412-8
Digital Multimeters-Testing Procedures		1413-6
Electrical/Electronic Fundamentals Workbook		1414-4
Electrical/Electronic Fundamentals Workbook Instructor's Guide		1415-2

Practical and content-filled.

TROUBLESHOOTING ELECTRICAL/ ELECTRONIC SYSTEMS

By Glen A. Mazur, Thomas E. Proctor

The heavily-illustrated text and workbook cover topics from residential wiring systems to industrial controls. Activities provide hands-on experience in solving troubleshooting problems.

476 pages, heavily illustrated, soft cover.

	ISBN
Text	1775-5
Instructor's Guide	1776-3
Workbook	1778-X
Instructor's Guide	1779-8
Transparencies	1777-1

TROUBLESHOOTING ELECTRIC MOTORS
2nd Edition

By Glen A. Mazur, Thomas E. Proctor

This text/workbook covers step-by-step procedures for troubleshooting motors and starters. Motor sizing and selection are presented.

Applications present additional information required when servicing motors. Activities provide hands-on experience.

299 pages, 126 illustrations, soft cover.

	ISBN
Text	1765-8
Instructor's Guide	1766-6
Transparencies	1764-X

ELECTRICAL/ELECTRONIC SYSTEMS

By Glen A. Mazur, Thomas E. Proctor

Electrical/Electronic Systems offers a comprehensive introduction to the concepts of electrical and electronic systems. All elements of residential, commercial, and industrial systems are presented in an easy-to-understand format. Contains full-color photographs from over 65 trade manufacturers and CAD-drawn illustrations.

490 pages, 701 illustrations, hard cover.

Electrical/Electronic Systems Workbook 16 Tech-Cheks contain questions based on the information presented in the corresponding chapter of the textbook. The Worksheets apply electrical and electronic theory to practical problems.

	ISBN
Text	1772-0
Workbook	1773-9
Instructor's Guide	1774-7
Transparencies	1771-2

42 full-color transparencies.

ELECTRICAL PRINCIPLES AND PRACTICES

By Glen A. Mazur, Peter A. Zurlis

Electrical Principles and Practices provides a comprehensive, content-filled introduction to basic electrical theory, circuit fundamentals, and practical wiring techniques. The text is based on typical residential, commercial, and industrial systems and represents the broad range of uses for electricity found in different industries. The *Workbook* contains questions based on the corresponding textbook chapter content, and the worksheets provide an opportunity to solve real-world problems.

460 pages, 574 illustrations, hard cover.

Heavily-illustrated, easy-to-read format.

	ISBN
Text	1755-0
Workbook	1756-9
Instructor's Guide	1757-7